BASIC ANATOMY AND
PHYSIOLOGY FOR RADIOGRAPHERS

Basic Anatomy and Physiology for Radiographers

M.R.E. DEAN
MB, BChir, DMRD
Consultant Radiologist
the Royal Shrewsbury Hospital

T.E.T. WEST
MD, FRCP
Consultant Physician
the Royal Shrewsbury Hospital

THIRD EDITION

OXFORD

BLACKWELL SCIENTIFIC PUBLICATIONS

LONDON EDINBURGH BOSTON
MELBOURNE PARIS BERLIN VIENNA

© 1970, 1975, 1987 by
Blackwell Scientific Publications
Editorial Offices:
Osney Mead, Oxford OX2 0EL
25 John Street, London WC1N 2BL
23 Ainslie Place, Edinburgh EH3 6AJ
238 Main Street, Cambridge,
   Massachusetts 02142, USA
54 University Street, Carlton,
   Victoria 3053, Australia

Other Editorial Offices:
Librairie Arnette SA
1, rue de Lille
75007 Paris
France

Blackwell Wissenschafts-Verlag GmbH
Düsseldorfer Str. 38
D-10707 Berlin
Germany

Blackwell MZV
Feldgasse 13
A-1238 Wien
Austria

First published 1970
Second edition 1975
Reprinted 1978, 1980
Third edition 1987
Reprinted in monochrome 1991, 1994

Set by Setrite Typesetters Ltd
Hong Kong
Printed in Great Britain
at The Alden Press, Oxford

DISTRIBUTORS

Marston Book Services Ltd
PO Box 87
Oxford OX2 0DT
(*Orders*: Tel: 0865 791155
       Fax: 0865 791927
       Telex: 837515)

USA
Blackwell Scientific Publications, Inc.
238 Main Street
Cambridge, MA 02142
(*Orders*: Tel: 800 759-6102
       617 876-7000)

Canada
Times Mirror Professional Publishing, Ltd
130 Flaska Drive
Markham, Ontario L6G 1B8
(*Orders*: Tel: 800 268-4178
       416 470-6739)

Australia
Blackwell Scientific Publications Pty Ltd
54 University Street
Carlton, Victoria 3053
(*Orders*: Tel: 03 347-5552)

British Library
Cataloguing in Publication Data

Dean, M.R.E.
Basic anatomy and physiology for
radiographers. — 3rd ed.
1. Anatomy, Human   2. Radiography,
   Medical   3. Physiology
I. Title   II. West, T.E.T.
611′.0024616   QM23.2

ISBN 0-632-00913-6

Contents

Preface to the third edition

The third edition of this book has given us the opportunity to update the physiology sections and to include some clinical applications where relevant. The chapter on the lymphatic system has been expanded and the chapter on the endocrine system has been rewritten. The scope of the chapter on general pathology has been considerably widened to include a brief but general classification of disease. The first impression of the third edition included some coloured diagrams, but it has unfortunately proved necessary to reproduce these in monochrome in later impressions in order to keep the price at a reasonable level for students.

We are grateful to Dr Norman Mitchell for his advice on the chapter on general pathology, to Mr Nicholas West for his help with the artwork and to Mrs Margaret Keeling-Roberts for her meticulous help with the time-consuming task of reading the proofs.

Preface to the first edition

This book is primarily intended for the student radiographer who is preparing for Part I of the M.S.R. examination. Particular emphasis has, therefore, been given to the chapters on the skeleton, the joints and the lymphatic system. It is important that the student should have available a half-skeleton to study while reading the section on the skeleton. A brief chapter on basic pathology has been included, but diseases of individual organs have only been mentioned when they aid the understanding of the physiology of the organ under discussion. Other diseases are best discussed at tutorials when X-ray films can be used for illustration.

I am grateful to Professor Harold Ellis for permission to reproduce illustrations 15, 17–19, 23–25, 31, 51, 66–69, 84, 95, 105, 122, 128, 129, 149–151, 153, 155, 167, 180, 181, 184, 187–191, 194, 203, 214, 217, 237, 243, 255*.

I am grateful to Mrs Adrienne Finch for her advice on the preparation of the manuscript, to Mrs Mary Woolley for her invaluable secretrial help and to Mrs Barbara Smedley, Mr D. Pollock, Dr P. McQuade and Miss J. Kitson for their assistance in correcting the proofs.

*The respective illustrations in the third edition being: 4.4, 4.6–4.8, 4.12–4.14, 4.21, 6.12, 7.6–7.9, 9.5, 10.3, 10.14, 10.31, 10.37, 10.38, 11.7–11.9, 11.11, 11.13, 11.26, 13.1, 13.2, 13.5, 13.8–13.12, 14.2, 14.10, 15.2, 15.5, 16.18, 17.3 and 12.7.

CHAPTER 1

General introduction

The cell

The simplest form of animal life is composed of a single unit called a cell. These unicellular animals are capable of ordered activity and they possess all those properties which distinguish living from dead matter.

The properties of animal cells

Respiration. An animal cell needs oxygen for combustion with food substances to provide the energy necessary for its activity. Respiration is the process by which oxygen is taken up from the surroundings and is used by the cell for the provision of energy. This combustion of oxygen with food substances produces carbon dioxide which is released into the surroundings of the cell.

Metabolism. Metabolism is the term applied to the processes by which food substances are used by the cell for: (a) growth and repair, (b) combustion with oxygen for the provision of energy and (c) the elaboration of secretions that the cell may produce.

Excretion. Waste products are formed during the processes of metabolism and these are removed from the cell by the processes of excretion.

Growth. The cell is able to increase in size by producing new cellular constituents from ingested foodstuffs.

Movement. Animals possess the ability to move in an organized manner.

Irritability. Irritability is the ability to respond to an external stimulus such as a change in the external environment.

Reproduction. Animals are able to reproduce their species.

More complex animals are composed of large numbers of cells which differ widely in appearance and function, but these cells all have certain characteristics in common. In these more complex animals the cells become specialized, or differentiated, to form tissues and organs. These tissues and organs in turn form systems each of which perform a specific function for the animal as a whole. The study of human anatomy is therefore best begun with the description of a typical animal cell.

A typical animal cell

All cells consist of a semifluid substance called *protoplasm*. Protoplasm is composed of water, which accounts for 70% of its volume, inorganic salts and large organic molecules, such as proteins, carbohydrates and fats. Each cell is surrounded by a membrane, called the *cell membrane*. Within the cell is a dense rounded mass, called the *nucleus*, and this is surrounded by a membrane called the *nuclear membrane*. The protoplasm of the cell body, excluding the nucleus, is called the *cytoplasm* and the protoplasm of the nucleus is called the *nucleoplasm*.

The use of the electron microscope has revealed considerable detail of cell structure and a number of smaller structures, or organelles, have now been identified in both the cytoplasm and the nucleus.

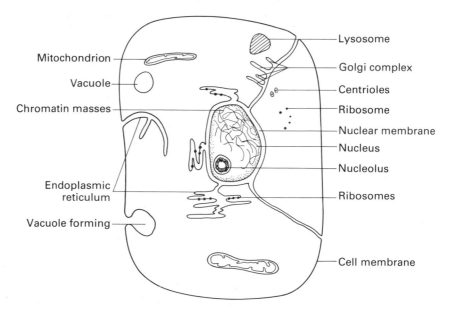

Fig. 1.1. A typical animal cell.

The cell membrane is a thin, flexible triple-layered membrane. The outer layer is composed of protein, with some carbohydrate present, the middle layer is composed of lipids and the inner layer of protein. This membrane is not only semipermeable (i.e. small molecules in solution can pass through the membrane into, or out of, the cell) but it can also be selectively permeable, only allowing entry to selected small molecules. Larger molecules can enter the cell by progressively indenting the cell membrane until it finally rounds off into the cell substance forming a vacuole (Fig. 1.2). Larger molecules can leave the cell by the reverse process.

Fig. 1.2. Vacuole formation: (a) a large molecule approaches the cell membrane; (b) and (c) the cell membrane is progressively indented; and (d) the cell membrane closes over the molecule which remains lying in a vacuole within the cytoplasm.

The endoplasmic reticulum has been demonstrated by electron microscopy. The complicated series of internal membranes form a series of fine channels throughout the cytoplasm, communicating with the cell membrane at the periphery and with the nuclear membrane at minute openings called the nuclear pores. This system of channels divides the cytoplasm into two compartments, that lying within the channels and that lying outside. The cytoplasm lying within the endoplasmic reticulum contains secretory products and that lying outside is a colloid suspension of proteins, carbohydrates, enzymes, ribosomes and ribonucleic acid (RNA). This ribonucleic acid regulates the synthesis of protein by the cell and the ribosomes are the site of protein synthesis.

The Golgi complex is a collection of fine membranous tubules and it usually lies close to the nucleus. This structure is especially prominent in secretory cells and it is thought to be concerned with the final elaboration of cell secretions and with the transport of cell secretions to the exterior of the cell. These protein secretions lie within the cytoplasm in membranous sacs called secretion granules. The Golgi complex is also responsible for the production of lysosomes.

The lysosomes are minute membrane-surrounded spheres which contain enzymes. They are capable of digesting ingested food particles and they act as the digestive system for the cell. The lysosomes of certain cells, such as the white cells of the blood (p. 26), are capable of digesting ingested bacteria and thus they have an important role to play in the body's defences against infection.

The mitochondria are elongated double-walled sacs. They are composed of a smooth outer membrane and a folded inner membrane. They are concerned with the release of energy by the combustion of organic foodstuffs with oxygen. They are particularly numerous in those cells which have high energy requirements, such as muscle cells.

The centrosome is a dense area of cytoplasm which contains two cylindrical bodies called centrioles. The centrosome lies near the cell nucleus. It is concerned with cell division and with the movement of the chromosomes during cell division.

The nucleus is surrounded by a double-layered membrane. The outer part is part of the endoplasmic reticulum and the inner part is perforated by small nuclear pores by which the nucleus communicates with the cytoplasm.

The chromosomes make up most of the mass of the nucleus, and, in periods between cell division, are visible only as clumps called the *chromatin masses*. During cell division, however, these chromatin masses become visible as individual threads. There are 46 chromosomes present in all human cells apart from the ova and the spermatozoa (p. 7). The chromosomes are composed of a protein support and a very large, coiled molecule of deoxyribonucleic acid (DNA). The chromosomes carry the genes which are ultimately responsible for heredity. Each gene is a portion of the large coiled molecule of DNA and it carries the particular characteristics of the parent cell to the two daughter cells during the process of cell division.

The nucleolus is a dense rounded structure found within the nucleus. It is mainly composed of ribonucleic acid and it is thought to produce the ribosomes which then pass to the cytoplasm and control the process of protein synthesis by the cell.

The cell cycle

In the human body, cells grow and divide at widely differing rates. Some cells, for example the neurons (nerve cells), actually lose the ability to divide altogether. Thus the rate of synthesis of new protein and new DNA occurs at different rates in different groups of cells. Even within a particular type of cell there are periods of activity and inactivity. These periods of cyclical activity are called the cell cycle. The period between cell divisions is called interphase, and interphase is divided into a first gap period called G_1, a period of synthesis called S-phase, and a second gap period called G_2. Protein synthesis occurs throughout interphase but DNA synthesis only occurs in the S-phase. The length of the G_1-phase varies with the type of tissue, whereas the length of the S- and G_2-phases are relatively constant irrespective of the type of tissue to which the cell belongs. Thus in epithelial, or surface, tissues where there is a lot of wear and tear, G_1 may only last a few hours but in other tissues it may last for some years. The S-phase usually lasts for 6 or 7 hours and the second gap phase, G_2, usually lasts 5 hours. The synthesis of new DNA, therefore, which occurs in the S-phase, is a fairly fixed prelude to the occurrence of cell division. Cell division or mitosis usually takes about 1 hour and the cell then re-enters interphase.

Mitosis, or cell division, is divided, for descriptive purposes, into four phases and it involves two distinct processes, the division of the nucleus and the chromosomes, and the division of the cytoplasm. During the S-phase of interphase new DNA is synthesised so that as mitosis commences the nucleus contains double the amount of DNA although the number of chromosomes is still only 46.

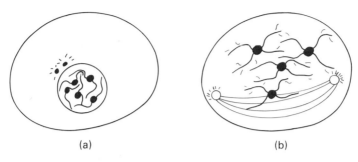

Fig. 1.3. (a) Early prophase, (b) late prophase.

Prophase is the first stage in cell division. During prophase the chromosomes separate from the chromatin masses and become visible as individual threads. They are seen as pairs of chromatids which are joined at a point called the *centromere*. Outside the nucleus, the centrioles start to move apart to opposite sides of the cell, but they remain attached to each other by minute thread-like microtubules called the *central spindle*. The nucleolus then disappears and the nuclear membrane disintegrates to release the chromosomes.

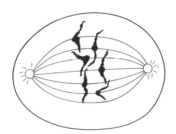

Fig. 1.4. Metaphase.

Metaphase is the second phase of cell division. With the disappearance of the nucleus the spindle moves into the centre of the cell. The chromosomes, which become shorter and thicker, then approach the equator of the spindle and become attached to the spindle by their centromeres.

Anaphase is the third phase of cell division. The chromosomes divide into two daughter chromosomes and these begin to move apart, along the spindle, to opposite poles of the cell. Thus there are now 46 chromosomes in each pole of the cell. During anaphase a slight constriction appears in the cell membrane, around the waist of the cell, heralding the start of cytoplasmic division.

Telophase is the final phase of cell division. The chromosomes again lengthen and start to merge into chromatin masses. The nuclear membrane reforms and the nucleolus reappears. The mitochondria and the organelles

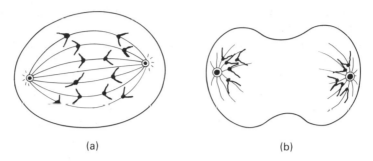

Fig. 1.5. (a) Early anaphase, (b) late anaphase.

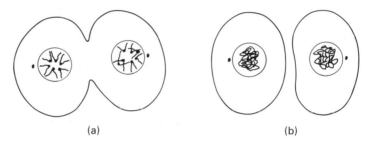

Fig. 1.6. (a) Early telophase, (b) late telophase.

are evenly divided into either side of the cytoplasm as the central constriction of the cytoplasm deepens. The spindle disintegrates and the central constriction finally divides the cell into two daughter cells.

Cells which are undergoing division are particularly sensitive to the effects of ionizing radiation. The effects of radiation on normal cells and tissues are discussed in Chapter 3.

The reproduction of the individual

Reproduction of a single cell takes place by this process of mitosis. Reproduction of the individual, however, involves more complex changes. The whole of the human body is derived from a single cell, called the *zygote*, which is formed by the fusion of two germ cells, the ovum from the female and the spermatozoon from the male.

The development of the germ cells

The female germ cell, or *ovum*, is developed in the ovary, and the male germ cell, or *spermatozoon*, is developed in the testis. The development of the germ cells involves a complicated type of cell division called *meiosis*. During meiosis the chromosomes of the parent cell in the ovary or testis are divided between the daughter germ cells so that each germ cell contains only half the number of chromosomes which are present in the parent cell.

There are normally 23 pairs of chromosomes (46 in total) in a typical human cell. This is called the *diploid* number. After meiosis the germ cells contain 23 single chromosomes and this is called the *haploid* number.

In the typical female cell there are 23 identical pairs of chromosomes, and one of these pairs is called the sex, or X chromosomes. In a typical female cell there are therefore 22 pairs of chromosomes and two X chromosomes. In a typical male cell, however, one of the sex chromosomes is unequal and is called the Y chromosome. Thus in the typical male cell there are 22 pairs of chromosomes plus one X chromosome and one Y chromosome. After meiosis all female germ cells contain one X chromosome, but the male germ cells contain either one X chromosome or one Y chromosome. When a male germ cell fuses with a female germ cell to produce the zygote, the chromosomes in the germ cells combine in the nucleus of the zygote to restore the number of chromosomes to 46. If a male germ cell containing an X chromosome fuses with the ovum, the child will be female (X,X), but if the male germ cell contains a Y chromosome, then the child will be male (X,Y).

Following fusion of the two germ cells the zygote has, therefore, derived half its chromosomes from the mother and half from the father.

The cells of the mother's ovary, however, contained 46 chromosomes of which 23 were derived from her mother and 23 from her father. The cells of the testis of the father similarly contain 46 chromosomes of which 23 were derived from each of his parents. During meiosis, however, the chromosome pairs become very closely applied to each other and exchange segments or genes. There is therefore a random redistribution of genetic material between the chromosomes of the parent cells before they become reduced to the haploid number (23) in the germ cells. When the zygote is formed by the fusion of the ovum and spermatozoon the chromosomes are therefore not only derived half from the mother and half from the father, but they also contain a random mix of genes from the maternal and paternal grandparents. It is this process of exchange of genes which is important in genetic variation within families.

The formation of the embryo

An ovum is released from the ovary of the female at approximately the middle of each menstrual cycle. This ovum passes into the uterine tube and moves slowly towards the uterus. If intercourse occurs within about 24 hours of ovulation, a spermatozoon may fuse with the ovum. This usually

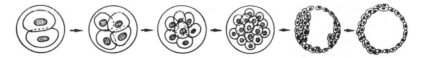

Fig. 1.7. The development of the blastocyst.

occurs in the lateral part of the uterine tube. This process of fusion is called fertilization, and further development of the ovum can only occur if this process takes place. Once fertilization has occurred, the new cell, or zygote, starts a rapid series of mitotic divisions to form a berry-like mass of cells called the *morula*. While these divisions are taking place, the zygote is gradually passing along the uterine tube towards the uterus which it reaches about 72 hours after fertilization.

Fluid soon starts to accumulate between the cells of the morula until a cyst-like structure, called the *blastocyst*, is formed (Fig. 1.7). One pole of this cyst, however, is thickened by a mass of cells which protrude into the cavity of the cyst. This is the *formative mass* which develops into the embryo. The walls of the cyst make up the *trophoblast* from which the membranes that protect, and supply nourishment to, the embryo, are developed.

At about the sixth day after fertilization the cells of the trophoblast adhere to the uterine lining, usually on the posterior wall above the middle of the uterus. The trophoblast then gradually burrows into the uterine mucosa.

The cells of the formative mass become differentiated into three layers, called the germ layers, and each of these layers gives rise to specific tissues.

The *ectoderm*, or outer layer, forms:
1   the skin, hair, nails and the lining of all the glands that open onto the surface of the skin;
2   practically the whole of the nervous system;
3   the epithelium lining the nose, the sinuses, the roof of the mouth, the gums and the cheeks, and the lower part of the anal canal.

The *endoderm*, or inner layer, forms:
1   the epithelial lining of the alimentary tract;
2   the epithelial lining of all the glands which open into the alimentary tract, apart from the salivary glands, but including the liver and the pancreas;
3   the epithelial lining of the respiratory tract and the air sacs of the lungs;
4   the epithelial lining of the thyroid, the parathyroid and the thymus glands;
5   the epithelial lining of the urinary bladder and most of the urethra.

The *mesoderm*, or middle layer, forms the remaining organs and tissues and this includes:

1   all the connective and skeletal tissues;
2   the skeletal muscles and the muscles of the viscera;
3   the blood, the blood vessels and the lymphatic vessels;
4   most of the urinary system with the exception of the lining of most of the urinary bladder, the prostate and the urethra.

During the third week after fertilization, the differentiation of these primary germ layers into tissues and organs starts, and, by the end of the eighth week the embryo is of recognizable human form. From this time the developing embryo is called a *fetus*. In the subsequent weeks, growth and maturation continues and birth, or parturition, usually occurs 38 weeks after fertilization.

The duration of pregnancy is usually dated from the start of the previous menstruation, and since fertilization usually occurs 14 days after this, a full pregnancy is said to last 40 weeks.

The formation of the placenta

The ovum is a large cell and it possesses a large quantity of cytoplasm in relation to the size of its nucleus. The cytoplasm contains a store of nourishment which is sufficient for the zygote during the early stages of its development. While the trophoblast is embedding in the uterine wall, nourishment is obtained from the uterine glands and from that part of the uterine mucosa which is destroyed as embedding proceeds. Subsequently nourishment is absorbed from the maternal tissues and blood. This source of nourishment becomes insufficient as growth of the embryo proceeds, and a structure called the *placenta* develops to provide the required supply (Fig. 1.8).

At the beginning of the process of embedding, the portion of the trophoblast which lies over the formative mass first adheres to the uterine mucosa. This area of the trophoblast becomes thickened, and spaces develop within it. The trophoblast invades and digests maternal tissues and blood vessels as embedding proceeds, and the spaces thus formed become filled with maternal blood. Ultimately the blastocyst becomes completely enveloped in the uterine mucosa, and the thickening of the trophoblast extends from the area above the formative mass to involve its entire surface. The spaces within the trophoblast gradually extend until a sponge-like network is formed. The portions of trophoblast between the spaces are called villi.

While these changes are occurring, a cavity called the *amniotic cavity*,

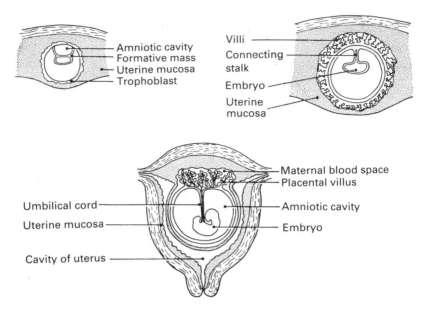

Fig 1.8. The development of the placenta.

develops between the formative mass and the trophoblast. Fluid accumulates in the amniotic cavity so that it gradually expands to surround the formative mass, which only remains connected to the trophoblast by a stalk of mesoderm called the connecting stalk.

Mesoderm from the connecting stalk then extends into the villi of the trophoblast and this mesoderm becomes differentiated into blood vessels as the embryonic circulation develops. These blood vessels are connected to the embryonic circulation by two arteries and one vein, which pass through the connecting stalk which is now called the *umbilical cord*. As growth proceeds, the villi around the attachment of the umbilical cord develop to form the placenta and the villi over the rest of the surface of the trophoblast disappear. The fetal and maternal blood are thus brought into close proximity by the placenta but they are separated by a layer of endothelial cells. However, maternal food substances and oxygen can pass to the fetal blood, and fetal waste products can pass into the maternal circulation.

The mature placenta is a disc-like structure about 228 mm in diameter and it is connected to the fetus by the umbilical cord which contains two umbilical arteries and one umbilical vein.

CHAPTER 2

The tissues

During development of the embryo, groups of cells become differentiated into tissues and the tissues in turn form organs. Groups of organs form systems, e.g. the digestive system and the circulatory system. There are, however, five basic tissues which are found in varying proportions in the organs of the body: *epithelial tissue, connective tissue, sclerous (skeletal) tissue, muscular tissue* and *nervous tissue*.

Epithelial tissue

Epithelial tissue is a lining tissue which is found in different forms on the external surface of the body and lining the cavities of the body. It is derived from all three of the embryonic germ layers (p. 8). Most *glands* are also of epithelial origin since they are usually formed as diverticula, or 'in-growths', of a lining epithelium.

Epithelial tissue functions as a protective layer, as a secretory or absorptive layer and in some cases as a sensory layer. It is composed of single or multiple layers of cells which have between them an intercellular ground substance, or *matrix*, which is composed of glycoproteins. The cells rest on a *basement membrane* which gives additional support to the tissue.

Since it is a covering tissue it is frequently subject to considerable wear and tear, which damages the surface cells, and it therefore has marked ability to regenerate and replace damaged cells.

Epithelial tissue may be divided into two main types: (a) *unilayered (simple) epithelium* which is composed of a single layer of cells resting on a basement membrane, and (b) *multilayered (compound) epithelium* which is composed of several layers of cells.

Unilayered (simple) epithelium

There are six types of simple epithelium:
1 **Squamous epithelium** is composed of a single layer of flat cells which fit together rather like paving stones. The cytoplasm is frequently so thin that the nuclei actually bulge onto the surface of the tissue. It is found lining: (a) the alveoli of the lungs, (b) parts of the renal tubules, (c) the serous cavities of the body (the pleura, the pericardium and the peritoneum)

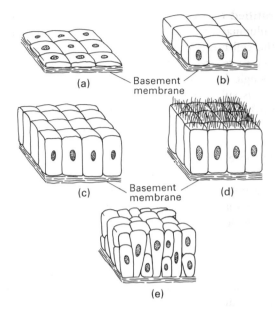

Fig. 2.1. Unilayered (simple) epithelium: (a) squamous; (b) cuboidal; (c) columnar; (d) ciliated columnar; and (e) pseudostratified.

where it is called *mesothelium*, and (d) the heart, the blood vessels and the lymphatic vessels, where it is called *endothelium*.

2   Cuboidal epithelium is composed of cells which are polygonal when viewed from above and are square when viewed from the side. Cuboidal cells are frequently seen in glands where they are sometimes able to change into columnar cells at times of high secretory activity.

3   Columnar epithelium is composed of cylindrical cells which are polygonal when viewed from above. The surface of the cells frequently possesses fine processes called *microvilli*. These are found in the small intestine, where they serve to greatly increase the surface area available for the absorption of foodstuffs, in the gall bladder, and in the proximal and distal convoluted tubules of the kidney, where they are called the brush-border.

In the respiratory tract, the uterine tubes and parts of the testis another type of process, called a *cilium* (plural *cilia*), is seen arising from the surface of the columnar epithelium. These cilia are capable of a sweeping motion which, in the respiratory tract, serves to move particles of dust towards the larynx. In the uterine tube they help to move the ovum towards the uterus.

Some columnar epithelial cells are able to produce mucus and are called *goblet cells*. They are frequently found in the lining of the respiratory and intestinal tracts. The mucus is produced within the cell body and is released from its surface giving the cell a goblet appearance.

4 Pseudostratified epithelium is composed of a single layer of some-what twisted columnar epithelial cells in which the nuclei lie at different levels within the cell body. When seen in a vertically cut section the cells can appear to lie in several layers but close inspection shows that there is indeed only a single layer. Columnar epithelium may itself assume these appearances if it is subjected to lateral compression.

5 Sensory epithelium is composed of cells which are capable of initiating or transmitting nervous impulses. It is found in the olfactory area of the nose, the taste buds of the tongue, and the inner ear. These epithelia will be described in more detail in Chapter 19.

6 Myoepitheliocytes are epithelial cells which are capable of contraction in the same way as muscle cells. They are present in the secretory areas and ducts of some glands where their action forces secretions along the ducts of the glands towards the exterior.

Multilayered (compound) epithelium

Multilayered epithelium is found in sites where there is surface friction. The superficial cells are constantly damaged or displaced by surface erosion but are continuously replaced by cell division in the deeper layers. There are two main types of multilayered epithelium:

1 Stratified squamous epithelium is composed of layers of epithelial cells which vary considerably in shape from the deep to the superficial layers. The deep cells, which are attached to the basment membrane, are columnar in shape. The more superficial layers of cells become increasingly flattened until the most superficial layers consist merely of flattened scales, called *squames*. The deep layers of cells undergo constant cell division and, as they move towards the surface, they change their structure until, on the surface, they are finally worn away by friction. There are two subtypes of stratified squamous epithelium, *keratinizing epithelium*, which is found on dry surfaces, and *non-keratinizing epithelium* which is found where there is surface moisture. In keratinizing epithelium, the cells, as they become more superficial, lose their nuclei and become filled with the protein, keratin, which provides a waterproof surface. In non-keratinizing epithelium the superficial cells retain their nuclei even though they are completely flattened and keratin is not laid down.

Keratinizing stratified squamous epithelium is is found in the skin (Chapter 19, p. 486) and in part of the mouth.

Non-keratinizing stratified squamous epithelium is found in the mucous membrane of the mouth, the lower part of the pharynx, the oesophagus, the vagina and part of the cervix uteri.

2 Urothelium (transitional epithelium) forms the lining of most of the urinary tract. It lines the collecting ducts of the kidneys, the calyces of

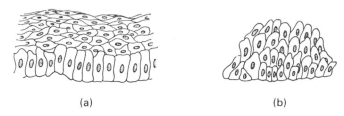

(a)   (b)

Fig. 2.2. Mulitlayered epithelium; (a) squamous; and (b) transitional (urothelium).

the kidney, the pelvis of the ureter, the ureter, the urinary bladder and the urethra. It is composed of rounded cells arranged in layers which are five or six cells thick. The cells are tightly bound together at anchorage points called desmosomes and, when the part of the urinary tract which they line is distended, the cells become flattened but still remain closely attached to each other. Because there is little surface friction, cell division in the basal layers is slow, but if the surface is damaged, the basal cells are capable of rapid regeneration of the superficial layers.

Glands

Glands are composed of epithelial cells which produce secretions. There are two main types of glands found in the human body.

1   Exocrine glands lie near the epithelial surface from which they are derived and either discharge their secretions directly onto that surface, or discharge their secretions into a fine channel, or duct, which opens onto that epithelial surface. Exocrine glands may be either unicellular or multicellular. Unicellular glands, such as the goblet cells, are single celled glands which are incorporated into an epithelial surface and discharge their secretions directly onto that surface. Multicellular glands are composed of epithelial cells which are depressed below the epithelial surface from which they are derived and they discharge their secretions onto the surface via a duct. If the duct is single and unbranched, the gland is called a simple gland, and if the duct divides into branches it is called a compound gland. Exocrine glands are further defined by the shape of the secretory portion. If the secretory portion is tubular it is called a tubular gland, if it is flask-shaped it is an alveolar gland and if it is composed of a rounded sac it is an acinar gland (Fig. 2.3). Exocrine glands usually have an abundant blood supply and the epithelial cells are usually supported by strands of connective tissue called septa. The salivary and mammary glands contain myoepitheliocytes which contract in response to stimulation by nervous impulses and thus expel their secretions.

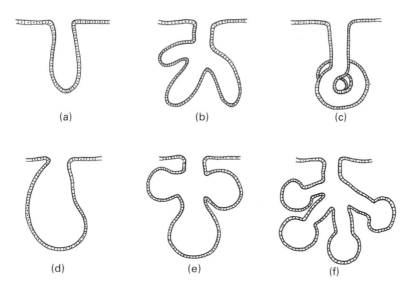

Fig. 2.3. Exocrine glands: (a) simple tubular; (b) compound tubular; (c) coiled tubular; (d) simple alveolar; (e) compound alveolar; and (f) compound acinar.

2   Endocrine glands do not have a duct and were originally referred to as ductless glands. They are usually composed of groups of cells which lie close to networks of blood vessels and their secretions are released directly into these blood vessels. These secretions are therefore able to exert an influence on cells in parts of the body which are remote from the glands' location. Endocrine glands are discussed in detail in Chapter 17.

Connective tissue

Connective tissue is, as its name suggests, a binding or supporting tissue. It is primarily developed from the mesoderm of the embryo. The cells of connective tissue are considerably more widely spaced than those of epithelial tissue and there is a corresponding increase in the intercellular ground substance, or matrix, which is predominantly secreted by the cells themselves. The matrix frequently contains fibres which determine the predominant characteristics of the tissue. Connective tissue can be divided into ordinary connective tissue, and specialized connective tissue, the former being found throughout the body, and the latter in bone and cartilage.

Connective tissues have both a structural function and a more specific defensive function with some of the cells being involved in the body's defences against infection and physical damage.

The cells of connective tissue

There are six types of cells which are usually present in connective tissue: *fibroblasts, fat cells, pigment cells, macrophages, lymphocytes* and *plasma cells*, and *mast cells*. Cells which are normally only present in the blood, such as eosinophil and neutrophil leucocytes, may also be present.

1 Fibroblasts are large irregularly shaped cells with branching processes. They form the white collagen fibres, which are usually present in the tissue and to which they usually adhere, and they also form reticulin fibres and yellow elastic fibres. They are particularly active in forming new fibres during the process of wound repair.

2 Fat cells are found in most connective tissues but are particularly numerous in adipose tissue. They consist of a thin circular rim of cytoplasm, which contains the nucleus, and the centre of the cell is occupied by a large globule of fat. During starvation this fat globule may disappear to leave the cell with a rather stellate (star-shaped) appearance.

3 Pigment cells, or chromatophores, are found in the connective tissue of the skin, and in the choroid and retina of the eye. They are star-shaped cells which contain numerous black granules called melanin granules.

4 Macrophages, or histiocytes, are found in large numbers in connective tissue. They may be seen attached to connective tissue fibres, or found moving freely within the tissue. When attached to fibres they are rather irregular in shape, but the freely moving cells are rounded. Macrophages have the power of phagocytosis, that is to say they can take up particles of foreign material, dead cells or bacteria into their cytoplasm. They form part of the macrophage system (p. 336) and they are therefore important in the body's defences against infection.

5 Lymphocytes and **plasma cells** are only found in connective tissues in significant numbers in areas of infection or injury. Lymphocytes are small cells (5−6 µm in diameter) and they are capable of producing antibodies (p. 45). Lymphocytes can enlarge to produce plasma cells which can also produce antibodies which they either store within their cytoplasm or which they release into the circulation.

6 Mast cells are round or oval cells which contain numerous granules in their cytoplasm. They are found particularly in loose connective tissue and they are concerned with the response of the body to inflammation (p. 36).

The matrix of connective tissue

The matrix of connective tissue is composed of the intercellular ground substance and three types of connective tissue fibres.

1 The ground substance is a rather viscous gel containing water, complex carbohydrates, and protein molecules. It is predominantly secreted by the fibroblasts.

2 Collagen fibres, or white connective tissue fibres, are the most numerous fibres present in ordinary connective tissue. They are elongated, straight, or slightly curved fibres which give tensile strength to the tissue. They frequently have fibroblasts attached to their surface.

3 Elastin fibres, or yellow elastic fibres, are yellowish in colour. They are usually thinner and shorter than white collagen fibres and they are frequently branched. They have excellent properties of elastic recoil.

4 Reticulin fibres are fine branching fibres which are found forming the supporting tissue of glands, the kidneys, lymph nodes and the spleen. They are also present in basement membranes.

Types of connective tissue

Connective tissue may be divided into *ordinary connective tissue* and *specialized connective tissue*. The former can be further divided into *irregular connective tissue*, and *regular connective tissue*, whilst the latter makes up *cartilage* and *bone*.

Ordinary connective tissue

1 Irregular connective tissue is of three types:

a) *Loose connective tissue* (areolar tissue) is the most widespread form of connective tissue. It is composed of a loose network of white collagen fibres and elastin fibres, embedded in a soft matrix which contains widespread cells. It is a binding tissue which has considerable elasticity that allows relative freedom of movement to the structures which it surrounds. It is found: (a) under the skin in areas where there is no subcutaneous fat, (b) surrounding muscles, blood and lymphatic vessels and nerves, (c) forming the submucous coat in the alimentary tract, and (d) binding together the component parts of various organs and glands.

b) *Dense irregular connective tissue* is composed of thick bundles of collagen fibres, some fibroblasts and relatively little ground substance. It has considerable tensile strength and is found: (a) forming the connective sheaths of muscles, (b) in the dermis of the skin, (c) forming the capsule of various organs, (d) forming the periosteum of bone, and (e) forming the sclera of the eye.

c) *Adipose* (*fatty tissue*) is predominantly composed of fat cells surrounded by vascular connective tissue and fibrous septa which divide the tissue into lobules. The fat cells are round or, when very closely packed, polygonal. In some regions, such as the soles of the feet, the palms of the hand and around the kidneys, adipose tissue has a mechanical function, whereas in other regions it simply provides a fat store. The distribution of adipose tissue seems to be genetically predetermined and the differences in subcutaneous fat distribution in the male and female are characteristic. If food

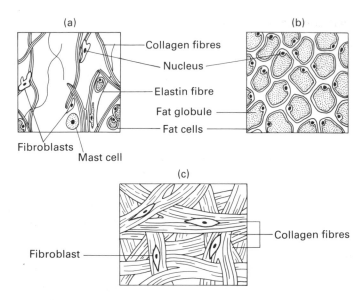

Fig. 2.4. Irregular connective tissue: (a) loose irregular (areolar) connective tissue; (b) adipose tissue; and (c) dense irregular connective tissue.

intake is reduced, for example during slimming, the fat stores are consumed to provide energy with a resultant loss in weight. The fat is initially lost from those areas in which it has no mechanical function but, with prolonged starvation, it is eventually lost from these areas as well.

2   Regular connective tissue is composed of regularly arranged sheets or bundles of connective tissue fibres which are so organized as to give the tissue a directional stability. Tendons, ligaments and fascial sheets are examples of regular connective tissue. In these tissues the fibres are almost exclusively white collagen fibres which, while to some extent interweaving, are nevertheless arranged in the direction of pull to which the tissue is subjected. Tendons, for example, are composed of interweaving bundles of collagen fibres which are arranged in the long axis of the direction of pull of the muscle to which they are attached. Bundles of white collagen fibres contain inactive fibroblasts between the fibres but active fibroblasts are found around the periphery of the bundle and these are thought to be able to form new fibres following injury.

In some sites, such as the vocal cords and the ligamenta flava of the spine (p. 199), the fibres which are present are predominantly elastin fibres. However even in ligaments and fasciae which are almost exclusively composed of collagen fibres, there are usually some elastin fibres present.

Specialized connective tissue

There are two specialized connective tissues, cartilage and bone. They differ from other connective tissues in that they possess a solid intercellular matrix. The structure of bone will be described in Chapter 5 (p. 84), but the structure of cartilage will be described here.

Cartilage

Cartilage is a firm tissue composed of a solid matrix in which there are a series of spaces, or *lacunae*, which contain groups of cartilage cells called chondrocytes. The firm matrix may contain connective tissue fibres and it is thus classified into three types: *hyaline cartilage*, which contains few fibres, *white fibrocartilage*, which contains numerous collagenous fibres and *yellow (elastic) fibrocartilage*, which contains yellow elastic fibres.

1   Hyaline cartilage is a firm, bluish, translucent tissue which has considerable elasticity. It is composed of a solid, apparently structureless, matrix containing lacunae in which the chondrocytes lie in groups of two or more. The chondrocytes secrete the matrix, which is composed of complex polysaccharides, proteins and water. There are relatively few blood vessels and these are usually found only on the surface of the cartilage. Metabolites actually pass through the matrix to reach the chondrocytes.

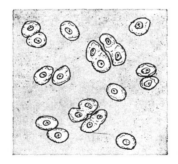

Fig. 2.5. Hyaline cartilage.

Hyaline cartilage is found: (a) forming the framework of the larynx, the trachea and the bronchi, (b) covering the articular surfaces of bones in synovial joints (p. 192), (c) as the costal cartilages which unite the ribs to the sternum, and (d) as a precursor of bone during the development of most of the skeleton.

2   White fibrocartilage is composed of bundles of white fibrous tissue and fibroblasts embedded in a cartilaginous matrix which contains scattered groups of chondrocytes. It is a dense whitish tissue which, whilst very tough, shows considerable elasticity and is able to resist both pressure and friction. It is found: (a) between the bodies of the vertebrae as the intervertebral discs, (b) deepening the sockets of the shoulder and hip joints, and (c) as articular discs between the articular surfaces of bones in certain synovial joints.

3   **Yellow (elastic) fibrocartilage** is composed of a network of yellow elastic fibres embedded in a firm cartilaginous matrix which contains normal chondrocytes. It is found: (a) forming the pinna of the ear (p. 497), (b) forming the epiglottis, and (c) forming the corniculate cartilages and the tips of the arytenoid cartilages in the larynx.

Muscular tissue

Muscular tissue is composed of muscle cells which are cells that have the ability to contract in an organized fashion. It may be divided into three types: *skeletal muscle*, which is capable of fast organized contraction, and is usually associated with the skeleton and voluntary movement; *cardiac muscle* which is found only in the heart, and *smooth muscle* which is found in the walls of the alimentary tract, the blood vessels and the urinary and genital tracts.

1   **Skeletal muscle** is composed of bundles of elongated muscle cells, or fibres, which vary in length from a few millimeters in small muscles, to as much as 30 cm in muscles of the limbs. Each bundle of muscle fibres is surrounded by loose connective tissue, which contains the blood vessels that supply the cells, and the whole muscle is surrounded by a sheath of dense connective tissue. Because it is usually under the control of conscious will, skeletal muscle is often referred to as voluntary muscle.

The muscle cells are circular, or polygonal, in cross-section and they have a diameter of 10–100 μm, the diameter varying with the individual muscle of which the cells are part. Each cell is surrounded by a cell membrane called the *sarcolemma*, beneath which lie the nuclei of the cell. There may be as many as several hundred nuclei in a single muscle cell. The cytoplasm of the cells has been shown to be composed of individual minute threads, or *myofibrils*, which are about 1 μm in diameter. The myofibrils possess regular, transverse dark and light bands, or striations, and skeletal muscle is, therefore, often called striped muscle. Projections of the cell membrane, or sarcolemma, cross the middle of the light bands and are seen as thin dark lines. These divide each myofibril into units which are about 2.5 μm in length and each of these units is called a *sarcomere*. The light and dark striations are intimately connected with the ability of the cell to contract. It is now known that each myofibril is itself composed of thin actin myofilaments and thicker myosin myofilaments which run in a longitudinal manner. The light bands, or stripes, are composed only of actin myofilaments and the dark bands are composed of actin and myosin filaments which are overlapping. Contraction of the cell occurs when the myosin myofilaments draw the actin myofilaments deeper between them, thus decreasing the overall length of the cell.

Each skeletal muscle has one or more nerves supplying it. These nerves contain both motor and sensory fibres. Within the main muscle the motor fibres divide into a number of branches each of which terminates on a single muscle fibre at a structure called the *motor end plate*. The arrival of a nervous impulse at the motor end plate causes contraction of the muscle fibre. Each individual nerve fibre may have between 10 and 2000 motor end plates at its terminations. In muscles where fine control is important, such as the muscles of the eye, each nerve fibre supplies relatively few muscle fibres, but where strong contraction is more important than fine control, then large numbers of muscle fibres are supplied by each nerve fibre.

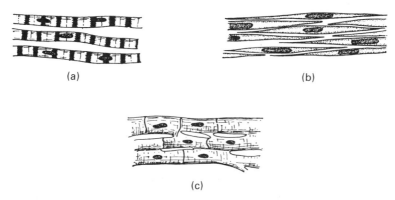

(a)   (b)

(c)

Fig. 2.6. Muscular tissue: (a) skeletal muscle; (b) smooth muscle; and (c) cardiac muscle.

2   **Cardiac muscle**, like skeletal muscle, is composed of striated cells but there are considerable functional and structural differences. Cardiac muscle is composed of tracts of cardiac muscle cells which are about 80 μm in length and 15 μm in diameter. Each cell contains a single nucleus which lies centrally. The ends of cardiac muscle cells frequently divide into several branches and these branches are closely applied to adjacent cells. Individual tracts, which are bound together by connective tissue, vary considerably in diameter and may have from only a few cells to several hundred cells across its diameter. The cells themselves are surrounded by loose connective tissue which contains reticulin fibres, blood vessels and nerves.

Cardiac muscle cells have transverse striations, but these are less marked than those of skeletal muscle. Like skeletal muscle cells, the cells of cardiac muscle are composed of overlapping myofilaments of actin and myosin, but the myofilaments are not grouped into myofibrils. Individual cardiac muscle cells are joined at structures called *intercalated discs*. Each cell has the power of rhythmic contraction and the structure of the intercalated disc allows this contraction to spread across the whole heart in a coordinated manner. The nerve supply to the heart simply modifies the rate of these contractions.

3   Smooth muscle is composed of spindle-shaped cells which vary in length and diameter according to the organ of which the muscle is a part. The cells, or *myocytes*, have a single, centrally placed nucleus which is elongated when the cell is relaxed, but becomes oval or circular during cell contraction. These spindle-shaped cells are arranged with their long axes parallel to the direction of contraction and are usually gathered into small bundles. Between the bundles there are loose connective tissue septa which contain blood vessels and nerves. The cytoplasm of smooth muscle cells does not have the striped appearance of skeletal and cardiac muscle, and smooth muscle is, therefore, often called unstriped muscle. The cytoplasm is nevertheless composed of obliquely arranged myosin myofilaments which are surrounded by actin myofilaments but the regular overlapping zones, which are seen in skeletal and cardiac muscle, are absent.

Smooth muscle cells have the power of spontaneous rhythmic contraction. The cells have a relatively poor nerve supply and the nervous impulses simply modify this rate of contraction. The smooth muscle, however, of the iris of the eye and the walls of the larger arteries, possesses a dense nerve plexus and in these sites the smooth muscle cells depend on the nervous supply to initiate contraction.

Nervous tissue

The structure of nervous tissue will be described in Chapter 16 (p. 415).

CHAPTER 3

General pathology

Whereas human anatomy is the study of the structure of the normal human body, and physiology is the study of its function, pathology is the study of the cause of disease and of the way in which the normal structure and function of the body are altered by it.

The human body is composed of a multiplicity of cells gathered together in their differing forms to make up tissues and organs each of which performs a specific function. In disease the structure and function of a number of these cells are altered and an understanding of these changes is essential to an understanding of the total effects of the disease. The study of pathology is, therefore, best begun by a consideration of those changes which occur in the cells and tissues in certain basic disease processes.

Diseases may be broadly classified into two main groups: *congenital* and *acquired diseases*.

Congenital disease

Congenital means literally 'born with', but although many of these diseases are evident at birth, in others the signs or symptoms may not become apparent until later life. The term congenital is often used to embrace two groups of diseases: those due to chromosomal or genetic disorders, and those acquired congenital diseases which have no inherited or familial basis but which are, for reasons known or unknown, acquired in the mother's womb.

Chromosomal and genetic disorders. Within the nucleus of each human cell are 46 chromosomes and, arranged along these chromosomes are the units of heredity, called genes. The chromosomes are arranged in pairs which are identified as pairs 1 to 23. As we have already seen, the entire human body is developed from a single cell, called the zygote, which is formed by the fusion of a male spermatozoon with a female ovum. The ovum and spermatozoon are produced by a process of cell division called meiosis, and during this process the number of chromosomes in the cell is halved from 46 to 23. When the ovum and spermatozoon fuse, the total number of chromosomes is restored to 46.

Certain diseases are caused by abnormalities which arise during the process of meiosis. Thus part of a chromosome may be absent from one of the daughter cells, or, alternatively, the daughter cell may possess one extra or one less chromosome. *Down's Syndrome* (*mongolism*) is characterized by the presence of an extra chromosome in the 21st pair. The disease is therefore sometimes called *trisomy* 21.

In genetic disorders one or more of the genes, inherited from one or both parents, carries the factor which causes the disease, and so passes it from generation to generation. The gene responsible may be dominant, so that it produces the disease even though a corresponding normal gene has been inherited from the other parent, or it may be recessive and the disease is not apparent unless both parents pass the abnormal gene to the child. Some abnormal recessive genes are linked to the X-chromosome and so may produce the disease only in males because there is no matching gene on the Y-chromosome. *Haemophilia* and a type of *muscular dystrophy* are caused by such sex-linked recessive genes.

Diseases, such as *diabetes mellitus* and certain cancers, are known to run in families, that is they have a familial tendency, but in these diseases the mode of inheritance is not fully understood.

Acquired congenital diseases. Certain congenital diseases are not related to chromosomal or genetic abnormalities, but appear to be acquired while the fetus is in the uterus. The cause of many of these diseases is still unknown but in others the disease can be related to viral infections or to the use of a particular drug by the mother during pregnancy. The damage to the fetus is usually most severe if this occurs during the first three months of pregnancy. Thus maternal infection by *rubella virus* (*german measles*) can lead to a number of congenital abnormalities (in particular to congenital heart disease), and the tragic effects of the drug *thalidomide* in producing severe congenital abnormalities, are by now well known. Other drugs have also been implicated as possible causes of congenital abnormalities and there are a number of drugs which are specifically listed as unsafe to be taken during pregnancy.

Exposure to radiation during pregnancy may also lead to congenital abnormalities and most X-ray departments will now only routinely X-ray females of child-bearing age during the 10 days following the onset of menstruation. Since ovulation does not normally occur for 14 days after the onset of menstruation it was hoped that this practice would avoid the risk of exposing the fetus to radiation during early pregnancy.

Acquired disease

Acquired diseases are those diseases which occur after birth due to some external influence; they may be caused by: *injury, infection, disorders of immunity, endocrine disorders, metabolic disorders,* and *circulatory disorders.*

Injury

Injury to the body may be caused by direct physical violence (trauma) as, for example, in a road traffic accident or a sporting injury, but it may also be caused by other physical agents such as heat, cold or radiation. Injury may also be caused by chemical agents, such as strong acids or alkalis, by a variety of chemical poisons, such as strychnine or cyanide, but in these days the commonest causes of chemical injury are probably drugs and alcohol.

The response of the body to injury involves a number of different mechanisms all of which combine in an attempt to limit the damage and to effect repair. Depending on the extent of the damage, these mechanisms may simply involve local changes at the site of damage or they may involve more general changes in the body as a whole.

Local effects of injury

Inflammation

The most basic local reaction which occurs in response to injury is inflammation. Acute inflammation is seen in its simplest form following a clean cut to the skin in which case the inflammatory response leads directly to repair. When the damage is more extensive, or complicated by infecting micro-organisms, the response is more complex and the outcome of the response may be considerably modified.

Inflammation is characterized by *redness, warmth, swelling* and *pain*. Depending on the site and severity of the injury, there is a variable loss of function in the part involved.

Following a clean cut there is dilatation of the small blood vessels in the damaged area, blood flow is increased and this produces the characteristic local warmth.

The blood flow through the damaged area is initially fast and there is exudation of white blood cells and protein from the capillaries into the damaged tissues. The capillaries and venules are not normally permeable to proteins but the increased pressure of blood in the adjacent arterioles increases the filtration rate of fluid from the capillaries. In addition, release of endogenous substances from the damaged tissue causes changes in the walls of the capillaries and venules which allows proteins to escape into the tissues. This increase in tissue fluid is responsible for the localized swelling. Stretching of the tissues by the increased fluid, and release of endogenous substances from the tissues, causes pain.

Although the blood flow is initially fast, it soon slows down and may almost cease. This is partly due to loss of fluid from the blood into the tissues, and partly to the blocking of capillaries by large numbers of white cells which stick to the walls of the capillaries prior to migrating into the damaged tissue. These white cells are mainly *polymorphonuclear leucocytes* (p. 266).

The plasma which passes into the damaged tissues contains a protein called *fibrinogen* which coagulates to form strands of fibrin in an attempt to wall off the damaged area. The white cells which migrate into the area are polymorphonuclear leucocytes, which are capable of independent amoeboid movement. They surround, and by phagocytosis, take up the dead cells and foreign matter which has entered the tissues at the site of damage, into their own cytoplasm. The polymorphonuclear leucocytes can also phagocytose invading micro-organisms if the injury is complicated by infection.

When the area of damage is limited and there is no complicating infection, the healing process starts immediately. Initially new capillaries grow into the damaged area and fibroblasts move out alongside the capillaries. The fibroblasts produce connective tissue fibres which bridge the damaged area. This combination of capillaries and fibrous tissue is called *granulation tissue*. At the same time that this granulation tissue is filling in the damaged area, the cells of the epidermis are gradually growing across the surface to close the gap and so complete the healing process.

When the area of damage is small there may be no residual sign of the injury. When the area of damage is larger the fibrous tissue which has filled the gap contracts, and a visible scar results.

General effects of injury

Shock

The main initial reaction to a moderate or severe injury is called shock. Shocked patients are cold, grey and sweaty. They have a rapid, weak pulse and a low blood pressure.

Shock is the name given to a series of changes which occur in the body as a result of impaired circulation which is secondary to a sudden severe fall in the cardiac output. The changes which occur in shock involve mechanisms which attempt to restore circulation to the tissues and, in particular, to those tissues which are vital to survival, namely the heart and the central nervous system.

Although shock frequently follows injury there are other causes which will be considered here as many of the mechanisms are similar.

Shock may thus be caused by: (a) a decrease in the blood volume (*surgical shock*), (b) acute heart failure (*cardiogenic shock*), and (c) severe infections (*infective shock syndrome*).

1   **Surgical shock** occurs when there is a reduction in blood volume due

to bleeding or loss of fluid due to extensive burns, or severe diarrhoea and vomiting. It is also called *hypovolaemic shock*. The loss of blood volume may be sudden, as in an *acute haemorrhage*, or more prolonged, following *severe burns* or *acute gastroenteritis*. A normal healthy adult can lose 10% of the circulating blood volume (about 500 ml) without any adverse reaction. With loss of 25% of the blood volume, however, it would take the body about 36 hours to restore the circulation and with loss of 50% of the blood volume, death would almost certainly result unless the volume of blood was restored by intravenous transfusion.

When the blood volume is reduced due to haemorrhage, loss of fluid from the surface due to burns or loss of fluid into the gastrointestinal tract with acute gastroenteritis, there is a reduction in the volume of blood returning to the right side of the heart via the venous system. The output of the heart therefore falls and there is a consequent drop in the blood pressure. The body responds to this by constriction of the veins and of the very small arteries (arterioles) in the tissues (vasoconstriction) and this effectively increases the volume of blood available to the heart. This vasoconstriction is due to the release of vasoconstrictive substances from the kidneys and the adrenal glands. The vasoconstriction produces a rise in the blood pressure, in spite of the loss of blood volume, but the circulation to the tissues remains impaired, so the signs of shock remain, namely a cold sweaty skin. The arteries to the heart and the central nervous system do not react to the vasoconstrictive substances, so that, although the tissue circulation generally is impaired, the circulation to the heart, brain and kidneys is effectively maintained.

If less than 25% of the blood volume has been lost, fluid from the tissues passes into the blood stream, via the capillaries, and the circulating blood volume is slowly restored. Any further loss of blood, however, would be critical and restoration of the blood volume by transfusion of either whole blood or intravenous fluid would become vital. Following an acute injury intravenous fluids are given initially in order to increase the volume of circulating blood but blood should be given as soon as the patient's blood group is known (see Chapter 11).

If the impaired circulation due to loss of blood volume is allowed to persist untreated, then, although all the tissues are affected, the impairment of circulation to the heart, the lungs, the kidneys or the brain, may lead to death of the patient.

2 **Cardiogenic shock** can be produced by an acute heart failure, which may follow a *myocardial infarct* (a heart attack). The damage to the cardiac muscle that occurs in this condition, produces a marked impairment of the cardiac output and a consequent fall in the level of the blood pressure. However, although the tissue circulation is impaired, there is no overall loss in the available blood volume, and the administration of intravenous

fluid may well simply overload the already impaired heart. Nevertheless the signs of shock, and the reaction to shock, in this condition are similar to those of shock due to loss of blood volume.

3   Infective shock syndrome occurs when bacteria or pus are present in the circulation. It is found in patients with septicaemia (infection in the blood stream) or severe localized infections, such as peritonitis or infected burns. The symptoms are frequently similar to those of surgical shock but they are more difficult to treat. The infecting organism which is reponsible for infective shock syndrome is commonly a Gram-negative bacterium (p. 35) and it is known that these bacteria, when they die, produce endotoxins which are thought to be responsible for the symptoms of infective shock syndrome.

Bone marrow response

When an injury is accompanied by significant haemorrhage the bone marrow responds by increased production of red blood cells. This occurs even if the haemorrhage is treated by blood transfusion.

Renal damage

A prolonged state of low blood pressure following an injury causes a fall in the output of urine by the kidney due to reduced renal filtration. This may lead to a complete failure of urine production (*anuria*) and it may lead to renal failure caused by damage to the kidney and this is called acute tubular necrosis. When there is extensive damage to muscular tissue, for example in crush injuries, the breakdown of muscle cells releases myoglobins, which particularly predispose to *acute tubular necrosis*.

Special types of injury

Burns

Burns may be caused by direct heat, by cold, by electricity, by chemicals or by radiation. Burns are usually classified into four types in order of severity:

1   *First degree burns* in which there is only localized erythema, or reddening, of the skin.

2   *Second degree burns* in which there is vesicle formation in the epidermis (blistering).

3   *Third degree burns* in which there is destruction of the epidermis and the upper layers of the dermis. This may lead to damage of the hair follicles and the sweat glands.

4   *Fourth degree burns* in which there is destruction of the whole thickness of the skin, that is, both the dermis and the epidermis. There may also be damage to underlying tissues, such as muscle or bone.

In the last three categories of burns there is coagulation of proteins and damage to the blood vessels of the skin with considerable loss of fluid from the surface of the involved area. This is accompanied by an inflammatory reaction subsequent to healing. If the epidermis is destroyed over a significant area then skin grafting is necessary, both to obtain adequate skin replacement and to avoid severe scarring.

Low temperature injuries

The damage caused by low temperatures depends on whether the injury is caused by temperatures just above or below freezing point.

Exposure of a part of the body to a continuous low temperature without freezing, results in damage to blood vessels and consequent tissue damage due to an impaired blood supply. The part of the body involved initially becomes white and blotchy but, on gentle warming may become red and swollen as the blood supply is restored. With prolonged exposure there may be thrombosis of the blood vessels and gangrene may follow. This type of injury, when it involves the feet, is called *trench foot*.

Exposure of part of the body to freezing temperatures may result in *frostbite*. In this type of injury actual crystals of ice form in the tissues. There is thrombosis of blood vessels and tissue death. Warming produces no improvement and gangrene occurs.

General exposure of the body to low temperatures with a consequent lowering of the body temperature is called *hypothermia*. There is depression of the vital centres in the brain and, if the exposure is prolonged, death may ensue.

Radiation injury

All types of radiation damage the body to a greater or lesser extent. The type of damage produced depends on two factors, the penetrative power, or energy, of the radiation and the total amount of radiation to which the body is exposed. Radiation of low penetrative power predominantly damages the more superficial tissues whereas highly penetrating radiation, such as gamma radiation or high voltage X-rays, damages the deeper tissues. When radiation passes through a cell, electrons may be displaced from the molecules which make up the constituent parts of the cell, i.e. ionization occurs. This ionization may have no visible effect on the cell, but, if one of the molecules of an important part of the cell is involved, then the function of the cell may be disrupted.

The most important cell function to be disrupted is the process of cell division, or mitosis. Depending on the dose of radiation to which a cell is exposed one of three outcomes may be involved: (a) the cell may be killed outright, (b) the cell may die at the next attempt at cell division and (c) the cell may lose the ability to divide but otherwise continue as before. The first and second effects are the basis of tumour treatment by radiotherapy, but the second effect is by far the most important.

All cells in the human body, whether normal or abnormal, can be damaged by radiation, but, since it is mitosis which is predominantly affected, it is those tissues in which the cells are actively dividing which are the first to show the effects of radiation. Thus the skin, the cells of the bone marrow which are producing new blood cells, the reproductive cells of the ovaries and testes, and the lining of the alimentary tract are the most sensitive to radiation. Bone, muscle, fibrous tissue and the cells of the nervous system are less sensitive.

The effects of radiation injury are divided into early and late effects.

1   *Early effects of radiation.* The early effects of radiation are those of a typical inflammatory response with an increase in the vascularity, tissue swelling and white cell infiltration. This may be followed by a normal healing process if the damage is limited. With larger doses of radiation there may be tissue necrosis and considerable scarring with the healing process.

2   *Late effects of radiation.* Radiation may lead to damage to blood vessels with impairment of the blood supply to the irradiated area. This impairment of the blood supply can cause tissue death as a delayed feature, with chronic ulceration and gross fibrosis. Radiation may also induce malignant tumours such as leukaemia, or tumours of the bone or skin. These may appear years after exposure to radiation. There is in addition a genetic risk with irradiation of the gonads and this could lead to mutations, or congenital abnormalities, in later generations.

The effects of radiation on specific tissues varies depending on the nature of the tissue involved.

1   **The skin**. The first noticeable change in the skin following exposure to radiation is reddening, or erythema. This may appear between a few days up to two weeks after exposure. This is a typical inflammation with subsequent healing and is seen in almost all patients who receive radiotherapy. Larger doses of radiation, however, may produce more damage and this can be divided into *acute* and *chronic changes*.

Acute changes may be divided, rather like burns, into four degrees of damage according to the dose of radiation received by the skin. With first degree damage there is erythema, and damage to the hair follicles with loss of hair. This usually occurs about 18 days after irradiation, but the hairs will subsequently grow again two or three months later. With second degree damage there is a bright red erythema, loss of hair due to hair follicle damage, and desquamation, or flaking, of the skin. The hair may not grow again and, following healing, there may be a permanent change in pigmentation of the skin. With third degree damage there may be blistering, or vesicle formation, and these blisters may burst to leave ulcerated areas which exude serum. There is damage to the sweat glands and permanent loss of hair. Healing can occur fairly rapidly, but damage to the hair follicles and the sweat glands is permanent. Fourth degree damage only occurs as a result of radiation overdoses. There is destruction of the epidermis, the dermis and sometimes the underlying tissues. Whole areas of skin may desquamate, and healing, if it occurs, is slow and often requires skin grafting.

Chronic changes following radiation to the skin include permanent changes in pigmentation, dilatation of capillaries in the skin, called telangectasia, late ulcer formation and the development of malignant tumours. Following radiation the skin may heal as a very thin atrophic area which is easily damaged, or there may be thickening of the skin due to gross fibrosis.

2 Haemopoetic tissue. The cells of bone marrow which produce blood cells and the cells of lymphoid tissue are particularly sensitive to radiation. Conventional radiotherapy, depending on the area of the body which is being treated, frequently leads to a fall in the total white cell count (*leucopenia*) and to a fall in the blood platelets (*thrombocytopenia*). This is due to damage of the parent cells in the bone marrow. Treatment is, therefore, always accompanied by regular white cell counts. The red blood cells, due to their longer life span, are less affected, but large doses of radiation may lead to cessation of all bone marrow activity (*aplastic anaemia*). Exposure to radiation may also, in the long term, cause *leukaemia*.

3 The testes and ovaries. In the female, irradiation of the ovaries may cause the cessation of ovulation and, with larger doses, the ovaries may be so damaged that hormonal activity also ceases. This cessation is often called an 'artificial menopause' and in the past this was sometimes done deliberately. In these days this only occurs as an incidental effect of radiation for a pelvic malignancy.

In the male, irradiation of the testes causes temporary or permanent cessation of spermatozoon production.

In both the male and the female, however, there is the likelihood of genetic damage and the possibility of congenital abnormalities in the next generation.

4 Mucous membranes. Mucous membranes are more sensitive to radiation than is the skin. The reactions of the two are otherwise very similar. Radiation to mucous membranes is followed by superficial ulceration, serous exudation and the cessation of mucous secretion. Ulceration may lead to bleeding and thus to death. The late effects of radiation to mucous membranes include fibrosis, and consequent obstruction of the bowel, adhesions and fistula formation.

5 The eye. Radiation to the eye may, in the immediate phase, cause conjunctival ulceration as the conjunctiva is more sensitive than the skin. The cornea can also be damaged and may become infected leading to corneal ulceration and consequently some loss of vision. Large doses of radiation may damage the lens which develops a *cataract*.

6 The kidney. Radiation to the kidney can cause an increase in the blood pressure due to damage to the renal arteries. It may also cause *nephritis*, and acute or chronic renal failure.

7 The brain and spinal cord. Although nervous tissue is relatively resistant to radiation, damage to the blood vessels supplying the brain and spinal cord can lead to death of nervous tissue. In the spinal cord the result may be a *paraplegia* (paralysis of the lower half of the body) or a *tetraplegia* (paralysis of all four limbs).

8 Bone and cartilage. Large doses of radiation to bone or cartilage may cause necrosis due to damage to the supplying blood vessels. In addition, malignant tumours (*sarcomas*) may develop some years after exposure to the radiation.

9 The lungs. The lungs can be damaged by radiation in much the same way as the skin or mucous membranes. Inflammatory reactions can occur and may be followed by *pulmonary fibrosis* (scarring following healing of the inlammation).

Diseases caused by infection

Infection can be defined as the presence in the body of a micro-organism at a site where it is not normally found and where its muliplying causes an infective disease. These micro-organisms may be bacteria, viruses, protozoa or fungi. Infection can also be caused by worms.

The transmission of infection

Infection can reach the body in five main ways:

1 *By direct contact.* Infection can be transmitted to the skin or mucous membranes by direct contact with another individual who is infected with the micro-organism. The venereal (sexually transmitted) diseases are transmitted in this way.

2 *By indirect contact.* Infections can be transmitted to the body by inanimate objects which have previously been contaminated by a person who is infected. Blankets, clothing, and unsterilized surgical instruments can transmit infection from one person to another.

3 *Through the air.* Infections of the upper respiratory tract are often transmitted in this way. During coughing and sneezing, micro-organisms are sprayed into the air in droplets, and with close contact these droplets may be inhaled and transmit the micro-organism. Other droplets, however, land on clothing or handkerchiefs and become dried. When the clothing or handkerchiefs are rubbed or shaken micro-organisms are emitted into the air on small particles of fluff or dust, and they may remain suspended in the air for long periods.

4 *By ingestion.* The swallowing of infected food or drink is a common way for organisms that infect the gastrointestinal tract to reach the body. Food may be contaminated by contact with an infected person or animal. Sometimes bacteria, which have contaminated food, may be killed by cooking and their toxin alone may be ingested. This may cause a short self-limiting type of food poisoning.

5 *By insect or animal bites.* Infection may be transmitted to man by other living organisms. Malaria is transmitted by mosquitoes and the rabies virus by the bite of an infected dog.

Bacteria

Bacteria are minute living organisms which are very widespread in nature. They are present in soil, water and in the air, and they also live in relationship with man and animals. Although a large number of bacteria are known, relatively few are harmful (or *pathogenic*) to man, and indeed many are beneficial.

Bacteria of medical importance are of the order of 1−5 μm in length (1 μm equals 1/1000th of a millimetre). They vary considerably in shape and are often named accordingly (Fig. 3.1). *Cocci* are spherical in shape; *bacilli* are rod-shaped; *vibrios* are curved rods; and *spirochaetes* are elongated spiral threads.

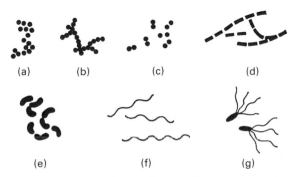

(a)   (b)   (c)   (d)

(e)   (f)   (g)

Fig. 3.1. Some types of bacteria: (a) staphylococci; (b) streptococci; (c) pneumococci; (d) bacilli; (e) vibrios; (f) spirochaetes; and (g) flagellated bacteria.

Some bacteria can propel themselves in a fluid medium by means of protoplasmic processes called *flagella* (singular flagellum).

A bacterium possesses a firm cell wall which surrounds the cytoplasmic contents. The bacterial nucleus is the site of the cell's genetic information and it consists of a single molecule of double-stranded deoxyribonucleic acid (DNA) which is coiled and tightly packed. This is the bacterial chromosome which transfers the characteristics of the parent cell to the daughter cells when it reproduces.

Bacteria usually reproduce by simple division, but under certain circumstances transfer of genetic material between bacteria can occur. They are capable of very rapid reproduction, and under ideal conditions can divide every 20 min so that in four hours one bacterium can reproduce to form 4000 daughter bacteria. Fortunately ideal conditions do not continue, and conditions become adverse, so that the bacteria may not survive.

Under adverse conditions of altered temperature or water shortage some bacteria can produce *spores*. A spore is a multilayered coat surrounding a cytoplasmic core which enables the bacterium to survive until conditions again become favourable. Spores are particularly resistant to heat and chemicals which would normally kill bacteria.

The cultivation and identification of bacteria

Bacteria can be cultured in the laboratory by providing suitable conditions for growth. Cultivation allows the bacteria to be studied and identified. The requirements for the growth of bacteria vary with the species. Thus some bacteria will grow only in the presence of oxygen, and are called *aerobic bacteria*, others will grow only in the absence of oxygen, and are called *anaerobic bacteria*, and some will grow in the presence or absence of oxygen and these are called *facultative anaerobic bacteria*.

The food requirements of bacteria also vary; some may require only water and simple salts, whereas others require complex food substances of animal origin such as amino acids, fats and carbohydrates. The temperature, humidity and acidity of the environment are also important. Most bacteria that cause disease in man grow best at body temperature, but other bacteria will grow at much higher or lower temperatures.

Bacteria are usually cultured in the laboratory in broth or on a solid agar medium to which amino acids, meat extracts or blood have been added (agar is an extract of seaweed). The temperature is controlled by the use of an *incubator*.

The identification of a bacterium is made by finely spreading a source of the bacterium over the required medium. This is incubated for a period of 24–48 hours. Some bacteria, such as the bacterium responsible for tuberculosis, require much longer periods of incubation. Individual bacteria grow to produce little mounds, called colonies, and the shape, form and colour of the colony is a step in the identification of the bacterium. Some of the colony is spread onto a microscope slide and examined after it has been stained with various chemical stains. The reaction of the bacterium to certain stains is a further step in the identification process. For example, bacteria can be divided into *Gram-positive* and *Gram-negative* groups according to their ability to take up colour with the Gram stain.

When the stained bacteria are examined under the microscope their shape is noted. The spherical bacteria, or cocci, can be divided into groups by the way in which they lie. Thus *streptococci* tend to lie in chains, *staphylococci* in clusters and *pneumococci* in pairs (Fig. 3.1). Frequently further tests, such as the ability of the bacterium to ferment sugar, or the reaction of the bacterium to specific antisera, are required to complete the identification.

Infection

In man there are always bacteria on the skin, in the nose and in the gastrointestinal tract. These normal inhabitants do not cause disease in these sites, and disease may indeed result if they are removed. Some people, who are called *carriers*, may be inhabited by bacteria that are pathogenic to other people, although they themselves are perfectly fit. Thus carriers of typhoid have the bacterium living in their gall bladder, and they can communicate this bacterium to others in whom it will cause typhoid fever.

Bacteria that are capable of causing disease are called *pathogens*. Infection is the reaction of the body to the invading organism. The reaction may be insignificant or it may be severe, according to the number and type of bacteria that have gained access, and to the resistance of the body to infection.

Virulence

The virulence of a bacterium is its ability to multiply in the tissues and damage the body. Thus an organism of low virulence may merely provoke a mild reaction, which may pass unnoticed, while an organism which is very virulent may cause a severe reaction and even death. The virulence of bacteria varies from species to species and may differ between different strains of the same species.

Toxins

An important factor in the virulence of a bacterium is its ability to produce substances called toxins. Some toxins exert their effects from within the bacterium, but others are released into the tissues. They may exert their effects locally, causing destruction of tissue, or they may be absorbed into the blood stream and damage tissues remote from the site of infection.

The defences of the body against infection

The body has certain defence mechanisms that act to prevent, localize and combat infection.

The surface defences. The skin acts as a mechanical barrier that prevents bacteria reaching the underlying tissues. Sweat contains substances that can kill bacteria on the surface of the skin. The mucous membranes of the body also act as a mechanical barrier to infection. Certain secretions, such as tears, nasal secretions, saliva and the prostatic fluid can destroy bacteria. The gastrointestinal tract is further protected by the high concentration of hydrochloric acid in the gastric juice, which can destroy most bacteria that are ingested with the food.

The internal defences:

1 **Inflammation**. The local reaction of tissues to bacterial infection is inflammation. The initial stages of the inflammatory reaction which occurs in infection are the same as those which occur in response to injury. Thus there are local changes of redness, heat, swelling and pain, and, depending on the site and severity of the infection there is some degree of impaired function. The effect of inflammation is to limit the bacterium infecting the tissues, to remove dead cells and to repair the damaged area. Inflammation occurs locally at the site of the infection, but its effects may produce changes in the body generally.

In the infected area, plasma and white blood cells leave the capillaries as an inflammatory exudate. The plasma contains the protein fibrinogen which coagulates to form fibrin strands in an attempt to isolate the infected area. In acute infections the cells which enter the tissues are polymorphonuclear leucocytes which phagocytose dead bacteria and dead tissue.

The general symptoms of inflammation may be slight or they may be severe. There is frequently a rise in body temperature, a fast pulse, loss of appetite and constipation. Toxins released from the infected area may cause delirium or even coma. If white cells are being damaged in large numbers by the infection, the bone marrow responds by producing an increased number of these cells and there is often an increase in the total number of white cells circulating in the blood stream. This is called a *leucocytosis*.

2 **The formation of antibodies**. One of the body's most important internal defences against infection is the formation of antibodies.

Certain diseases, such as chickenpox or measles usually only occur once in a person's lifetime, even though they may repeatedly come into close contact with a person suffering from the disease. A person who has suffered from one of these diseases is thus said to be immune to a further attack. This immunity is due to proteins, called *antibodies*, which circulate in the blood of the immune person. These antibodies are produced by the cells of the *macrophage system* (p. 336) together with certain lymphocytes when a bacterium, a virus or certain other foreign material, usually a protein, enters the body. The toxins of bacteria, being proteins, can also provoke the formation of antibodies, which are then called *antitoxins*. Substances which provoke the formation of antibodies are called *antigens*.

Antibodies have a close molecular relationship to the protein on the surface of the virus or bacterium which initiated their production. When the bacterium or virus again comes into contact with the appropriate antibody, because of this molecular resemblance, the bacteria or viruses are *agglutinated*, that is gathered together in clumps, and are thus rendered harmless. Antibodies are therefore specific and only produce agglutination of the bacterium or virus that initiated their production.

Immunity can therefore be acquired naturally following attacks of certain infections, but it can also be produced artificially in two ways, *actively* or *passively*.

a) **Active immunity** can be produced artificially by injecting into the body dead or live organisms which have been treated in such a way that they cannot cause disease. Similarly a toxin can be treated in such a way that it is powerless to cause damage but is still capable, on injection into the body, of provoking the production of the appropriate antitoxin.

b) **Passive immunity** is produced by injecting into the body the serum of an animal or of another man that is known to be immune to the disease. The antibodies in this serum will produce a temporary immunity, but this immunity is immediate, whereas active immunity takes some weeks to produce. The injection of foreign proteins, however, may cause undesirable side effects and passive immunity should only be conferred if the patient has been exposed to a potentially virulent infection to which he or she is not already actively immune.

The spread of infection

If the inflammatory reaction fails to contain an infection locally then spread may occur in a number of ways:

1 *Cellulitis.* Infection may spread throughout the spaces between the surrounding cells and tissues so that the adjacent areas become inflamed.
2 *Lymphatic spread.* Bacteria may pass into the local lymphatic vessels and be transported to the nearest lymph nodes. A secondary focus of infection may then occur in the lymph nodes; this is called *lymphadenitis*.

The lymphatic vessels leading to these nodes may also become inflamed and may be visible as red streaks leading away from the site of infection; this is called *lymphangitis*.

3 *Blood spread.* Bacteria may pass into the local blood vessels and multiply in the blood stream; this is called *septicaemia*. The bacteria may be deposited into capillaries in other tissues to form new foci of infection and this is called *pyaemia*.

The outcome of inflammation

1 *Repair.* If the infection is contained locally, with no complications, repair of the damaged area follows quickly. New capillaries grow out from the edge of the damaged area and fibroblasts appear alongside these new capillaries forming granulation tissue. If the damage is slight there is little fibrous tissue formed and the surface epithelium grows over the damaged area with little evidence of scarring. With greater areas of damage more fibrous tissue is formed to fill the defect and visible scarring will result.

Repair depends, however, on the type of tissue that has been damaged. In most circumstances, bone and connective tissue have a good ability to regenerate. The liver, thyroid, pancreas and kidneys can regenerate under favourable circumstances, but the cells of the central nervous system and voluntary muscle cells cannot regenerate.

2 *Suppuration.* If an invading micro-organism is more virulent, and the tissues and white cells are more severely damaged, then *pus* may form. Pus is composed of dead white cells and bacteria together with damaged tissue which has died and liquified. If the infected area is superficial then the pus merely seeps away from the surface, but if the inflammation occurs in deeper tissues pus may accumulate to form an *abscess*. The pus in the abscess sometimes tracks through the tissues to the surface forming a *sinus*. A sinus which communicates with a hollow organ, such as the bowel, is called a *fistula*.

3 *Ulceration.* With surface infections, an area of damaged tissue may die and become separated from the surrounding tissue, leaving an underlying cavity which is slow to heal. These cavities are called *ulcers*. Eventually granulation tissue will usually fill these cavities and healing occurs but with some scarring.

Types of inflammation

Inflammation varies both in rate of onset and in rate of repair and it can be divided into three types.

1 *Acute inflammation* which has a rapid course. The cells which accumulate in the infected area are polymorphonuclear leucocytes.

2 *Subacute inflammation* is an intermediate type between acute and chronic inflammation and both polymorphonuclear leucocytes and lymphocytes are present in the infected area. Also present are cells called plasma cells, which resemble lymphocytes but which are only found in the tissues and not in the circulating blood. They are concerned with local antibody formation.

3 *Chronic inflammation* occurs either as a prolonged later stage of acute inflammation or as a reaction to a less virulent bacterium which does not stimulate the body's defences so actively. The cells that accumulate in an area of chronic inflammation are usually lymphocytes and plasma cells. There is often considerable formation of fibrous tissue in areas of chronic inflammation.

The prevention and treatment of infection

The prevention of infection is as important as the treatment of an infection once it has become established.

1   Sterilization. An object is said to be sterile when it is freed from all living organisms, including spores. To prevent infection it is important that objects, such as syringes, needles and surgical instruments, should be sterile before use. Similarly dressings for wounds should be sterile. Sterilization of an object can be achieved by either *physical* or *chemical* means, and the method selected depends on the resistance to damage of the object to be sterilized.

a)   **Physical sterilization.** *Heating* is the most commonly used method of physical sterilization. Dry heat is usually used for sterilizing instruments in laboratories. Moist heat is more effective than dry heat at a given temperature. Boiling for 5 min will kill most bacteria but will not kill all spores. Sterilization by moist heat is carried out most effectively by autoclaving. In an *autoclave* the objects to be sterilized are subjected to steam under pressure as this raises the boiling point temperature. In a modern high vacuum autoclave, objects to be sterilized are subjected to steam in the absence of air, at a pressure of 220.6 kN m^{-2} for 3 min. This will effectively kill all bacteria and spores. (At this pressure the temperature produced is $135.3°C$).

Radiation. Certain heat sensitive objects are sterilized by exposure to gamma rays. This method is used commercially for sterilizing disposable equipment but it is too expensive for use in hospitals.

b)   **Chemical sterilization.** Chemical sterilization is performed by the use of substances called disinfectants. Disinfectants kill bacteria, but certain of them also damage human tissue and their use is limited. Chemical sterilization of instruments is not carried out unless sterilization by physical means damages the instrument. The most commonly used disinfectants are iodine, alcohol, cetrimide, Hibitane, formaldehyde and ethylene oxide vapour.

Iodine in an alcoholic solution is frequently used for disinfecting the skin. Iodine stains the skin and thus it is possible to see the area that has been sterilized. Some people, however, are sensitive to iodine, and their skin can be quite severely damaged by its application. Iodine has the advantage that it kills spores. *Providone-iodine* is a detergent preparation which may be safely used on the skins of patients who are sensitive to iodine.

Alcohol, in a 70% solution as surgical spirit, is a disinfectant which is frequently used for sterilizing the skin, but it does not kill spores.

Cetrimide (CTAB) is a detergent that is used for cleaning wounds and the skin.

Hibitane (chlorhexidine) is occasionally used for disinfecting instruments that cannot be sterilized by heat. It is also useful for cleaning the skin.

Formaldehyde is occasionally used for sterilizing instruments, and it is used in the form of tablets which release formaldehyde vapour into the enclosed space in which the instruments arc stored. *Glutaraldehyde* is used to sterilize instruments which would be damaged by heat. Some people, however, develop skin sensitivity to glutaraldehyde.

Ethylene oxide vapour may be used to sterilize heat-sensitive equipment such as arterial catheters, heart-lung machines and respirators.

2 Asepsis during surgery. In surgical operations, where there is an open wound, it is important not only that the instruments should be sterile, but that bacteria should not be allowed to come into contact with the wound from the air. Thus aseptic techniques are used. The operating room should be dustproof and washed throughout with disinfectants as frequently as is practical. All persons entering the theatre should wear a face mask to prevent droplet infection. They should also wear sterile gowns, shoes and caps. The surgeon performing the operation, and any staff handling instruments, should scrub their hands thoroughly and should wear sterile gloves. The skin of the patient, in the area of the operation, should be thoroughly cleansed with disinfectant and the surrounding area should be covered with sterile towels. After the operation the area of the wound is covered with a sterile dressing before the patient leaves the theatre.

3 Isolation. Patients with certain infectious diseases have to be isolated in order to prevent the spread of infection. This is done by moving the patient into a single room and staff of the hospital who come into contact with the patient must take stringent precautions to avoid carrying the disease to other patients.

4 The treatment of infection. Once an infection is established in the body it is often necessary to take measures to aid the body's defences to limit the infection and kill the bacteria. Apart from the antisyphilitic drugs it is only in the last 50 years that agents which have antibacterial activity have become available. These antibacterial substances are of two types:

Chemotherapeutic agents, which are chemicals that kill or restrict the growth of bacteria without damaging the patient's tissues. The sulphonamides are chemotherapeutic agents.

Antibiotics, which are substances produced by living organisms, such as moulds, are toxic to bacteria. Penicillin was the first such antibiotic to be of practical value.

The chemotherapeutic agent, or antibiotic, to be used depends on the site of the infection and the species of the invading bacteria. No antibacterial agent is effective against every type of bacterium. Prolonged or inadequate use of these agents may allow a species of bacterium to become resistant to their action.

Some common bacteria which are pathogenic to man

Staphylococci are frequently found infecting wounds. They are, unfortunately, normal inhabitants of the skin, only occasionally causing disease, such as boils or impetigo, but they are readily transferred to wounds or surgical incisions.

Streptococcus pyogenes is a common source of infection in the throat or tonsils and it may cause sepsis in burns or wounds. *Streptococcus faecalis* is occasionally found in infections of the urinary tract and it may infect wounds.

Clostridia. Clostridium tetani is the causative organism of tetanus. It gains entry to the body through cuts and scratches, but, being anaerobic, it only establishes itself if the wound is dirty or if there is dead tissue present. From a local infection it produces a powerful toxin which can be lethal. *Clostridium welchii* is another anaerobic bacterium; it causes gas-gangrene and it is also a normal inhabitant of the gut. Both these types of clostridium are inhabitants of manure and thus of soil.

Salmonellae and *Shigellae* are organisms that cause gastrointestinal infections. Salmonellae are primarily pathogens of animals and so food poisoning may be caused by their presence in food, where they produce a toxin which causes diarrhoea and vomiting when the food is eaten. Certain types of salmonellae are responsible for enteric fever. Shigellae are the causative organisms of dysentery.

Escherichia coli is found in large numbers in the healthy gut, but some types can cause severe diarrhoea in babies and young children, and they may also cause travellers' diarrhoea in adults.

Haemophilus influenzae is a frequent causative organism of chest infections and meningitis.

Mycobacterium tuberculosis is the bacterium responsible for tuberculosis. It gains entry to the body through either the respiratory tract or the gastrointestinal tract. It usually reaches the lungs via infected sputum from a person suffering from the diease. This organism, which grows slowly, may also invade bones and joints. Cows are subject to tuberculosis, and the bacterium can reach the human gastrointestinal tract via infected milk. Pasteurization, however, has virtually eradicated this type of infection.

Treponema pallidum is a spirochaete and it is the causative organism of syphilis. It is normally transmitted by the sexual organs. Unless treated, a generalized infection can eventually occur which involves almost every tissue of the body, but its most damaging effects are produced in the arteries and the brain.

Viruses

Viruses are very much smaller than bacteria and are normally invisible through the ordinary light microscope. They may be regarded as intermediate between living organisms and inanimate objects. Many viruses may be crystallized and yet can reproduce themselves, although only when they are within living cells. They consist essentially of nucleic acids, which are the substances forming the genetic material of living cells. When a virus infects a cell it actually enters the cell and uses the cell's metabolic machinery to grow and reproduce, to the detriment of the cell.

Unlike bacteria, viruses cannot be cultivated on artificial media but have to be grown in the laboratory in living cells. This is done by growing the virus in cell cultures, which are living cells that are themselves artificially maintained, by infecting laboratory animals or by growing the virus in fertilized hen's eggs.

Viruses may be classified according to the tissues that they infect.

Viruses that infect the central nervous system

Poliomyelitis is caused by a virus that attacks the spinal cord, causing paralysis.

Rabies is caused by a virus that is usually transmitted from an infected dog. The virus travels to the brain and eventually produces convulsions and death.

Viruses that infect the respiratory tract

Influenza is caused by a virus that infects the respiratory tract and also produces muscular pains and a high temperature.

The common cold is caused by a virus that infects the nose, the throat and the larynx.

Pneumonia. Some cases of pneumonia, particularly in children, have been shown to be due to viruses.

Viruses that infect the skin

Smallpox, chicken pox, Herpes simplex and *measles* are all caused by viruses which enter the body through the respiratory tract and pass into the blood stream before settling in the skin where they cause a rash.

Miscellaneous viruses

Mumps is caused by a virus that infects the parotid salivary glands. The mumps virus sometimes infects other glands such as the pancreas and the testis.

Viral hepatitis (infective jaundice) is caused by a virus that infects the liver and produces jaundice.

Infectious mononucleosis (glandular fever) is due to a virus which causes enlarged lymph nodes, a high temperature and the appearance in the blood of characteristic mononuclear cells, called glandular fever cells.

Protozoa

A number of infections are caused by protozoa, which are small unicellular parasites. Most protozoan infections are beyond the scope of this book, as they rarely occur in this country, but two such infections merit mention.

Malaria can occur in most parts of the world but it is commonest in tropical or subtropical climates. It is transmitted to man by the bite of an infected mosquito. In man it produces a disease characterized by severe attacks of fever, by anaemia and by enlargement of the liver, the spleen and the lymph nodes. Certain forms of malaria can be fatal. Malaria can be seen in this country in people who have recently returned from a malarial area.

Toxoplasmosis also has a world-wide distribution. It is common in domestic animals, and human infection is usually caused by infected cats. The infection may cause no apparent symptoms or there may be generalized enlargement of the lymph nodes together with bouts of fever. The virus may also involve the heart and lungs, or the eyes. Infection of mothers during pregnancy can lead to severe fetal abnormalities or to stillbirths.

Worms

Infections due to worms are most common in the Middle and Far East and in Africa. Three infections caused by worms, however, deserve mention.

Threadworms infect the gastrointestinal tract and are most common in children. The female worms emerge from the anus to deposit their eggs and they may cause intense anal irritation.

Tapeworms also infect the gastrointestinal tract in man and they gain access if undercooked infected meat is eaten. They produce diarrhoea, anaemia and weight loss.

Hydatid disease is transmitted by sheep, cows or pigs. It is relatively uncommon in England but does occasionally occur in Wales. The characteristic lesions are cysts in the liver, but these can also occur in the lungs and kidneys.

Yeasts and fungi

Most yeasts and fungi are only weakly invasive and are most frequently found infecting the surface of the skin (e.g. *athlete's foot*). *Moniliasis* (thrush) is the most commonly encountered, producing mild infections of the skin of the axillae or groins, of the mouth, or of the vagina. In debilitated patients, or patients who have been on long-term oral antibiotics, moniliasis may involve the whole of the oesophagus producing a characteristic appearance on a barium examination.

Disorders of immunity

Immunity is the name given to the ability of the body to recognize, isolate and reject foreign proteins and certain other large foreign molecules, which are called antigens. The reaction to antigens is called the immune response and this may be defined as the ability of the body to recognize foreign material and to respond to that material. As has already been seen earlier in this chapter, the body responds to the presence of invading bacteria, bacterial toxins or viruses by producing antibodies, but the production of antibodies is only a part of the activities of the immune system. There are two types of immunity recognized, *humoral immunity* and *cell-mediated immunity*. Humoral immunity involves the production of antibodies which circulate in the blood stream and cell-mediated immunity is concerned with phagocytic activity. The lymphocytes (p. 313) of the body are intimately involved in the immune process. Two types of lymphocytes, called B-lymphocytes and T-lymphocytes, which are not distinguishable by microscopic examination, are involved in the process.

The *T-lymphocytes*, which form about 80% of the total blood lymphocytes, depend on a factor produced by the *thymus* to become antigenically active or competent. These T-cells are involved in cell-mediated immunity and they produce factors which mobilize macrophages to the site of infection or to the site of reaction with an antigen.

The *B-lymphocytes* form 20% of the total blood lymphocytes and they depend upon a gut-produced factor to become immunologically competent. In birds this gut factor is provided by a structure called the *bursa of Fabricius* and thus the name B-cells or 'bursa-cells'. It is not clear where this factor originates in man but it may well be in the lymphoid tissue in the gut. B-lymphocytes, when antigenically stimulated can proliferate to produce plasma cells which in turn produce antibodies in the form of large protein molecules known as *immunoglobulins* (Ig).

There are various classes of immunoglobulins. The majority of circulating antibody (85%) is in the form of IgG. IgG has the smallest molecular size and is able to cross the placenta to confer passive immunity to the fetus. IgM accounts for 5−10% of plasma antibody and appears early in the immune response. IgA accounts for 1−4% of the plasma antibody. It is produced by the mucosa associated lymphoid tissue and is secreted onto the mucosal surface of the respiratory, gastrointestinal and urogenital tracts. It is also present in breast milk. IgE is involved in hypersensitivity reactions (see below).

Humoral immunity

The B-lymphocytes are thus involved in humoral immunity. When an antigen enters the body it is taken up by the cells of the macrophage system. The B-lymphocytes are then stimulated, either by the antigenically primed macrophages or directly by the antigen itself. When antigenically stimulated, the B-lymphocytes divide to produce two types of cell, *plasma cells* and *memory cells*. The plasma cells produce large quantities of the appropriate antibody. These antibodies are called immunoglobulins as they form part of the globulin group of plasma proteins in the blood. The memory cells are so-called as they retain the identity of the antigen so that if a further invasion of the same antigen occurs, the response of antibody production is more rapid. The first exposure to an antigen initiates the so-called *primary response* in which there is a delay in the production of antibodies. The *secondary response* occurs with a second exposure to the same antigen and because of the memory cells the response is very much more rapid and larger quantities of antibody are produced.

Cell-mediated immunity

The T-lymphocytes are concerned with cell-mediated immunity. When there is exposure to an antigen, phagocytes activate the T-lymphocytes which divide to produce large numbers of lymphocytes which pass in the blood stream to the area of the antigen which provoked their formation. These lymphocytes infiltrate the area of the antigen and they attract macrophages, in the form of mononuclear cells, from the blood, and these mononuclear cells phagocytose the antigen. Certain lymphocytes, called *K-cells* or *killer cells*, are also recognized and they can destroy cells which are coated with antibody. Cell-mediated immunity is important in the protection of the body against tuberculosis and certain viral infections, and it is also involved in certain fungal infections and in the reaction of the body to material which is grafted from another person. In some infections, however, both humoral and cell-mediated immune responses are involved.

Hypersensitivity

Normally the immune response is a protective mechanism, but some individuals who have been exposed to a particular antigen may, on a second exposure to that antigen, have a harmful reaction called an *allergic reaction*. They are thus said to be hypersensitive to that antigen. During the reaction there are clear indications that the immune system has been activated. If the reaction is immediate it is the humoral system which is responsible and this is called *immediate hypersensitivity*. The reaction may be relatively mild, as in the reaction to grass pollen in patients with hay fever, but it may be considerably more serious or even fatal.

The injection of contrast media in the X-ray department may, if the patient is hypersensitive (e.g. to iodine), provoke severe reactions during which there is severe bronchospasm (asthma), oedema of the larynx or the lungs, and circulatory collapse which may lead to cardiac arrest.

A more delayed type of reaction may occur in response to the administration of certain drugs or to the ingestion of some foods in food allergies. This type of hypersensitivity, called *delayed hypersensitivty*, is due to cell-mediated immunity. The reaction may cause fever, lymph node enlargement, skin rashes and joint pains.

Tissue grafting

The ability to recognize foreign protein assumes great importance in the surgical practice of tissue grafting. Skin may be grafted from one part of a patient to another part and, if the blood supply is adequate, it will grow normally. If skin is grafted from one patient to another, the skin graft is, within a few days, infiltrated with lymphocytes and the graft dies. This is due to the fact that the graft is recognized as foreign protein and the lymphocytes cause the death, or rejection, of the graft.

For tissue grafting from one person to another to be successful, it is neccessary to suppress the body's immune reactions by the use of drugs. It is now possible to tissue-type people in the same way as blood groups can be determined. An exact match of tissue types, however, cannot be obtained, except in the case of identical twins. Thus, even with a compatible tissue match, such operations as renal transplants still require the use of drugs to suppress the immune response. Even so, a number of renal transplants still suffer rejection by the recipient.

Auto-immune diseases

Since the body has the ability to recognize foreign proteins it follows that it must also have the ability to recognize its own proteins, against which it does not produce an immune response. There are, however, a number of diseases which are known to be, or are thought to be, caused by the body producing antibodies against its own proteins. These diseases are called auto-immune diseases. The reasons for the development of this auto-immunity are not yet fully understood but the number of diseases which are thought to be of this nature are increasing. The diseases which are thought to be due to autoimmunity are:

Hashimoto's thyroiditis in which the thyroid gland becomes enlarged and infiltrated with lymphocytes. This results in abnormalities in the production of thyroid hormone and goitre.

Some *Type I insulin dependent diabetes* is known to be caused by auto-antibodies to the cells of the Islets of Langerhans in the pancreas which are responsible for the production of insulin. This disease is unusual among the auto-immune diseases in that it occurs in children and young adults.

Blood disorders. Some forms of *haemolytic anaemia* and *thrombocytopenic purpura* are thought to be auto-immune in origin. In haemolytic anaemia the red blood cells are destroyed more rapidly than normal resulting in anaemia and sometimes jaundice due to breakdown of the released haemoglobin which increases the amount of bile pigments in the blood. Thrombocytopenic purpura is caused by lack of circulating platelets in the blood resulting in a tendency to bleed. There may be a purpuric rash due to small haemorrhages into the skin and there may also be haemorrhages into the gastrointestinal tract or into some joints.

Systemic lupus erythematosus is a complicated disease which involves many tissues of the body. Principally it involves the skin, the kidneys, the joints, the serous membranes and the heart. Its onset may be acute or chronic but it is a relapsing disease and death frequently occurs from renal failure.

Rheumatoid arthritis mainly involves the joints but it may also involve many other tissues. The joint lesions involve destruction of the articular cartilage and may result in gross deformities.

Ulcerative colitis. There is some evidence that ulcerative colitis, which is characterized by colonic ulceration with severe diarrhoea and colonic bleeding, may be auto-immune in origin.

Endocrine disorders

The endocrine glands are frequently called ductless glands, for they possess no ducts and their secretions, which are called hormones, are released directly into the blood stream. A hormone may be defined as a substance which is produced by a gland, is released into the blood stream and is carried to other tissues and organs on which it exerts its effects.

. Endocrine disorders are caused by overproduction or underproduction of the hormones of one or more of the endocrine glands. Since the diseases which result give valuable information as to the physiology of the gland or glands involved, these disorders will be described in the chapter on the endocrine system (Chapter 17).

Metabolic disorders

Metabolic disorders may be divided into extrinsic metabolic disorders, which are caused by dietary deficiencies, or, more often in Western Societies, by dietary excesses, and intrinsic metabolic disorders which are caused by defects in the body which prevent the normal absorption or metabolism of specific substances in spite of a balanced diet.

Extrinsic metabolic disorders

A balanced diet must contain an adequate quantity of carbohydrates, fats, proteins, vitamins, mineral salts and water. An inadequate daily intake of calories in the form of carbohydrates, fats and proteins leads to loss of weight and, ultimately to *starvation*. Starvation is rare in Western Societies, except in the elderly who may no longer bother to eat an adequate diet, and in the psychological condition of *anorexia nervosa* in which the patient develops an aversion to food. It is more common now to encounter an excessive intake of calories which leads to *obesity*, with an increased risk of heart disease, obesity induced diabetes mellitus, and other problems related to overweight.

An adequate intake of vitamins is also necessary for normal growth and the maintenance of health, but the disorders which result from vitamin deficiencies will be described in the chapter on the digestive system (Chapter 14).

Intrinsic metabolic disorders

There are a large number of disorders which are caused by intrinsic metabolic disorders. Some of these abnormalities are inherited and are called *inborn errors of metabolism*, some have a possible genetic basis and others occur sporadically. Many of them result from the lack of a specific enzyme. A description of most of these diseases is beyond the scope of this book, but one does merit description, namely gout.

Gout is a disease which is characterized by attacks of acutely painful arthritis, classically of the first metatarso—phalangeal joint, but other joints are frequently involved. The disease is due to inability to metabolize nucleic acids and an excess of uric acid results. Uric acid may be deposited as crystals in and aound the affected joint and these deposits may calcify and be visible radiologically. Patients with gout may develop uric acid stones in the kidneys.

Circulatory disorders

The circulatory system, which includes the heart, the blood vessels and the lymph vessels, is responsible for the transport of oxygen and foodstuffs to the tissues, and the removal of waste products of the cell's metabolism from the tissues to the exterior. Diseases of the circulatory system are the major cause of death and are therefore of great importance. Diseases of the blood, such as anaemia, are described in the chapter on the circulatory system (Chapter 11) and malignant tumours of the blood cells, such as leukaemia, are described in the last section of this chapter under the heading of tumours. Diseases of the heart, the blood vessels and the lymph vessels will be described here.

General diseases of the circulation

General diseases of the circulation are caused, in most cases, by specific disease of the heart, the blood vessels or the blood. They are amongst the most commonly encountered diseases in medical practice.

Oedema

Oedema is the abnormal accumulation of fluid in the tissues. It may be localized or generalized. Localized collections of fluid in the body cavities are usually given specific names, thus a *pleural effusion* is fluid in the pleural cavity and *ascites* is fluid in the peritoneum.

The causes of oedema are as follows:
1 *Increased venous pressure.* An increase of pressure in the veins draining the tissues will force blood out of the capillaries into the tissues. Local oedema may be caused by mechanical obstruction to a vein by exterior pressure, or by occlusion of a vein by a blood clot (*thrombosis*). Generalized oedema due to increased pressure in the vein occurs in heart failure when

the heart is unable to return to the tissues all the blood which passes to it from the superior and inferior venae cavae.

2 *Lymphatic obstruction.* Obstruction to the lymph vessels in a region of the body will give rise to localized oedema. Lymphatic obstruction may rarely occur as a congenital abnormality, but acquired lymphatic obstruction may occur as a sequel to infection of the lymph nodes or due to infiltration of lymph vessels and lymph nodes by malignant tumours.

3 *Decreased osmotic pressure of the blood.* The plasma proteins, which cannot pass through the walls of capillaries, are mainly responsible for maintaining the osmotic pressure of the blood. In conditions in which there is a reduction in the plasma proteins, oedema can occur due to reduced osmotic pressure. A fall in the plasma protein levels can occur in severe starvation, malabsorption from the intestines or loss of protein through the kidneys in renal failure.

4 *Increased capillary permeability.* Damage to the capillary walls can lead to loss of both fluid and plasma proteins into the tissues. In cardiac failure not only is there an increase in the venous pressure, but also a decrease in the oxygen supplies to the tissues (*hypoxia*). This tissue anoxia causes increased capillary permeability which contributes to the generalized oedema. Local increases in capillary permeability may be caused by infections, burns or allergic reactions.

Congestion

Congestion is a term used to describe an increase in the amount of blood in the tissues and it is often a precursor to oedema. Congestion is most commonly encountered in the lungs in early left ventricular failure, and in right ventricular failure there may be congestion of the liver.

Thrombosis

Thrombosis may be defined as the formation of a blood clot within an artery, a vein or the heart. Thrombosis may be caused by:

1 *Slowing of the blood flow*, and this may be due to varicose veins, partial venous obstruction from external pressure, or prolonged immobility, as may occur during and after a surgical operation.

2 *Damage to the blood vessels*, which may be caused in arteries by atheroma, in the heart by damage to the wall caused by a myocardial infarct, or in arteries and veins by mechanical trauma or inflammation.

3 *Disorders of the blood*, such as polycythaemia or leukaemia, in which there is a total increase in the number of blood corpuscles and an increased viscosity due to changes in the plasma proteins.

Embolism

An embolus is a undissolved mass which circulates in the blood stream until it impacts in a blood vessel which is too small to allow it free passage. The most common type of embolus is a thrombus. Thrombosis of a vessel may be followed by detachment of all or part of the thrombus from the vessel walls forming an embolus. Emboli can arise from both arteries and veins.

Venous thrombosis occurs most commonly in the lower limbs and, slightly less commonly, in the veins of the pelvis. Emboli released from these sites usually impact in the pulmonary arteries and are called pulmonary emboli.

Arterial thrombosis may occur in the heart due to damage to the wall of the heart, or due to an abnormal cardiac rhythm, as in atrial fibrillation when no coordinated atrial contraction occurs. Thrombosis may also occur in arteries secondary to atheroma (see below). Detachment of such a thrombus results in an arterial embolus which then obstructs a more distal vessel. The most common sites for arterial emboli are the arteries of the brain and the arteries of the lower limb.

Fat embolism occurs when globules of fat pass into the blood vessels following fractures of bones. Either immediately following the fracture, or during manipulation or pinning, fragments of bone marrow and fat may be forced into the local veins and these emboli usually impact in the lungs, although fat emboli may reach the brain. Fat emboli can be fatal.

Infarction

An infarct is a localized area of tissue damage or death caused by occlusion of, or insufficiency of, the arterial supply to the area. Most infarcts are the result of atheroma or arterial embolism. Sudden death may be caused by infarcts of the cardiac muscle (a heart attack), or of the brain (a stroke), but an embolus into a pulmonary artery from a leg vein is also a cause.

Gangrene

Gangrene may be defined as the death of an area of tissue. It is most commonly caused by arterial occlusion either by an arterial embolus or by atheroma which may be accompanied by arterial thrombosis. Gangrene is most commonly encountered in the lower limbs secondary to occlusion by atheroma, thrombosis or embolism of the arteries of supply.

Diseases of the arteries

The arteries carry the blood from the heart to the tissues. The exact structure of the arteries varies with the size of the artery but they are all composed of three layers: an inner layer, called the tunica intima, which consists of a single layer of endothelial cells supported by a basement membrane; a middle layer, called the tunica media, which consists of yellow elastic fibres and circular smooth muscle cells, and an outer layer, called the tunica adventitia, which is composed of fibrous tissue. Diseases of the arteries are usually divided into three groups, degenerative diseases, inflammatory diseases and disorders of arterial contraction.

1 Degenerative arterial disease.

a) *Atherosclerosis* is a degenerative condition which involves the aorta, the large arteries and some medium sized arteries, in particular the coronary and cerebral arteries and the arteries of the lower limbs. It is extremely common and almost every person over the age of 30 is affected to a variable degree. Atherosclerosis consists of plaques of thickening of the tunica intima which are yellowish in colour and are composed of fat deposits (lipids) with some associated fibrosis. These plaques may coalesce to narrow the lumen of the arteries and lead to actual obliteration of the arterial lumen. In the coronary arteries this causes damage to the cardiac muscle, or *ischaemic heart disease*, leading to 'heart attacks', and in the cerebral arteries to *cerebral infarction*, or 'strokes'. In the arteries of the lower limbs, atheroma may cause pain on walking, (intermittent claudication), and in a more advanced form, *gangrene* of the toes or foot may result. The exact cause of atheroma is not known, but excess of fats in the diet, lack of exercise and cigarette smoking are all important in accelerating the disease. Heredity is also an important factor.

b) *Arteriosclerosis* is a degenerative disease in which the smooth muscle and yellow elastic fibres in the tunica media are gradually replaced by fibrous tissue. This occurs to a greater or lesser extent with advancing age. It is accelerated in patients with hypertension but is usually only of major importance when it involves the arteries of the kidney.

c) *Aneurysms.* A *true aneurysm* is a localized enlargement of the lumen of an artery and is either fusiform, and involves the whole circumference of the artery, or saccular, when it is an asymmetrical bulging which does not involve the whole circumference. A *dissecting aneurysm* is a term used to describe a condition in which the wall of an artery splits, and blood tracts within the wall of the artery instead of through the lumen. The vast majority of aneurysms are secondary to atherosclerosis although small congenital aneurysms occur in the Circle of Willis in the brain. Aneurysms, if they are not treated surgically, will ultimately rupture leading to catastrophic haemorrhage and consequent sudden death.

d) *Hypertension* is a disease in which there is a sustained rise in the blood pressure. There is no full agreement as to the definition of what constitutes a raised blood pressure, but a systolic pressure about 140 mm Hg and diastolic pressure above 90 mm Hg is a good working definition. Hypertension is included in this section because all the arteries in a patient with hypertension develop changes of arteriosclerosis and there are accelerated changes of atheroma. In 85% of cases there is no discernible cause for the hypertension and this is referred to as primary hypertension. In 15% of cases a recognizable cause is discovered and in the majority of patients the cause is disease of the kidneys, but some rare tumours of the adrenal gland can also give rise to hypertension.

Primary (essential) hypertension is commonest in males and it leads to a slow and progressive rise in the blood pressure. There is increasing arteriosclerosis particularly in the arteries of the kidneys. The rise in blood pressure may cause enlargement of the left ventricle of the heart, and if untreated, will ultimately lead to left ventricular failure. Strokes due to cerebral thrombosis or haemorrhage are common and atheroma of the coronary arteries increases the incidence of heart attacks. In many patients, however, the disease may remain symptomless for a considerable time. 10% of patients with primary hypertension develop the condition of *malignant hypertension* in which the arteriosclerotic changes in the arteries are particularly progressive and this is particularly noticeable in the arteries of the kidneys. If untreated, malignant hypertension leads rapidly to renal failure.

2 Inflammatory arterial disease.

a) *Syphilitic aortitis* is now a rare disease but it is still ocassionally seen. The disease usually starts in the ascending aorta and may lead to aneurysms, damage to the aortic valve and ischaemic heart disease due to involvement of the origins of the coronary arteries.

b) *Buerger's disease* is uncommon but it is an inflammatory disease of arteries which occurs almost exclusively in males. The disease is closely associated with heavy smoking. The arteries of the lower limbs are predominantly involved leading to intermittent claudication and ultimately to gangrene.

c) *Polyarteritis nodosa* is an inflammatory condition of arteries, of uncertain cause. There is localized infiltration of the walls of small arteries by polymorphonuclear leucocytes, and this inflammation may lead to thrombosis of the vessel or to the formation of small aneurysms. The disease particularly involves the arteries of the kidneys and the diagnosis is sometimes established by renal angiography, which may demonstrate the presence of tiny aneurysms.

3 Raynaud's disease is a disease which involves abnormal spasmodic contractions of the arteries to the fingers and occcasionally the toes. It appears to be an abnormal reaction to cold, but it may lead to arterial thrombosis. It predominantly affects females and may become evident in adolescence.

Diseases of the veins

The veins return the blood from the tissues to the heart. Like arteries their walls are composed of three layers but these walls are thin in comparison to the size of their lumen. The inner coat, called the tunica intima, is composed of a single layer of endothelial cells which are shorter and broader than those in the arteries. The middle layer, called the tunica media, is much thinner than the equivalent layer in an artery, and it is composed of connective tissue and yellow elastic fibres, although in some veins there are also some circular smooth muscle fibres present. The outer layer, called the tunica adventitia, is composed of connective tissue fibres. Most veins have, on their inner surface, valves which serve to prevent the backflow of blood. These valves consist of a fold of the tunica intima which is strengthened and thickened by connective tissues and yellow elastic fibres. They are semilunar in shape and allow blood to flow in one direction only, that is, towards the heart.

Varicose veins are abnormally tortuous veins and they are most commonly encountered in the superficial areas of the lower limbs. They are caused by incompetence of the valves in the lumen of the veins. There is a familial tendency in the condition but it also occurs in obese patients and in those whose occupations involve prolonged standing. Stasis of blood occurs in varicose veins and this may lead to thrombosis and to ulceration of the skin, particularly around the ankles.

Varicose veins sometimes occur in the rectum, where they are called *haemorrhoids*.

Venous thrombosis is a common condition in which there is occlusion of a vein, or veins, by blood clot. Venous thrombosis is usually divided into two types.

1 *Phlebothrombosis* is a condition in which the vein is occluded by a blood clot due to a relative decrease in bloodflow through the vein, or due to partial obstruction of the vein by pressure from the exterior. It can occur following surgical operations when the patient is relatively immobile.

2 *Thrombophlebitis* is a condition in which the lining of a vein is damaged by a mild inflammatory reaction which leads to occlusion of the vein by a blood clot.

Diseases of the heart

Diseases of the heart are the commonest cause of death in Western countries, and ischaemic heart disease, secondary to coronary artery disease, is the commonest of these diseases.

Cardiac failure is a condition in which the heart fails to maintain an adequate circulation to the tissues in spite of a sufficient return of blood to the heart from the veins. It may be caused by the following:

1 *Failure of the ventricular muscle*, and the commonest cause of this is ischaemic heart disease.

2 *Increased resistance* to the outflow of blood from the heart, as in hypertension.

3 *Increased requirement for blood flow*, which may occur when one of the heart valves is narrowed or leaks, but it can also occur due to reduced oxgen content of the blood associated with chronic lung disease, and in severe anaemia.

Acute cardiac failure is of sudden onset and occurs most commonly due to coronary artery occlusion and may lead to cardiogenic shock.

Chronic cardiac failure may occur either following partial recovery from coronary artery occlusion or as a result of disease of the valves of the heart. It leads to distension of the veins and accumulation of fluid in the tissues.

Left ventricular failure may occur following a coronary artery occlusion, in untreated hypertension, or in disease of the aortic valve. In the presence of a normally functioning right ventricle, fluid accumulates in the lungs, and this is frequently visible on X-rays as *pulmonary oedema*. Clinically it presents as acute or chronic breathlessness.

Right ventricular failure can occur due to pulmonary embolism, chronic disease of the lungs or disease of the mitral valve. It leads to venous congestion, oedema of the tissues, and enlargement of the liver.

Ischaemic heart disease

Ischaemic heart disease is caused by an inadequate blood supply to the heart muscle. It is almost entirely due to atheroma of the coronary arteries. It presents in five ways:

1 **Angina pectoris**, which is characterized by severe pain in the chest usually precipitated by exercise. It is the equivalent of intermittent claudication, which occurs in the lower limbs, and it is due to an inadequate blood supply to part of the heart muscle. It is relieved by rest or by drugs which dilate the coronary arteries.

2 **Myocardial infarction**, in which there is death of part of the ventricular muscle due to occlusion of a coronary artery. It is not relieved by rest or by drugs which dilate the coronary arteries.

3 **Cardiac arrest**, in which sudden death occurs due to total cessation of the heart's action following an occlusion of a coronary artery.

4 **Left ventricular failure** with acute breathlessness, secondary to a symptomless myocardial infarct or due to more diffuse damage to the cardiac muscle, secondary to coronary artery insufficiency.

5 Cardiac arrhythmias, or changes in the normal coordinated cardiac rhythm, due to damage to the conducting fibres in the heart by myocardial ischaemia.

Inflammatory diseases of the heart

Acute rheumatic fever is a disease which presents with acute joint pains and damage to the myocardium. It follows streptococcal sore throats and usually occurs in children and young adults. It has, however, become relatively rare in this country due to the use of antibiotics. Although death may occasionally occur due to myocardial damage, the most common and serious effect of the disease is damage to the heart valves.

Myocarditis is an inflammatory condition of the cardiac muscle which may be viral, bacterial or due to bacterial toxins. It can occasionally be fatal due to cardiac failure.

Infective endocarditis is a condition in which thrombi, containing the infecting micro-organism, form on the endocardium, or inner lining, of the heart, particularly on the endocardium of the heart valves. It is particularly liable to occur in patients who have pre-existing damage to the valves as a result of rheumatic fever. Infective endocarditis will worsen this damage. Energetic antibiotic therapy has greatly improved the treatment of the disease but some cases are still fatal.

Diseases of the valves of the heart

Mitral stenosis, or narrowing of the mitral valve, is usually due to rheumatic heart disease. The narrowing impairs filling of the left ventricle and leads to an increased workload requirement by the right ventricle. This may lead to pulmonary hypertension and right ventricular failure.

Mitral incompetence is a condition in which the mitral valve fails to close properly during contraction of the left ventricle with the result that blood is forced backwards into the left atrium during systole. The left and right ventricles become overloaded and may fail. Mitral incompetence may be caused by rheumatic heart disease or by ischaemic heart disease due to stretching of the ventricular muscle and consequently the mitral valve.

Aortic stenosis (narrowing of the aortic valve) may be due to rheumatic heart disease, but in this country the cause now is more frequently due to calcific aortic stenosis in which the valve becomes fibrosed, calcified and rigid. Aortic stenosis leads to left ventricular enlargement and ultimately to left ventricular failure.

Aortic incompetence is a condition in which the aortic valve allows blood to flow backwards into the left ventricle from the aorta following ventricular contraction. It leads to enlargement of the left ventricle and ultimately to left ventricular failure. The commonest cause is rheumatic heart disease.

Advances in surgical techniques and modern materials allow valvular disease of the heart to be treated by replacing the damaged valves with artificial valves, but this must be undertaken before damage to the heart and lungs is severe.

Tumours

The tissues of the body are continuously being repaired and replaced following injury or loss due to the normal processes of wear and tear. Generally, therefore, growth and repair are continually occurring according to the demands of the body. Certain organs and tissues can increase their size (*hypertrophy*), or they can increase their total number of cells (*hyperplasia*), if an increased demand is made upon them.

A tumour, however, is an abnormal mass of cells. The cells of the tumour start to grow, for reasons that are not yet understood, without regard for the needs of the body, and frequently they are detrimental to the body since they absorb nourishment from the blood at the expense of normal cells.

Tumours are often called *neoplasms* (new growths) and they are divided into two types, benign and malignant.

Benign tumours

Benign tumours grow slowly but progressively and they do not invade other tissues. They exert pressure on adjacent tissues but their growth may sometimes spontaneously cease and their size remain stationary. The cells of benign tumours usually closely resemble the cells of the tissue from which they have arisen. Their complete removal by surgery is not followed by recurrence. Usually they do not shorten the lifespan of the patient, but tumours in certain situations may be fatal because they interfere with the function of a vital organ. Benign tumours of the brain, for example, are frequently difficult, or impossible, to remove completely and their pressure, exerted within the rigid walls of the skull, may cause death by pressure on the brain. Similarly benign tumours of endocrine glands may secrete an excess of the hormone of the gland of origin, and death can occur from this excess unless the tumour is removed.

Malignant tumours

Malignant tumours grow progressively and, if untreated, they almost invariably cause death of the patient. They usually grow rapidly and the cells of which they are composed are less well differentiated than the cells of benign tumours. That is to say that they tend to resemble the cells of the embryonic tissue from which the organ of origin was developed. Malignant tumours frequently spread to other tissues and local removal of the tumour is often followed by recurrence, usually because it has already spread beyond the line of excision. Spread of malignant tumours occurs by the direct invasion of adjacent tissues and it also occurs by the formation of secondary tumours, called metastases, in organs that are remote from the site of the primary tumour. Metastases are most commonly found in the lymph nodes, the lungs, the liver, the bones, the brain, the adrenal glands or the kidneys. Spread to distant tissues occurs in three ways.

1 **Lymphatic spread**. Tumour cells grow into adjacent lymph vessels and become detached and carried in the lymph to the nearest lymph nodes where secondary growth may occur. The secondary growth may ultimately destroy the lymph node, and tumour cells may again escape into the lymph vessels to be carried onwards through the lymphatic system.

2 **Vascular spread**. Tumour cells may invade the adjacent capillaries and veins and thus gain access to the blood stream. These tumour cells then become lodged in capillary networks in other organs forming secondary growths. Tumour cells that enter the portal vein become lodged in the liver and the liver is, therefore, the commonest site for metastases from malignant growths of the alimentary tract. Tumour cells from other sites enter systemic veins and tend to form metastases in the lungs. Some tumour cells, however, may pass through the pulmonary capillaries and escape into the arterial circulation to set up metastases in other organs. Malignant cells from primary lung tumours may enter the pulmonary veins and thus pass into the systemic circulation to produce metastases in other tissues.

3 **Spread by implantation**. Tumour cells from tumours which arise close to a serous membrane may become detached and implanted on other areas of the serous membrane. A carcinoma of the ovary may spread in this manner throughout the peritoneal cavity and a carcinoma of the lung may spread over the surface of the pleura.

The local invasiveness and the metastatic spread of tumours make their complete removal difficult. The rate of growth of malignant tumours varies considerably but, in general, the less well differentiated the cells the faster the rate of growth.

Malignant tumours may intially cause no symptoms but with progressive growth, loss of weight, muscle wasting and anaemia are common. This is called *malignant cachexia*. Death from malignant tumours may occur in a number of ways, but complications, such as pneumonia, are a common cause. Tumours within the alimentary tract can cause death by obstruction or, more rarely, death from liver failure due to metastases in the liver. Tumours of the brain, or metastases within the brain, may cause death due to pressure on the brain.

The classification of tumours

Tumours are most conveniently classified according to their tissue of origin.

Tumours of epithelial origin

Benign tumours of epithelial origin are either papillomas or adenomas.

A *papilloma* arises from the surface of the skin or a mucous membrane and is elevated from the surface of origin. It may be connected to the surface of origin by a broad base or by a narrow stalk (pedunculated) which contains blood vessels. In the gastrointestinal tract, papillomas are frequently called polyps. Papillomas may be further classified according to the type of epithelium from which they arise. Thus a squamous cell papilloma arises from squamous epithelium, such as the surface of the skin, a transitional cell papilloma arises in the urinary tract and a columnar cell papilloma arises in the gastrintestinal tract.

An *adenoma* arises from glandular epithelium and most commonly occurs in the intestine, the ovary, the breast, the pancreas or one of the endocrine glands. An adenoma may produce the same secretion as the tissue from which it arose and when this origin is an endocrine gland the excess of hormone may itself produce a disease.

Malignant tumours of epithelial origin are carcinomas and they are by far the most common of all malignant tumours. Carcinomas differ considerably in their rate of growth and in their degree of differentiation. They spread both by local invasion and by metastases. They are further classified according to the epithelial tissue of origin.

A *squamous cell carcinoma* arises from squamous epithelium and thus occurs on the surface of the skin, the mouth and tongue or in the oesophagus and the cervix uteri. They do, however, also occur in the bronchus and in the urinary tract.

A *transitional cell carcinoma* arises from transitional epithelium and thus occurs in the bladder or the pelvis of the ureter.

An *adenocarcinoma* arises from glandular tissue and the commonest sites of origin are the breast, the stomach, the intestines, the kidneys, the uterus, the pancreas and the thyroid gland.

A *choriocarcinoma* is a rare tumour which arises from trophoblastic (placental) tissue of a fetus. They usually invade the maternal blood vessels and spread rapidly throughout the maternal tissues.

A *schirrous carcinoma* is a term which is used to describe a hard carcinoma which is surrounded by a dense fibrous reaction. It describes the clinical features of the growth and is not a reference to a tissue of origin. It is most commonly encountered in the breast or in the stomach.

Tumours of connective tissue origin

Benign tumours of connective tissue origin are named according to the tissue from which they arise, and they closely resemble that tissue. Thus a *fibroma* arises from fibrous tissue, a *lipoma* from fatty tissue, a *neurofibroma* from the connective tissue sheaths of nerves, an *osteoma* from bone and a *chondroma* from cartilage.

A **malignant tumour of connective tissue** origin is called a sarcoma. Many sarcomas are highly malignant and the cells are so poorly differentiated that it may be impossible to determine the exact tissue of origin. Like benign connective tissue tumours, they take their names from the tissue of origin. Thus an *osteosarcoma* arises from bone and a *fibrosarcoma* from fibrous tissue. Although sarcomas are rare tumours, they most commonly occur in children and young adults. They sometimes arise in a pre-existing benign connective tissue tumour. They are generally rapidly growing tumours which metastasize into the blood stream. If the tumour is so malignant that its tissue of origin cannot be recognized it is usually called an *anaplastic* tumour. Some tumours are so poorly differentiated that it may not even be possible to decide whether they are sarcomas or carcinomas.

Tumours of lymphoid tissue, of haemopoietic tissue and of the macrophage system

The cells of the haemopoietic system, which are the cells in the bone marrow responsible for the production of the blood cells, the cells of lymphoid tissue, and the cells of the macrophage system, are all inter-related and are widely spread throughout the body. They give rise to a number of malignant tumours which are classified in a variety of ways by different authorities. Because of the widespread distribution of these cells, the neoplasms do not present as a tumour mass but appear to involve a number of different sites at the same time. The majority of the malignant tumours, however, arise in the bone marrow or in lymph nodes.

Leukaemias are neoplastic diseases of the bone marrow cells which produce the leucocytes. They are divided into acute and chronic types.

The *acute leukaemias* are *acute lymphoblastic leukaemia*, which mainly occurs in children, and *acute myeloblastic leukaemia* which mainly occurs in adults. Acute myeloblastic leukaemia arises from the cells in the bone marrow which will give rise to new polymorphonuclear leucocytes. Acute lymphoblastic leukaemia arises from the cells in the bone marrow which will give rise to new lymphoctes. Both are characterized by infiltration and destruction of the bone marrow by the leukaemic tissue, with marrow failure, anaemia and infection from increased liability to infection due to a lack of normal white cells in the peripheral blood.

The *chronic leukaemias* are *chronic myeloid leukaemia* and *chronic lymphatic leukaemia*, both of which are more common in the middle-aged and elderly. In both diseases there is a massive increase in the number of white cells in the peripheral blood, and many of these are 'primitive' or only poorly differentiated. In chronic myeloid leukaemia these cells are predominantly neutrophil polymorphonuclear leucocytes, and in chronic lymphatic leukaemia they are predominantly lymphocytes. In both diseases there is gradual replacement of the bone marrow by leukaemia tissue.

Polycythaemia vera is a neoplastic infiltration of the bone marrow by the cells which produce the new erythrocytes. Unlike the leukaemias, however, the bone marrow is not destroyed and the disease is characterized by a marked increase in the total number of red blood cells in the peripheral blood. The neoplasm is relatively chronic and death is usually due to thrombosis secondary to the increased number of circulating red blood cells.

Multiple myeloma is a malignant tumour of the plasma cells which form part of the macrophage system, but the tumour remains predominantly in the bone marrow. It is described in more detail in the chapter on the lymphatic system (Chapter 12).

Malignant tumours of lymphoid tissue are divided into Hodgkin's Disease and non-Hodgkin's lymphomas. These are described in the chapter on the lymphatic system (Chapter 12).

Tumours of naevus cell origin

Naevus cells produce the pigment which impairts colour to the skin and the pupils. A **benign tumour of naevus cells** is referred to as a '*mole*'. A *malignant melanoma* is a tumour of naevus cell origin and it may arise from a pre-existing mole. They are highly malignant and frequently metastasize by both blood and lymphatic spread.

Tumours of the central nervous system

Tumours of the nerve cells of the nervous system are rare, but when they do arise they are called *neuroblastomas*. Most cerebral tumours arise from glial cells, which are the connective tissue cells of the central nervous system, and the tumours are called *gliomas*. Tumours may also arise from the meninges and are called *meningiomas*. Cerebral tumours usually cause death by pressure on, or destruction of, the brain. Most of the tumours are benign but some may be locally invasive. Metastases from cerebral tumours are rare. The commonest intracranial tumour, however, is a metastasis from a tumour outside the cranial cavity.

Tumours of mixed tissue origin

Tumours of mixed tissue origin are tumours which contain cells of several different types of tissue. The commonest are called *teratomas* and they occur most frequently in the ovary or the testis. A **benign teratoma** is often called a *dermoid cyst* as it contains cystic spaces. Within the tumour there is frequently found recognizable hair, teeth and bone, and the last two may be visible on radiographs. A **malignant teratoma** is usually solid and, although some of them may contain recognizable tissues such as teeth or bone, most of them are composed of malignant cells with few of the cells having an obvious tissue of origin.

In children a tumour of the kidney of mixed origin, called a **Wilm's tumour,** is not infrequently encountered. This is composed of embryonic cells with some recognizable renal tubules amongst them.

Tumours of endothelial origin

Tumours of endothelial origin arise from the endothelial cells lining the blood and lymphatic vessels. The **benign tumours** are called *haemangiomas* and *lymphangiomas*, and the **malignant** *tumours* are called *haemangiosarcomas* and *lymphangiosarcomas.*

CHAPTER 4

Surface and regional anatomy

The anatomy of the human body can be described either regionally, that is by a description of all the structures of a given region, or systematically by a description of the organs that make up an entire system, such as the circulatory system or the nervous system. In the subsequent chapters the systematic approach has been adopted. In this chapter, therefore, the contents of the thorax, abdomen and pelvis are described in order to help the understanding of subsequent chapters.

To the student radiographer knowledge of the position and surface markings of organs is of great importance, therefore important surface landmarks of the body, and the relationship of some of the organs to these landmarks are also described.

The anatomical position

During the study of the human body it is assumed to be in the erect position with the arms by the side and the palms of the hands facing forwards (Fig. 4.1). This position is called the *anatomical position*. During descriptions of the relative positions of organs it must be remembered that the terms used apply to the anatomical position. It is important to have a clear understanding of the terms that are used in relating structures: *anterior*; towards the front of the body: *posterior*; towards the back of the body: *lateral*; the side farthest away from the midline of the body: *medial*; the side nearest the midline of the body: *superior*; towards the upper end of the body: *inferior*; towards the lower end of the body: *dorsal*; towards the back of the body: *ventral*; towards the front of the body: *palmar*; belonging to the palm of the hand: *plantar*; belonging to the sole of the foot: *dorsum*; a term that is used to refer to the back of the hand or the upper surface of the foot: *efferent*; a structure that passes away from an organ; and *afferent*; a structure that passes towards an organ.

During descriptions of the limbs the term *proximal* means nearest the trunk and the term *distal* means furthest away from the trunk.

The term *median plane* means a plane that passes from front to back through the midline of the body. *Sagittal planes* are planes that pass parallel to the median plane. The meaning of the term *horizontal plane* needs no explanation. The term *coronal plane* applies to a plane that passes vertically

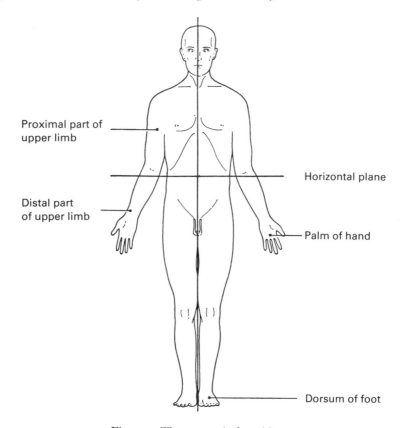

Proximal part of
upper limb

Horizontal plane

Distal part
of upper limb

Palm of hand

Dorsum of foot

Fig. 4.1. The anatomical position.

at right angles to the median plane. These planes are illustrated in Figs. 4.1
and 4.2.

The body can be divided, for descriptive purposes, into the *head, neck,
thorax* or *chest, abdomen, pelvis, upper limbs* and *lower limbs*.

The head

The skeleton of the head is formed by the *skull* which is composed of an
upper box-like portion, the cavity of which is called the *cranial cavity*, and a
lower irregular portion, which forms the skeleton of the face.

The cranial cavity contains the brain with its investing membranes,
which are called the *meninges*, and the blood vessels that supply the intracranial contents.

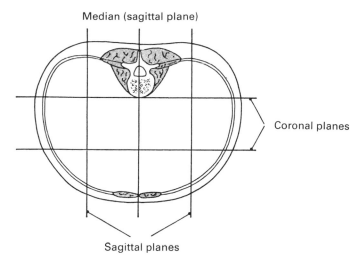

Fig. 4.2. Horizontal section through the trunk to illustrate the coronal and sagittal planes.

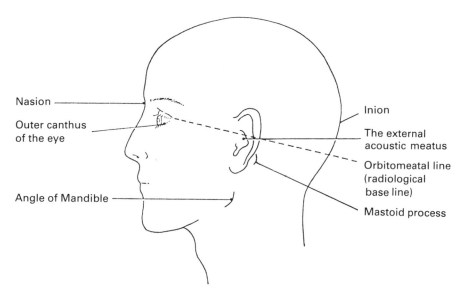

Fig. 4.3. Landmarks of the head.

Important surface landmarks of the head

The nasion is the depression at the root of the nose and is an important surface landmark in diagnostic radiography.

The inion lies at the back of the head in the midline. It is a prominence that is also called the external occipital protuberance.

The inner canthus is the medial angle of the eye, and *the outer canthus* is the lateral angle of the eye.

The orbitomeatal line is the line drawn between the outer canthus of the eye and the external acoustic meatus. This is referred to as the radiological base line in diagnostic radiography of the skull.

The sella turcica (pituitary fossa) lies 2.5 cm above and in front of the external acoustic meatus.

The neck

The neck extends between the base of the skull and the upper opening of the thoracic cavity.

Important surface landmarks of the neck

The following structures can be felt in the midline of the anterior surface of the neck.

The hyoid bone lies at the level of the 3rd cervical vertebra.

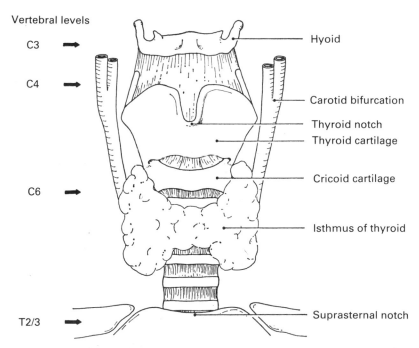

Vertebral levels

C3 ➡

C4 ➡

C6 ➡

T2/3 ➡

Hyoid

Carotid bifurcation

Thyroid notch
Thyroid cartilage

Cricoid cartilage

Isthmus of thyroid

Suprasternal notch

Fig. 4.4. Landmarks of the neck and their vertebral levels.

The notch of the thyroid cartilage of the larynx lies at the level of the 4th cervical vertebra.

The bifurcation of the common carotid artery lies at the level of the 4th cervical vertebra.

The cricoid cartilage of the larynx is continuous with the trachea at the level of the 6th cervical vertebra.

The suprasternal notch, between the inner ends of the clavicles, lies at the level of the junction of the 2nd and 3rd thoracic vertebrae.

Important structures in the neck

Posteriorly are the seven *cervical vertebrae* and their associated muscles.

In front of the cervical vertebrae is the *pharynx* which is continuous below, at the level of the 6th cervical vertebra, with the *oesophagus*.

In front of the lower part of the pharynx is the *larynx* which is continuous below, at the level of the 6th cervical vertebra, with the *trachea*.

On either side of the lower part of the larynx and the trachea is the *thyroid gland*. The left and right lobes of the thyroid gland are joined across the midline by the narrow *isthmus of the thyroid gland* which covers the 2nd and 3rd rings of the trachea. On the posterior aspect of the lobes of the thyroid gland are the *parathyroid glands*.

Lying posterolaterally to each lobe of the thyroid gland is the *common carotid artery* which is enclosed in a sheath of fibrous tissue together with the *internal jugular vein*, the *vagus nerve* and the *sympathetic trunk*, which is part of the autonomic nervous system.

In the root of the neck are the right and left *subclavian arteries*, the right

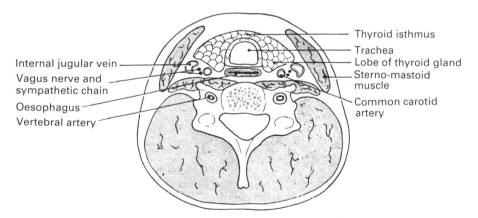

Fig. 4.5. Transverse section of the neck at the level of the lower part of the 6th cervical vertebra.

and left *subclavian veins*, the *thoracic duct*, the *brachial plexus* and the apical parts of the *lungs* with their investing *pleura*.

The thorax

Important surface landmarks of the thorax

The upper border of the manubrium sterni lies at the level of the junction of the 2nd and 3rd thoracic vertebrae.

The sternal angle (angle of Louis), which is the prominence of the junction of the manubrium sterni with the body of the sternum, lies at the level of the lower border of the 4th thoracic vertebra.

The junction of the xiphisternum with the body of the sternum lies at the level of the 9th thoracic vertebra.

On the lateral aspect of the thorax the *lower margin of the rib cage* lies at the level of the 3rd lumbar vertebra.

Posteriorly the *inferior angle of the scapula* can be palpated and lies at the level of the 8th thoracic vertebra.

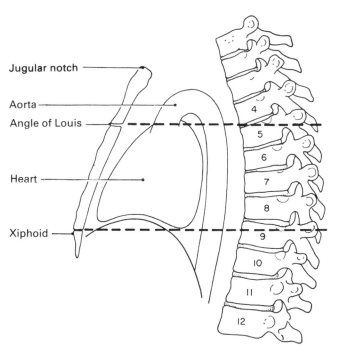

Fig. 4.6. Lateral view of the thorax to show the surface markings and their vertebral levels.

The boundaries of the thoracic cavity

The boundary between the neck and the thoracic cavity is called the *thoracic inlet*. Because of the downward slope of the 1st rib, the inlet is oblique, and its anterior margin lies at a lower level than its posterior margin. The inlet is bounded posteriorly by the 1st thoracic vertebra, anteriorly by the upper border of the manubrium sterni, and on each side by the 1st rib.

The anterior wall of the thoracic cavity is formed by the sternum, the costal cartilages and the anterior ends of the ribs with the intervening intercostal muscles.

The posterior wall of the thoracic cavity is formed by the thoracic vertebrae and the intervertebral discs.

The lateral walls are formed by the ribs and intercostal muscles.

The floor of the thoracic cavity is formed by the diaphragm, a musculotendinous structure, which separates the thoracic cavity from the abdominal cavity.

The cavity of the thorax is divided into three main parts, the *right* and *left pleural cavities* and *mediastinum*, which lies between them. The right and

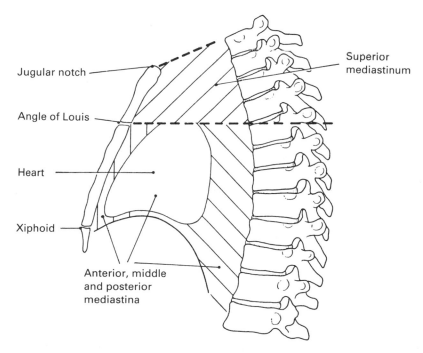

Fig. 4.7. The subdivisions of the mediastinum.

left pleural cavities contain the right and left lungs and their investing pleura.

The mediastinum

The mediastinum is divided, for descriptive purposes, into four regions.

The *superior mediastinum* lies above an arbitrary line drawn between the manubriosternal junction anteriorly and the lower border of the 4th thoracic vertebra posteriorly. Below this line lies the *inferior mediastinum*, which is divided into: *the anterior mediastinum* which lies in front of the heart and pericardium; *the middle mediastinum* which is formed by the heart and the pericardium; and *the posterior mediastinum* which lies behind the heart and the pericardium.

The superior mediastinum contains the *aortic arch* and its main branches, the brachiocephalic *veins* and part of the *superior vena cava*, the *trachea*, the oesophagus, the upper part of the *thoracic duct*, the *thymus gland* and the two vagus nerves, the *phrenic nerves* and the *left recurrent laryngeal nerve*. The relationship of the structures at the level of the body of the 4th thoracic vertebra is shown in Fig. 4.8.

The anterior mediastinum contains some loose *fatty tissue*, a few *lymph nodes* and the *internal thoracic (mammary) arteries*.

The middle mediastinum contains the *heart* and the *pericardium*, the *ascending aorta*, the *pulmonary arteries*, the *pulmonary veins*, the *bifurcation of the trachea*, the *phrenic nerves* and the *tracheobronchial lymph nodes*.

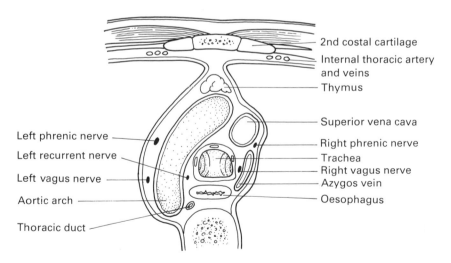

Fig. 4.8. Transverse section through the thorax at the level of the 4th thoracic vertebra to show the structures in the superior mediastinum.

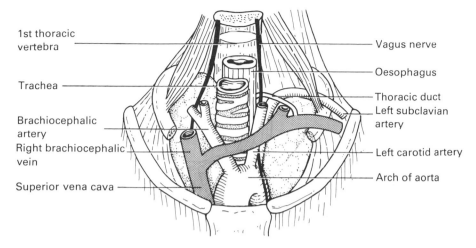

1st thoracic vertebra

Trachea

Brachiocephalic artery

Right brachiocephalic vein

Superior vena cava

Vagus nerve

Oesophagus

Thoracic duct

Left subclavian artery

Left carotid artery

Arch of aorta

Fig. 4.9. The main structures in the superior mediastinum. The manubrium sterni has been removed.

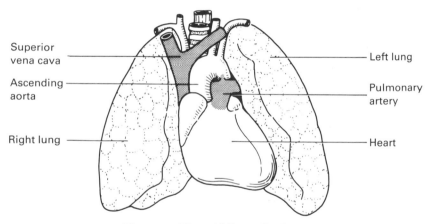

Superior vena cava

Ascending aorta

Right lung

Left lung

Pulmonary artery

Heart

Fig. 4.10. The middle mediastinum.

The posterior mediastinum contains the *descending thoracic aorta*, the *oesophagus*, the *azygos veins*, the *thoracic duct* and *lymph nodes*.

The surface markings of the heart

The *apex* of the heart lies in the 5th intercostal space about 9 cm to the left of the midline.

The *left border* of the heart extends upwards and to the right from the

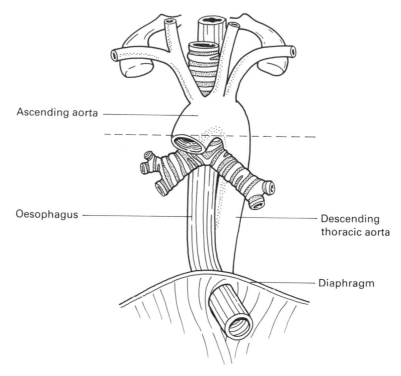

Ascending aorta

Oesophagus

Descending
thoracic aorta

Diaphragm

Fig. 4.11. The posterior mediastinum.

apex to end at the lower border of the 2nd left costal cartilage about 1 cm
from the left sternal edge.

The *right border* of the heart describes a gentle curve about 1 cm to the
right of the right border of the sternum from the 3rd to the 6th costal
cartilages.

The *lower border* of the heart extends from the lower margin of the right
border to the apex.

The surface markings of the lungs and pleura

The *apex* of the lung and its investing pleura extends upwards posteriorly to
a point about 3 cm above the middle of the clavicle. The apex of the lung,
however, does not rise above the level of the neck of the first rib which
runs obliquely downwards from back to front.

Anteriorly the edge of the pleura extends downwards and medially
behind the sternoclavicular joint to meet the opposite pleural edge in the
midline at the level of the sternal angle.

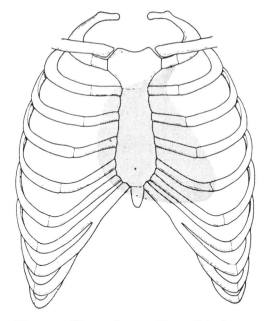

Fig. 4.12. The surface markings of the heart.

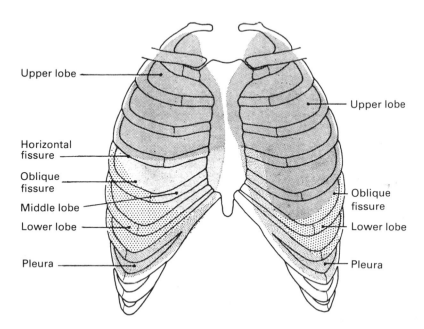

Fig. 4.13. The surface markings of the lungs and pleura—anterior view.

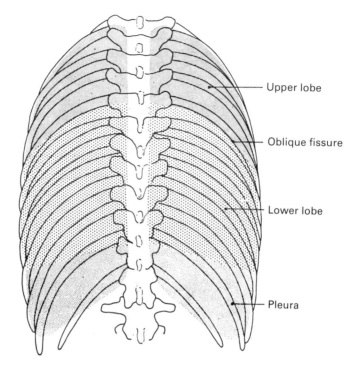

Upper lobe

Oblique fissure

Lower lobe

Pleura

Fig. 4.14. The surface markings of the lungs and pleura—posterior view.

The pleural edges then pass downwards towards the 6th costal cartilage and become continuous with the inferior margin of the pleura. The left lung does not extend as far towards the midline as the right lung, as the lower part of its anterior border is indented by the heart, forming the cardiac notch. The *inferior margin* of the pleura extends downwards and laterally from behind the 6th costal cartilage to cross the 8th rib in the line of the nipple, the 10th rib in the midaxillary line and the 12th rib at the lateral margin of the sacrospinalis muscle. The lung does not extend as far downward as the pleura and during normal breathing it lies about two rib spaces above the inferior margin of the pleura.

The abdomen

Important surface landmarks of the abdomen

The *junction of the xiphisternum* and the body of the sternum lies at the level of the 9th thoracic vertebra.

The *lower margin of the rib cage*, which is formed by the 10th rib, lies at

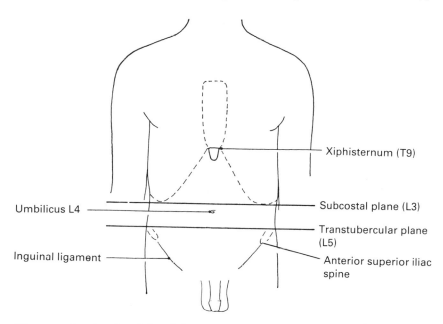

Fig. 4.15. Landmarks of the abdomen and their vertebral levels.

the level of the 3rd lumbar vertebra. A line drawn horizontally across the abdomen at this level is called the *subcostal plane*.

The *upper surface of the iliac crests* lies at the level of the upper border of the 5th lumbar vertebra. A horizontal line drawn across the abdomen at this level is called the *transtubercular plane.*

The *umbilicus* is somewhat inconstant in position but it usually lies at the level of the 4th lumbar vertebra.

The iliac crests end anteriorly at the *superior anterior iliac spines* and these are easily felt through the skin. The *inguinal ligament* extends downwards and medially from the anterior superior iliac spine to the pubic tubercle.

The regions of the abdomen

The abdomen is divided, for descriptive purposes, into nine regions by two vertical planes and two horizontal planes.

The transpyloric plane is a line drawn horizontally across the abdomen at the level of the 1st lumbar vertebra. This is approximately one hand's breadth below the junction of the body of the sternum and the xiphisternum. (The **subcostal plane**, which has already been described, is often used instead of the transpyloric plane as the upper horizontal plane for dividing the abdomen into these regions.)

The transtubercular plane has already been described.

The right and left lateral planes are vertical lines which run upwards from a point midway between the anterior superior iliac spine and the pubic tubercle. These lines, extended upwards over the thorax, are referred to as the *midclavicular lines*.

The names of the nine regions are shown in Fig. 4.16.

The boundaries of the abdominal cavity

The *anterior wall* of the abdominal cavity is formed by muscles.

The *posterior wall* is formed by the lumbar vertebral column, the psoas and quadratus lumborum muscles and the posterior parts of the iliac bones.

The *lateral walls* are formed by muscles above and by the iliac bones below.

The abdominal cavity is separated above from the thorax by the diaphragm. Below, it is continuous with the pelvic cavity through the pelvic inlet.

The contents of the abdominal cavity

The abdominal cavity contains most of the *alimentary canal*, the *liver*, the *gall bladder* and its ducts, the *pancreas*, the *kidneys* and *ureters*, the *spleen*, the *abdominal aorta* and most of its branches, the *inferior vena cava* and its branches, *lymph nodes*, *lymphatic vessels* and *nerves*.

The stomach is the expanded portion of the alimentary canal which is continuous below the diaphragm with the oesophagus. Its position varies according to the build of the individual and the volume of its contents, but two points are relatively fixed. One is the cardiac orifice, at which the oesophagus opens into the stomach, and the other is the pyloric orifice at which the stomach opens into the duodenum. The fundus, or upper part of the stomach, lies under the left dome of the diaphragm. The oesophagus opens into the stomach at the level of the body of the 11th thoracic vertebra about 5 cm to the left of the midline. The stomach opens into the duodenum through the pyloric canal, which lies at the level of the transpyloric plane about 5 cm to the right of the midline.

The duodenum is the first part of the small bowel and is relatively constant in position. It is continuous with the pyloric canal of the stomach and describes a C-shaped curve, the concavity of which faces to the left and surrounds the head of the pancreas. The duodenum is continuous with

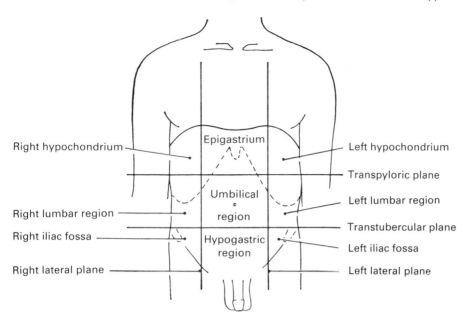

Fig. 4.16. The regions of the abdomen.

the jejunum at the level of the 2nd lumbar vertebra 3.5 cm to the left of the midline.

The small intestine is approximately 7 m in length. It is attached to the posterior abdominal wall by a fold of peritoneum. The upper portion of the small intestine is called the *jejunum* and the lower portion of the small intestine is called the *ileum*. The jejunum lies in the upper left part of the abdominal cavity and the ileum in the lower right part. The small intestine opens into the proximal part of the large intestine, called the caecum, in the right iliac fossa.

The large intestine is composed of the *caecum*, the *colon*, the *rectum* and the *anal canal*. The caecum lies in the right iliac fossa. It is continuous above with the colon. The first part of the colon, called the *ascending colon*, runs upwards from the caecum to the under surface of the liver where it turns to the left at the *hepatic flexure*. The hepatic flexure lies at about the level of the 2nd lumbar vertebra. The 2nd part of the colon, called the *transverse colon*, runs to the left across the abdominal cavity from the hepatic flexure. Close to the medial surface of the spleen it turns downwards at the *splenic flexure* and runs downwards on the left side of the abdominal cavity, as the *descending colon*, to enter the pelvis.

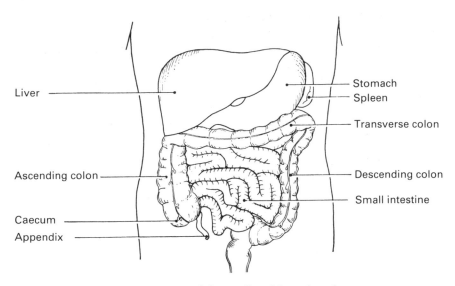

Fig. 4.17. General arrangement of the small and large bowel.

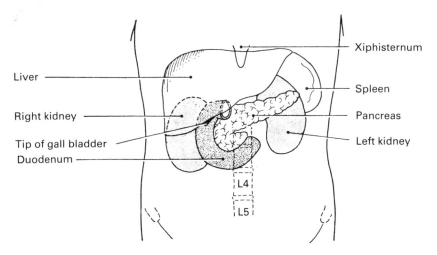

Fig. 4.18. Arrangement of the viscera on the posterior abdominal wall.

The liver lies in the upper part of the abdomen more on the right than the left. Its lower border can be outlined by a line which extends from the tip of the 10th rib on the right to the level of the space between the 5th and 6th left ribs about 7 cm to the left of the midline. A line drawn horizontally to the right from this point corresponds approximately to the upper border of the liver. The right border of the liver is related to the lower part of the right rib cage from which it is separated by the diaphragm.

The gall bladder lies on the undersurface of the right lobe of the liver and its tip usually projects from below the liver at the level of the 9th costal cartilage at the point where the lateral margin of the right rectus abdominis muscle crosses the costal margin.

The pancreas lies on the posterior abdominal wall and is composed of a head, neck and body. The neck of the pancreas lies at the level of the transpyloric plane in front of the vertebral column. From the neck, the head of the pancreas extends slightly downwards and to the right into the concavity of the duodenum, and the body of the pancreas runs slightly upwards and to the left towards the spleen.

The spleen lies in the upper left part of the abdomen, in contact with the under surface of the left dome of the diaphragm, which separates it from the 9th, 10th and 11th ribs.

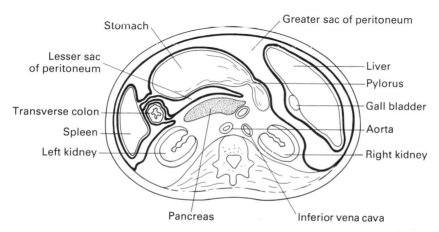

Fig. 4.19. Transverse section of the abdomen through the transpyloric plane (L1).

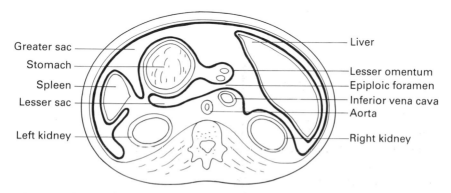

Fig. 4.20. Transverse section through the epiploic foramen at T12 to show the arrangement of the lesser and greater sacs of the peritoneum.

The kidneys lie on either side of the vertebral column on the posterior abdominal wall. The right kidney is usually about 1 cm lower in position than the left. The hilum of the right kidney lies just below the transpyloric plane and the hilum of the left kidney just above the transpyloric plane. The upper poles of the kidneys lie about 2.5 cm from the midline at the level of the spine of the 11th thoracic vertebra. The lower poles of the kidneys lie about 7 cm from the midline about 2.5 cm above the transtubercular plane.

The peritoneum is a large complex sac of serous membrane which lines the abdominal cavity and is reflected onto the abdominal viscera. In the male the peritoneal sac is completely closed but in the female it is open to the exterior via the uterine tubes, the uterus and vagina.

That part of the peritoneum which lines the walls of the abdomen is called the *parietal peritoneum* and that part which covers the abdominal viscera is the *visceral peritoneum*. The peritoneal cavity is composed of the *greater sac*, which is the main cavity of the peritoneum, and the *lesser sac*, which is a diverticulum of the main cavity and lies behind the stomach. The lesser sac of the peritoneum opens into the greater sac at an opening called the *epiploic foramen*.

The complex arrangement of the peritoneum will be better understood if it is remembered that the peritoneum is a closed sac and that the cavity of the sac contains no viscera. During the development of the abdominal viscera, some of the viscera gradually extend into the posterior wall of the sac, and thus become ensheathed by peritoneum. They remain connected to the posterior wall of the abdomen, however, by a double fold of peritoneum which contains vessels and nerves. Other viscera project only slightly from the posterior abdominal wall and are thus covered by peritoneum only on their anterior surfaces.

The mesentery. The jejunum and ileum are suspended from the posterior abdominal wall by a double fold of peritoneum called the mesentery. The mesentery arises from the posterior abdominal wall by a narrow base called the *root of the mesentery*. From the root of the mesentery, which is about 12 cm in length, the folds of the mesentery fan out to be attached to the small intestine along the whole of its 6 m.

The greater and lesser omenta. The stomach and the first part of the duodenum are suspended from the lower surface of the liver by a double fold of peritoneum called the lesser omentum. The two layers of this fold part to envelop the stomach and then reunite at the greater curvature of the stomach to form the greater omentum. The greater omentum hangs down in front of the abdominal viscera like a large apron, and is heavily loaded with fat. Near the lower part of the abdominal cavity the greater omentum turns upwards again to pass to the transverse colon.

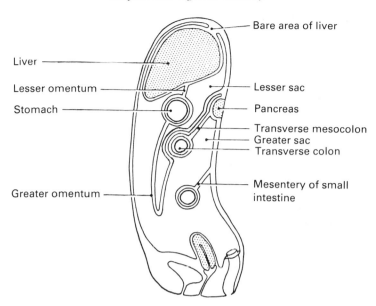

Fig. 4.21. The peritoneal cavity in longitudinal section.

The two layers of the greater omentum then pass in front of the transverse mesocolon.

The transverse mesocolon. The two layers of peritoneum which ensheathe the transverse colon reunite at its posterior border and then pass back to the posterior abdominal wall as the transverse mesocolon.

The lower part of the peritoneum is reflected over the pelvic viscera.

The pelvic cavity

The pelvic cavity lies within the walls of the true pelvis (p. 171), and it is continuous above with the abdominal cavity.

The boundaries of the pelvic cavity

The *anterior wall* of the pelvic cavity is formed by the pubic symphysis and the bodies of the public bones.

The *posterior wall* is formed by the anterior surfaces of the sacrum and the coccyx.

The *lateral walls* are formed mainly by the ischial bones and the internal obturator muscles.

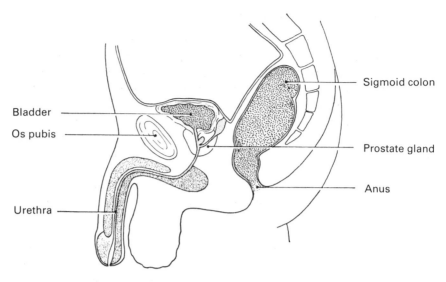

Fig. 4.22. The principal organs of the male pelvis.

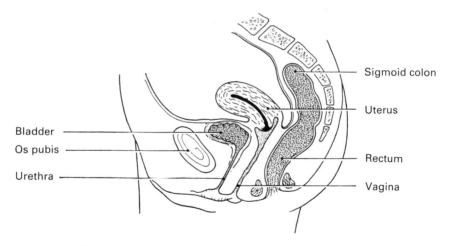

Fig. 4.23. The principal organs of the female pelvis.

Above, the pelvic cavity is continuous with the abdominal cavity at the pelvic inlet.

The *floor* of the pelvic cavity is formed by the levator ani and coccygeus muscles.

The contents of the pelvic cavity

The contents of the pelvic cavity differ in the male and the female.

The contents of the male pelvic cavity. The *bladder* lies anteriorly. At the base of the bladder is the *prostate gland* and behind the prostate gland are the *seminal vesicles*. Lying posteriorly to the bladder and the prostate gland are the *rectum* and the lower part of the *sigmoid colon*. Several loops of small intestine lie in the upper part of the pelvic cavity.

Also present in the pelvis are the *internal iliac vessels* and their branches, *lymph nodes* and *lymphatic vessels*, *nerves* and the lower portions of the two *ureters*.

The contents of the female pelvic cavity. The bladder lies anteriorly. Behind the bladder is the *vagina* and the *cervix uteri*. The *uterus* usually projects forwards from the cervix uteri to lie on the superior surface of the bladder. The *uterine tubes* extend laterally from the upper part of the uterus and are enclosed in a fold of peritoneum called the *broad ligament of the uterus*. At the lateral end of the uterine tubes are the *ovaries* which lie on the lateral pelvic wall. Behind the vagina and the cervix uteri lies the *rectum*. A pouch of peritoneum, called the *pouch of Douglas*, lies between the upper part of the vagina and the rectum.

The other contents of the female pelvic cavity are the same as in the male.

CHAPTER 5

The skeleton

The skeleton is made up of a large number of bones some of which are single and others paired. The bones vary considerably in size and shape, but they can be classified into five main groups.

Long bones consist of a roughly cylindrical shaft with an expanded portion at either end. This group includes most of the limb bones.

Short bones vary considerably in shape but they are very approximately cuboidal. They include the bones of the proximal portions of the hands and feet, i.e. the carpal and tarsal bones respectively.

Flat bones have a large surface area compared to their depth and include the bones of the skull vault and the ribs.

Irregular bones vary considerably in shape and are not included in any of the previous groups. They include the bones that make up the spinal column, i.e. the vertebrae, and some of the skull bones.

Sesamoid bones develop in tendons near joints. The most important bone in this group is the kneecap, or patella. There are many other sesamoid bones which are found particularly in the tendons of the hands and feet; some of these are regularly present, others are only found occasionally.

The structure of bone

Bone is a connective tissue and is one of the hardest tissues in the human body. It consists of an organic matrix in which inorganic salts are deposited. The rigidity of bone is provided by these inorganic salts, which constitute about 60 per cent of the total weight. They consist mainly of calcium and phosphate with some magnesium and carbonate. It is possible to remove the salts from a bone by soaking in dilute mineral acid. After treatment the bone is unaltered in shape but is very flexible and a long bone, for example, can be bent so that its oppposite ends touch.

A fibrous membrane, the *periosteum*, covers the outer surface of bones except for those areas that form moving joints, where the bone is covered with articular cartilage. The periosteum contains a network of blood vessels from which vessels penetrate into the underlying bone.

A cross-section of bone shows it to consist of two types:

1 *Compact bone*, which to the naked eye is dense and structureless, and which makes up the outer layer of a bone.

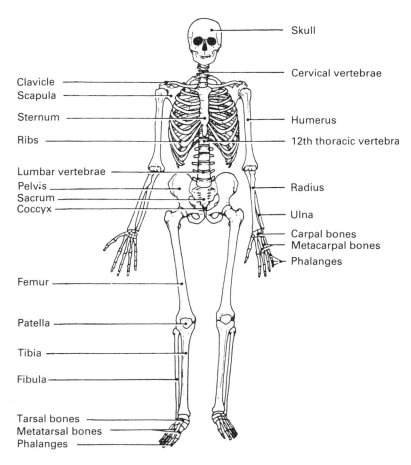

Fig. 5.1. The human skeleton.

2 *Spongy (cancellous) bone*, which consists of strands of bone called trabeculae, with intervening spaces that are visible to the naked eye.

The relative amount of each type of bone present varies from bone to bone, and from one area of a bone shaft to another, and depends on the degree of strength required. In the shaft of a long bone there is a thick outer layer of compact bone, but in an irregular or a short bone the layer of compact bone is relatively thin.

Compact bone

When compact bone is examined under the microscope it is seen to have a regular arrangement of units called *Haversian systems*.

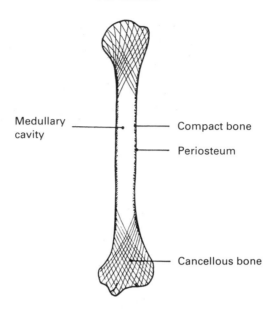

Fig. 5.2. Longitudinal section of a long bone.

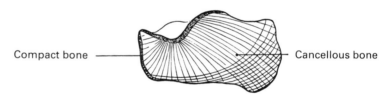

Fig. 5.3. Longitudinal section of a short bone.

Each Haversian system consists of:

1 a central canal, the *Haversian canal*, which contains vessels and nerves;

2 concentric rings of bone which surround the central canal and are called *lamellae*;

3 *lacunae*, which are the spaces between the lamellae and which contain bone cells;

4 *canaliculi*, which are minute channels which cross the lamellae and connect the lacunae. It is through these canaliculi that nutrient fluids are distributed from the vessels in the Haversian canal.

Between adjacent Haversian systems there are further lamellae called *interstitial lamellae* and the whole of the circumference of a bone is surrounded by further lamellae called *circumferential lamellae*.

Adjacent Haversian canals are connected by horizontal channels called

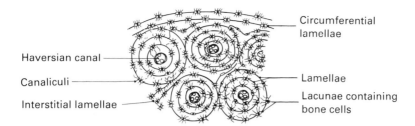

Circumferential lamellae

Haversian canal

Canaliculi

Interstitial lamellae

Lamellae

Lacunae containing bone cells

Fig. 5.4. Transverse section of compact bone.

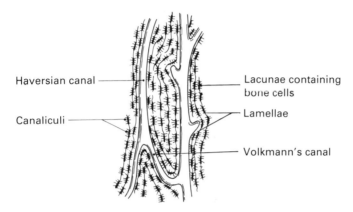

Haversian canal

Canaliculi

Lacunae containing bone cells

Lamellae

Volkmann's canal

Fig. 5.5. Longitudinal section of compact bone.

Volkmann's canals and it is through these that the vessels pass from one Haversian system to another.

Although compact bone looks structureless to the naked eye it can be seen that it contains spaces and the main difference between compact and spongy bone is the size of the spaces.

Spongy bone

Spongy bone contains lamellae which are similar in structure to those of compact bone but the spaces are larger and Haversian systems are only found in the large trabeculae. The bone receives its nourishment from surrounding vessels.

Bone marrow

The shaft of a long bone contains a central cavity called the medullary cavity. This cavity is filled with bone marrow which is also found between the trabeculae of spongy bone. At birth all the bone marrow produces blood cells and is called *red marrow*. By the time adult life is reached red marrow is only found in the bones of the skull, the clavicles, the scapulae, the vertebrae, the ribs, the pelvis and the upper ends of the humeri and femora. Elsewhere red marrow has been replaced by *yellow*, or fatty, *marrow*, which has little blood forming tissue present.

Functions of bone

1 Bone forms the framework of support for the soft tissues, and thus gives support to the weight of the body.
2 Bone provides levers on which the muscles can exert their pull to produce movement.
3 Bone provides protection for certain vital organs, such as the brain within the skull.
4 Bone is important for the production of blood cells since it contains the tissue responsible for this.
5 Bone acts as a store for calcium salts.

The development and growth of bones

During the development of the human body most of the bones are first modelled in cartilage, but a few of the bones, namely the clavicles and the bones of the skull vault, are first represented by membranes composed of mesoderm and have no cartilaginous stage. These cartilaginous or membranous precursors are later changed to bone by the process of ossification. Some of the bones are developed from a single centre of ossification; that is to say bone formation starts in one area of the precursor and extends until the whole bone is ossified. Other bones, including the long bones, are developed from more than one centre of ossification, so that at some stages of development the bone is divided by cartilaginous portions into several parts. The first centre to appear in any bone is called the *primary centre of ossification*, and the subsequent centres to appear are called the *secondary centres of ossification*. The stage of development at which the centres appear in any one bone is relatively constant, and it is thus possible to assess the age of a normal child by taking radiographs of certain areas of the skeleton.

The development and growth of a typical long bone

In a typical long bone the primary centre of ossification appears in the centre of the cartilage precursor at about the eighth week of intrauterine life and extends until the whole of the shaft is ossified. The shaft of a developing long bone is called the *diaphysis* and the ossification of the diaphysis is usually complete before birth.

At a later stage, either just before or after birth, secondary centres of ossification appear at the ends of the long bone. The end of a developing long bone is called the *epiphysis*. The secondary centre of ossification extends until the epiphysis is ossified but the epiphysis remains separated from the diaphysis by a cartilaginous plate called the *epiphyseal cartilage*. This epiphyseal cartilage persists until the growth of the bone is complete; ossification of the plate then occurs and the epiphysis and diaphysis fuse.

The increase in length of a developing long bone occurs by growth of the epiphyseal cartilage followed by ossification. The increase in diameter of a long bone is brought about by a layer of cells under the periosteum called *osteoblasts*, which produce new bone around the circumference of the developing bone. At the same time the medullary cavity increases in diameter due to bone absorption produced by cells called *osteoclasts*. The osteoblasts and osteoclasts are continuously active, producing new bone and absorbing old bone, so that a constant process of remodelling occurs until the bone assumes its adult size and shape.

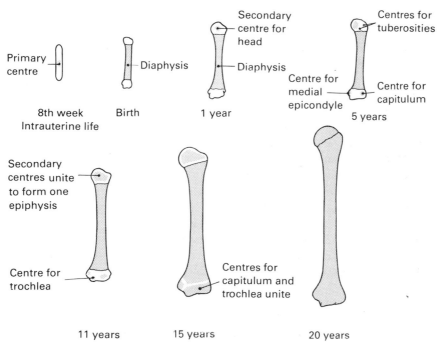

Fig. 5.6. The development of a typical long bone as illustrated by the development of the humerus.

Bone formation

Osteoblasts at the bone forming surface lay down a collagen matrix called *osteoid*. Vitamin D stimulates this process as well as the subsequent mineralization which involves the deposition of *hydroxyapatite*, a mineral containing calcium and phosphorous. Normally the processes of bone formation and osteoclastic bone resorption are balanced, but during childhood bone formation proceeds at the more rapid rate, and in a number of pathological conditions the balance between these two processes is disturbed.

Nutritional factors that affect bone growth

Bone is continually being destroyed and laid down again. This process is called remodelling. Because of the continuous synthesis−breakdown−resynthesis cycle, there is a need throughout life for the raw materials of bone. The following factors are needed for normal bone synthesis:

1 An adequate intake of *calcium* and *phosphorous*.
2 An adequate intake of, and normal metabolism of, *Vitamin D*.
3 Adequate supplies of other minerals, such as *magnesium* and *potassium*.
4 Adequate secretion of *gonadal steroid hormones, thyroxine* and *growth hormone* (see Chapter 17).

The daily requirements and sources of calcium and phosphorous are described in the chapter on the alimentary system. Calcium is actively absorbed in the upper small intestine under the influence of 1,25-dihydroxy-cholecalciferol, which is the most active metabolite of Vitamin D. Absorption is increased in the young, during pregnancy and during lactation. The presence of other foodstuffs in the intestine can also influence the rate of calcium absorption. After absorption calcium enters the *calcium pool* in the plasma and the extracellular fluid. This calcium pool is continuously added to by calcium absorption from the intestine and by calcium liberation from bone through osteoclastic activity. In contrast the calcium pool is depleted by bone formation, by urinary calcium excretion and by loss in the faeces of calcium secreted into the intestine. The body contains about 1 kg of calcium of which 99% is in the skeleton. The plasma calcium is kept within narrow normal limits in healthy individuals by a balance struck between the factors influencing the rate of gain or loss of calcium from the pool.

Most of the phosphorous in the body is found in the skeleton. The rate of phosphorous absorption does not appear to be so precisely controlled as that of calcium. Most dietary phosphorous is absorbed from the intestine and losses from the body are mainly in urine, since faeces contain virtually none of the element. Phosphorous filtered at the glomerulus is reabsorbed mainly in the proximal renal tubule, a process which is influenced by parathyroid hormone and Vitamin D.

Vitamin D metabolism will be described in the chapter on the alimentary system. The most active metabolite of vitamin D is 1,25-dihydroxycholecalciferol (1,25-DHCC) which is formed in the kidney from 25-hydroxycholecalciferol under the influence of parathyroid hormone. The rate of production of 1,25-DHCC is stimulated by parathyroid hormone and a low plasma phosphorous. Low plasma calcium concentrations, by stimulating increased secretion of parathyroid hormone, lead to the production of 1,25-DHCC.

1,25-DHCC has actions on the intestine, bone, kidney and muscle:

1 In the *small intestine*, 1,25-DHCC stimulates the synthesis of a calcium binding protein which enhances calcium absorption. It may also enhance phosphorous absorption by a similar means.

2 In *bone*, 1,25-DHCC increases calcium liberation independently of the action of parathyroid hormone, favours the deposition of calcium and also stimulates formation of the collagen matrix of bone.

3 In the *kidney*, 1,25-DHCC increases the reabsorption of calcium and phosphorous.

4 1,25-DHCC probably has a direct action on *muscle* increasing the power of muscular contraction.

The healing of fractures

When a bone is fractured a series of events occur at the site of the broken ends which, assuming that the area is immobilized, leads to healing of the fracture by new bone formation.

1 *Haematoma formation.* Following a fracture there is bleeding from damaged vessels and a blood clot is formed around the broken ends of the bone.

2 *Inflammation.* A normal inflammatory reaction occurs at the site, and the increased blood flow may be responsible for the loss of bone density around the bone ends which is frequently seen on plain radiographs of the area.

3 *Callus formation.* Following the inflammatory reaction, macrophages invade the blood clot and remove the debris. If there are detached fragments of bone these fragments undergo necrosis and are removed by the macrophages. As in other tissues, following inflammation, granulation tissue is formed. Capillaries and fibroblasts extend into the damaged area and the bone ends become reunited by fibrous tissue. Osteoblasts then begin the process of new bone formation. If the fracture is fully immobilized the osteoblasts form a mass of bone with no lamellar structure and this is called woven bone. It is at this stage that radiographs show the evidence of new bone formation and this is referred to as callus. If the fracture cannot be fully immobilized (e.g. a rib fracture), cartilage rather than woven bone is formed and this cartilage initially reunites the bone ends.

4 *Remodelling.* Over succeeding weeks the woven bone or cartilage is gradually coverted to lamellar bone and the remodelling process may be so effective that it is impossible on later radiographs to determine the original site of the fracture.

Osteoporosis

Osteoporosis is a condition in which the total bone mass is reduced, although the relative proportions of bone matrix and mineral salts are normal. Most people over the age of 50 have some degree of osteoporosis and in those over the age of 70 the condition is more obvious. Some women after the menopause are particularly affected. In general, women are more severely affected by osteoporosis than men. Those who are more physically active have less osteoporosis than those who are inactive.

The trabecular bone is particularly affected by osteoporosis and this results in weakening of the bones with an increased liability to fractures. These fractures most commonly occur in the vertebral bodies, the neck of the femur and the distal radius.

In addition, certain disease states predispose to osteoporosis. These include Cushing's syndrome, thyrotoxicosis and hypogonadism (see Chapter 17). Immobilization, even for a few weeks, leads to some loss of bone mass, and individual bones immobilized (e.g. after a fracture) can show quite definite osteoporotic changes. These changes are, however, usually only recognizable on plain radiographs when 50% of the bone mass has been lost.

Osteoporosis, because of fractures of the femoral neck, is an important cause of hospital admission in the elderly.

Osteomalacia

Osteomalacia is a bone disorder characterized by failure of calcification of osteoid due to deficiency of Vitamin D or to resistance to its action. Nutritional deficiency of Vitamin D causing osteomalacia (or rickets in children) is rare in the UK. However, immigrants from India and Pakistan have been found to have Vitamin D deficiency. This is probably due to a number of factors including dietary deficiency and decreased formation of Vitamin D in the skin because of skin pigmentation and diminished exposure to sunlight.

Osteomalacia in adults causes bone pains and muscle weakness which leads to particular difficulty in walking and climbing stairs. In children the condition is known as rickets and leads to bowing of the legs, thickening of the wrists, prominence of the costochondral junctions and bossing of the skull.

Paget's disease of bone

Paget's disease is characterized by disordered bone remodelling, followed by excessive bone formation. Such changes result in bones that are denser on radiographs, but are often weaker thus fracturing more easily. The axial skeleton is more commonly affected. The bones involved (including the skull) are larger, and nerves, even the spinal cord, may be compressed causing neurological symptoms. The cause of Paget's disease is unknown but its frequency increases with age and it is commoner in men.

CHAPTER 6

The skull

The skull is the name given to the bones which make up the skeleton of the head. The only bone of the skull which is moveable is the mandible, or lower jaw. The rest of the bones are firmly joined together at immoveable joints called sutures.

The skull is divided into an upper box-like portion called the *calvarium*, and a lower irregular portion which constitutes the *skeleton of the face*.

The calvarium contains the brain. It is made up of eight bones: one *frontal bone*; two *parietal bones*; two *temporal bones*; one *occipital bone*; one *sphenoid bone*; and one *ethmoid bone*.

The facial skeleton is made up of 14 bones: two *zygomatic bones*; two *maxillae*; two *nasal bones*; two *lacrimal bones*; one *vomer*; two *palatine bones*; two *inferior nasal conchae*; and one *mandible*.

These bones will be described individually later in this chapter, but it is important first to study the skull as a whole.

The skull viewed from above

The skull seen from above has a smooth rounded appearance and is oval in outline. Anteriorly is the *frontal bone*, laterally are the *left* and *right parietal bones* and posteriorly is the *occipital bone*. Three important sutures can be seen:

1   **The coronal suture** anteriorly which unites the frontal bone with the right and left parietal bones.
2   **The sagittal suture** in the midline which unites the two parietal bones.
3   **The lambdoid suture** posteriorly which unites the right and left parietal bones with the occipital bone.

The junction of the coronal suture and the sagittal suture is called the *bregma*. In the early months of life there is at this site a membrane covered gap called the *anterior fontanelle*.

The junction of the sagittal suture and the lambdoid suture is called the *lambda*, and the membrane covered gap found at this site in the early months of life is called the *posterior fontanelle*.

The skull viewed from the front

The skull viewed from the front is wider above than below. The upper portion is smooth and is made up of the *frontal bone*; the lower portion is

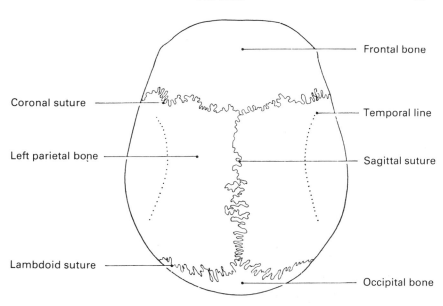

Fig. **6.1.** The skull seen from above.

irregular. On either side, below the frontal bone, are the orbital cavities, and above the medial half of these the frontal bone has a smooth rounded elevation called the *superciliary arch*. The superciliary arches are joined in the midline by a prominence called the *glabella*. Below the glabella the frontal bone recedes to form the root of the nose. This point is called the *nasion*.

The orbital openings are roughly quadrilateral. The upper margin is formed almost entirely by the frontal bone. This margin bears on its medial third a notch called the *supraorbital notch*. This is often completely enclosed with bone, however, and is then called the supraorbital foramen.

At the lateral side of the orbit the *zygomatic process of the frontal bone* projects downwards for a short distance to unite with the *frontal process of the zygomatic bone*. These two processes form the lateral margin of the orbit.

The inferior margin of the orbit is made up of the *zygomatic bone* laterally and the *maxilla* medially.

The medial margin is formed above by the *frontal bone* and below by a projection of the maxilla—the *frontal process of the maxilla*.

The anterior opening of the nasal cavities is roughly triangular. The apex and the upper part of the sides is formed by the *nasal bones*, and the base and the lower part of the sides is formed by the *maxillae*. The two nasal bones articulate with the frontal bone above at a point called the nasion. They articulate with each other in the midline and with the frontal process of the maxilla on either side. Where the two maxillae meet in the midline below the nasal openings there is a pointed process, the *anterior nasal spine.*

Below the orbits the two maxillae form a large part of the skeleton of the face. Laterally each maxilla bears a short process, the *zygomatic process*, which articulates with the zygoma. Just medial to this process is an opening which transmits vessels and nerves, and is called the *infraorbital foramen.* The two zygomatic bones form the prominence of the cheeks on either side. Below the zygomatic processes the maxilla is somewhat narrower and consists of the *alveolar margin* which bears the upper teeth.

The lowest part of the facial skeleton consists of the *mandible*, which carries the lower teeth.

The orbits

The orbits are roughly pyramidal in shape. Each orbit thus has a base, an apex, a roof, a floor, and lateral and medial walls.

The base is the orbital opening onto the face and this has already been described.

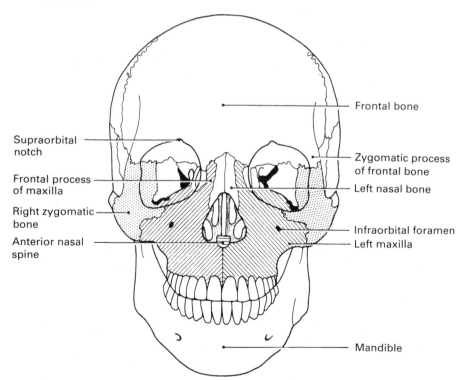

Fig. 6.2. The skull seen from the front.

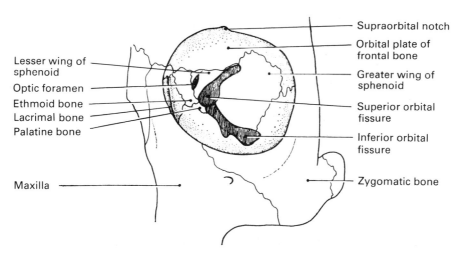

Fig. 6.3. The left orbit seen from the front.

The roof of the orbit is thin and is slightly concave from side to side. It is formed almost entirely by the *frontal bone*, but the *lesser wing of the sphenoid bone* forms a small area of the roof posteriorly. Anteriorly, near the lateral margin of the roof, is a hollow, called the *lacrimal fossa*, and this contains the lacrimal gland. At the junction of the roof and the posterior part of the medial wall is the *optic foramen* or canal. Through this foramen the optic nerve and the ophthalmic artery enter the orbit from the middle cranial fossa.

The medial wall of the orbit is also thin. It is formed from front to back by the *frontal process of the maxilla*, the *lacrimal bone*, the *orbital plate of the ethmoid* and a small portion of the *body of the sphenoid*. The medial wall slopes gently downwards and laterally from the roof to merge with the floor of the orbit. In the anterior part of the medial wall is a groove in which the lacrimal sac is found. Below this groove is a canal, the *nasolacrimal canal*, which runs downwards to enter the nasal cavity.

The floor of the orbit is triangular, and slopes upwards and medially to merge with the medial wall. The greater part of the floor is formed by the *maxilla*, but the *zygomatic bone* forms a small area anterolaterally and the *orbital process of the palatine bone* forms a very small area posteriorly. The floor of the orbit forms the roof of the *maxillary antrum* which lies immediately below. The posterior part of the floor is separated from the lateral wall by a narrow slit called the *inferior orbital fissure*. This fissure communicates with a space at the base of the skull called the *infratemporal fossa*.

The lateral wall of the orbit is somewhat thicker than the other walls. It is formed mainly by the anterior surface of the *greater wing of the sphenoid*, but a small area anteriorly is formed by the *zygomatic process of the frontal bone* and the *frontal process of the zygomatic bone*. Posteriorly the lateral wall and the roof are separated by a slit called the *superior orbital fissure*. This fissure transmits the oculomotor, the trochlear and the abducens nerves which supply the extrinsic muscles of the eye. This fissure widens out towards its medial end.

The apex of the orbit is the medial end of the superior orbital fissure.

The skull viewed from the side

When the skull is viewed from the side five of the bones which make up the calvarium can be distinguished.

Anteriorly is the *frontal bone* which articulates posteriorly with the *parietal bone* at the coronal suture.

The parietal bone articulates posteriorly with the *occipital bone* at the lambdoid suture, and inferiorly for most of its length with the *temporal bone*.

Anteriorly to the temporal bone a small area of the side of the skull vault is formed by the *greater wing of the sphenoid*.

The portion of the temporal bone which forms the side of the vault is called the *squamous* part of the temporal bone. Extending forwards from the posterior part of the squamous portion of the temporal bone is the *zygomatic process* which unites with the *temporal process of the zygomatic bone* to form the *zygomatic arch*.

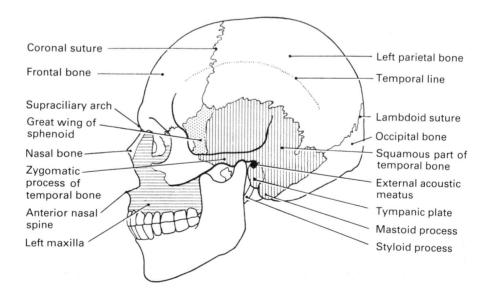

Fig. 6.4. The skull seen from the left side.

Just posterior to the root of the zygomatic process is the *external acoustic meatus*. The anterior and inferior margins of the meatus are formed by a plate of bone called the *tympanic plate*. The edge of this plate is roughened where the cartilaginous portion of the external acoustic canal is attached.

Behind the external acoustic meatus is the *mastoid* part of the temporal bone, and this projects downwards to form the *mastoid process*. In between the external acoustic meatus and the mastoid process is the *styloid process*, a slender process which points downwards and forwards. The styloid process is attached to the hyoid bone by the stylo-hyoid ligament.

The posterior border of the mastoid temporal bone articulates with the occipital bone.

Below the frontal bone anteriorly can be seen the *zygomatic bone*, the *maxilla* and the *mandible*.

When the side of the vault is examined, a curved line, which starts at the zygomatic process of the frontal bone and curves backwards over the frontal and parietal bones, can be seen. This line, called the *temporal line*, marks the upper border of a shallow hollow on the side of the skull called the *temporal fossa*. This fossa is limited anteriorly by the posterior surface of the frontal process of the zygoma and inferiorly by the upper surface of the zygomatic arch. The floor of this fossa gives rise to the temporalis muscle, the tendon of which passes deep to the zygomatic arch to be attached to the coronoid process of the mandible.

Below and medial to the anterior part of the temporal fossa is the *infratemporal fossa*, which is obscured when the mandible is in place. The infratemporal fossa is bounded anteriorly by the anterior part of the maxilla and above by the greater wing of the sphenoid and a small area of the squamous part of the temporal bone. Medially it is bounded by a projection from the base of the sphenoid bone called the *lateral pterygoid plate*. The infratemporal fossa communicates with the orbit through the *inferior orbital fissure*.

Extending upwards and medially from the infratemporal fossa is a narrow space between the posterior part of the maxilla and the sphenoid bone. This space, the *pterygopalatine fossa*, also communicates with the orbit through the inferior orbital fissure.

The skull viewed from below

The base of the skull is irregular and, for convenience of description, is divided into three areas:

1 *An anterior part* which extends from the front teeth to the posterior margin of the hard palate.

2 *A middle part* between the posterior margin of the hard palate and the anterior margin of the foramen magnum.

3 *A posterior part* behind the anterior margin of the foramen magnum.

The anterior part of the base of the skull

The anterior part of the base of the skull is formed by the *hard palate* and the *alveolar process of the maxillae* and is at a lower level than the rest of the base. The hard palate is bounded anteriorly and laterally by the horseshoe-shaped alveolar process of the maxillae which contains the upper teeth. In the adult skull these number 16, eight on each side.

The hard palate is formed in its anterior two-thirds by the palatine process of each maxilla and in its posterior third by the horizontal plates of the palatine bones. The palatine bones do not, however, form any part of the alveolar process.

In the midline of the hard palate is a bony ridge which marks the site of fusion between the two maxillae in its anterior two-thirds, and the horizontal plates of the palatine bones in its posterior third. Crossing this at right angles is a transverse line which marks the site of the suture between the maxillae and the palatine bones.

The margin of the hard palate is a free sharp edge which carries, in the midline, a backwardly directed spine called the *posterior nasal spine*.

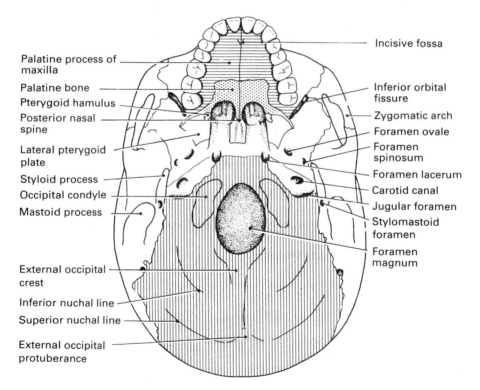

Fig. **6.5.** The base of the skull.

At the anterior end of the suture between the two sides of the hard palate is a shallow depression, the *incisive fossa*. Opening into the floor of this fossa are several small foramina which transmit vessels and nerves. Further small foramina for vessels and nerves are found at the posterolateral corners of the hard palate.

The middle part of the base of the skull

The middle part of the base of the skull extends from the posterior margin of the hard palate to the anterior margin of the foramen magnum.

In the midline anteriorly is the *vomer* which forms the posterior part of the nasal septum separating the two halves of the nasal cavity. Immediately behind this is a wide bar of bone which extends backwards to the anterior margin of the foramen magnum. This bar of bone consists of the *basal part of the sphenoid bone* anteriorly and this is directly continuous with the *basal part of the occipital bone* posteriorly.

At the posterior end of the alveolar process is the *pterygoid process of the sphenoid bone*. This protrudes downwards from the junction of the greater wing of the sphenoid with the body of the sphenoid and consists of two plates of bone; a medial plate which is narrow and points directly backwards, and a lateral plate which projects backwards and laterally. The space between the two plates is called the *pterygoid fossa*. The medial plate has, on its posterior border, a blunt projection called the *hamulus*.

Lateral to the pterygoid process is the *infratemporal fossa*. The *lateral pterygoid plate* forms the medial wall of the fossa, and the lower surface of the *greater wing of the sphenoid* forms most of the roof. Laterally the greater wing of the sphenoid articulates with the *squamous part of the temporal bone* which forms the rest of the roof of the fossa.

There are two important foramina in the infratemporal surface of the greater wing of the sphenoid. Medially, close to the roof of the pterygoid process, is a large oval foramen, the *foramen ovale*, which transmits a division of the trigeminal nerve from the cranial cavity (the mandibular division of the trigeminal nerve). Posterolaterally to this is a smaller, round foramen—the foramen spinosum—through which the middle meningeal artery enters the cranial cavity.

Between the posterior part of the greater wing of the sphenoid and the anterior part of the basi-occiput is an irregular wedge of bone, the *petrous* part of the temporal bone. Just behind the foramen spinosum there is a groove between the greater wing of the sphenoid and the petrous part of the temporal bone. This groove lodges the auditory tube, which is a communication between the pharynx and the middle ear. Between the apex of the petrous temporal and the body of the sphenoid is an irregular foramen called the *foramen lacerum*. Behind and lateral to the foramen lacerum there is a large circular foramen which is the opening of the *carotid canal*. The internal carotid artery runs upwards in this canal, which turns forwards and medially to enter the foramen lacerum, and the internal carotid artery then enters the skull through the upper part of the foramen lacerum. The lower part of the foramen lacerum is closed, in the living body, by cartilage.

Posteriorly to the carotid canal, between the petrous temporal bone and the basi-occiput, is the *jugular foramen*. Anteriorly to this the bone is hollowed out to form the *jugular fossa*. The jugular foramen contains several important structures. Anteriorly it transmits the inferior petrosal sinus, its middle portion transmits the glossopharyngeal, vagus and accessory nerves and its posterior portion transmits the internal jugular vein.

On the lateral side of the jugular foramen is the *styloid process*. Anterolaterally to the styloid process is a smooth hollow, the *articular fossa* of the temporomandibular joint. In front of the articular fossa is a rounded elevation called the *articular tubercle*, which is continuous on its lateral side with the root of the zygomatic arch.

The posterior part of the base of the skull

In the midline anteriorly is the *foramen magnum* which transmits the spinal cord and its coverings, the spinal accessory nerves and the vertebral arteries. Anteriorly the sides of the foramen magnum are overlapped by the *occipital condyles* which articulate with the superior articular facets of the 1st cervical (atlas) vertebra. The occipital condyle is pierced anteriorly by the *hypoglossal canal* which transmits the hypoglossal nerve.

Posterolaterally to the styloid process is a small foramen, the *stylomastoid foramen*, through which the facial nerve emerges from the temporal bone. Behind and lateral to this foramen the *mastoid process* projects downwards.

In the midline behind the foramen magnum is a ridge of bone, the *external occipital crest*, which runs backwards to a prominent process called the *external occipital protuberance*. Running horizontally on either side of the external occipital protuberance is a ridge of bone, the *superior nuchal line*. Halfway along the external occipital crest is another horizontal line, the *inferior nuchal line*. The bone in this region is roughened for the attachment of muscles and ligaments.

The interior of the skull

The interior of the skull can be examined by removing the top of the vault, the skull cap.

The interior of the skull cap

The interior of the skull cap is formed anteriorly by the *frontal bone*, on each side by the *left* and *right parietal bones* and posteriorly by a small portion of the *occipital bone*. The coronal and sagittal sutures and a small part of the lambdoid suture can usually be seen on the interior of the skull cap, but with advancing age they may be obliterated.

Anteriorly, in the midline of the frontal bone, is a small crest of bone called the *frontal crest*. Running backwards from this is a shallow groove, the *sagittal sulcus*, which runs in the median plane across the skull cap. In life this lodges a large venous channel, the superior sagittal sinus (p. 303). On either side of the sagittal sulcus there are a number of shallow depressions called granular pits.

Running upwards from the cut edge on either side of the skull cap are a number of shallow grooves which mark the position of the meningeal arteries and veins.

Wider grooves are present, particularly near the cut edges, which mark the position of the cerebral convolutions, but they are not well marked in the skull cap.

The interior of the base of the skull

The interior of the base of the skull is naturally subdivided into the *anterior, middle* and *posterior cranial fossae.*

The anterior cranial fossa is formed almost entirely by the *frontal bone* but the *lesser wings of the sphenoid* form the posterior edges of the floor and the *cribriform plate of the ethmoid* forms a small part of the floor in the midline.

The middle cranial fossa is formed by the *body* and *greater wings of the sphenoid bone* and by the *temporal bones*. Its posterior border is formed by the upper borders of the *petrous portions of the temporal bones*.

The posterior cranial fossa is formed mainly by the *occipital* and *temporal bones.*

The anterior cranial fossa

The anterior fossa is at a higher level than the middle cranial fossa. It is separated from the middle cranial fossa, in the midline, by the optic groove which is a groove on the body of the sphenoid bone between the two optic foramina. Laterally it is separated from the middle cranial fossa by the greater wings of the sphenoid and the temporal bones.

In the midline of the anterior cranial fossa is a sharp upward projection of bone, the *crista galli* of the ethmoid bone. The *falx cerebri*, a fold of dura

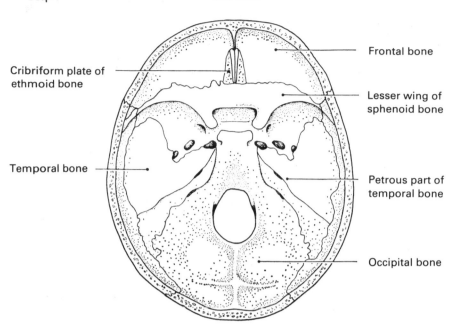

Cribriform plate of
ethmoid bone

Frontal bone

Lesser wing of
sphenoid bone

Temporal bone

Petrous part of
temporal bone

Occipital bone

Fig. 6.6. The interior of the base of the skull.

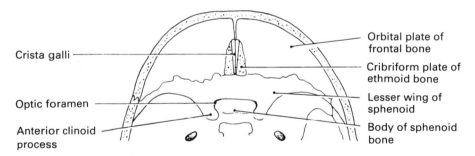

Crista galli

Optic foramen

Anterior clinoid
process

Orbital plate of
frontal bone

Cribriform plate of
ethmoid bone

Lesser wing of
sphenoid

Body of sphenoid
bone

Fig. 6.7. The anterior cranial fossa.

mater which lies between the two cerebral hemispheres, is attached to the
crista galli. Laterally to this, the floor of the anterior cranial fossa is formed
by a perforated plate of bone called the *cribriform plate of the ethmoid bone*.
Branches of the olfactory nerve enter the nasal cavity through these perfor-
ations in the cribriform plate. This fits into a groove between the two *orbital
plates of the frontal bone* which form the greater part of the floor of the
anterior cranial fossa.

The posterior part of the floor of the anterior cranial fossa is formed by the *body of the sphenoid* in the midline and the *lesser wings of the sphenoid* laterally. The lesser wing of the sphenoid is perforated, close to its junction with the body of the sphenoid, by the *optic foramen*. The medial ends of the two lesser wings bear processes which are backwardly directed and are called the *anterior clinoid processes*.

The middle cranial fossa

The middle cranial fossa has a central raised portion which extends from the optic groove anteriorly to the dorsum sellae posteriorly. On either side of this raised portion is a deep hollow.

The median raised portion of the middle cranial fossa is formed by the body of the sphenoid. Anteriorly is the groove between the two optic foramina, the *optic groove*, which in life contains the optic chiasma (p. 493). Behind this is a hollow called the *sella turcica*, in which the pituitary gland lies. The posterior wall of the sella turcica is formed by the *dorsum sellae*. On the upper surface of the dorsum sellae are two lateral projections called the *posterior clinoid processes*. Lateral to the sella turcica is a groove which contains the internal carotid artery.

The lateral portion of the middle cranial fossa is bounded anteriorly by the posterior border of the *lesser wing of the sphenoid*. The floor of the middle cranial fossa is formed from before backwards by the *greater wing of the sphenoid*, the *squamous portion of the temporal bone* and the upper surface of the *petrous portion of the temporal bone*.

Anteriorly, between the greater and lesser wings of the sphenoid, is the *superior orbital fissure* which communicates with the orbit. The oculomotor, trochlear and abducens nerves enter the orbit through this fissure. Just behind the medial end of this fissure there is a round foramen in the greater wing of the sphenoid, the *foramen rotundum*. The maxillary division of the trigeminal nerve passes through the foramen rotundum to reach the *pterygomaxillary* fissure.

Just behind and lateral to the foramen rotundum is the *foramen ovale* and posterior to this is the *foramen spinosum*. The mandibular division of the trigeminal nerve leaves the cranial cavity through the foramen ovale. The middle meningeal artery enters the skull through the foramen spinosum and running laterally from this foramen is the groove in which the artery lies.

Medial to the foramen ovale is the *foramen lacerum*, through the upper part of which the internal carotid artery enters the cranial cavity.

The posterior cranial fossa

The posterior cranial fossa is the largest of the three cranial fossae. It is bounded anteriorly by the dorsum sellae in the midline and by the upper

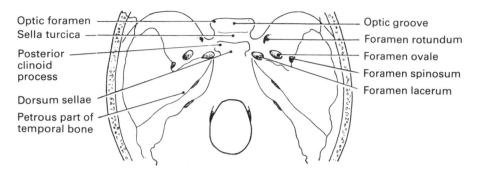

Optic foramen —
Sella turcica —
Posterior clinoid process —
Dorsum sellae —
Petrous part of temporal bone —

— Optic groove
— Foramen rotundum
— Foramen ovale
— Foramen spinosum
— Foramen lacerum

Fig. 6.8. The middle cranial fossa.

borders of the petrous temporal bone laterally. Posteriorly there is a bulge which corresponds with the external occipital protuberance on the outside of the skull and running laterally from this is a groove which contains a venous sinus—the transverse sinus. This groove marks the posterior and lateral limits of the posterior cranial fossa. The floor of the fossa is formed in the midline anteriorly by the *body of the sphenoid* and the *basal portion of the occipital bone*. Laterally the floor is formed by the *petrous* and *mastoid portions of the temporal bones* and the *squamous portions of the occipital bone*.

In the floor, in the midline, is the *foramen magnum*, the anterior portion of which is narrowed on either side by the occipital condyles. Above each condyle can be seen the opening of the *hypoglossal canal* through which the hypoglossal nerve leaves the cranial cavity.

The groove which marks the posterior and lateral limits of the posterior cranial fossa curves downwards at the anterior margin of the fossa to form an S-shaped groove. This contains the sigmoid sinus. At the anterior end of this groove is the opening of the *jugular foramen*. Above and medial to the jugular foramen is the *internal acoustic meatus*.

The nasal cavity

The nasal cavity is an irregularly-shaped cavity which lies above the floor of the mouth and below the anterior cranial fossa. It is divided into right and left halves by the *nasal septum*. The anterior part of the septum is formed in the living body by a cartilaginous plate so that when the skull is examined there is only a single anterior opening to the nasal cavity. Opening into the nasal cavity are the *paranasal sinuses*.

Each half of the nasal cavity has a roof, a floor, and lateral and medial walls.

The roof of the nasal cavity is horizontal in its central portion where it is formed by the *cribriform plate of the ethmoid*, but anteriorly and posteriorly

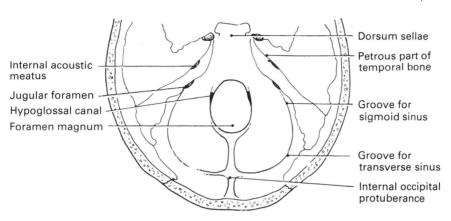

Internal acoustic meatus

Jugular foramen

Hypoglossal canal

Foramen magnum

Dorsum sellae

Petrous part of temporal bone

Groove for sigmoid sinus

Groove for transverse sinus

Internal occipital protuberance

Fig. 6.9. The posterior cranial fossa.

it slopes downwards. The anterior part of the roof is formed by the *nasal bones* and the posterior part by the *body of the sphenoid*.

The floor of the nasal cavity is formed by the upper surface of the *hard palate*, which separates the nasal cavity from the mouth.

The medial wall of the nasal cavity is formed by the bony septum which is composed of the *perpendicular plate of the ethmoid* above, and the *vomer* below.

The lateral wall of the nasal cavity is irregular due to the presence of three bony projections, the inferior, middle and superior nasal conchae. Each of these three projections forms the roof of a passage called the inferior, middle and superior sections of the meatus of the nose. The lateral wall is formed in front and below by the nasal surface of the *maxilla*, behind by the *perpendicular plate of the palatine bone* and above by the nasal surface of the *ethmoidal labyrinth*, which is a mass of air cells in the portion of the ethmoid bone which separates the orbit from the nasal cavity. The *lacrimal bone* forms a small part of the lateral wall between the maxilla and the ethmoid.

The inferior nasal concha is a curved plate of bone which projects into the nasal cavity and is a separate bone. The *middle* and *superior nasal conchae* are projections of the nasal surface of the ethmoidal labyrinth.

The nasolacrimal canal, a canal which descends from the orbit, opens into the inferior meatus of the nose.

The paranasal sinuses

The paranasal sinuses are cavities contained within certain of the bones of the skull. These cavities communicate with the nasal cavity and are lined

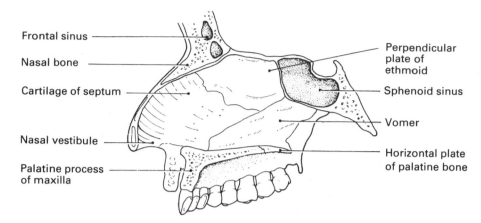

Frontal sinus

Nasal bone

Cartilage of septum

Nasal vestibule

Palatine process
of maxilla

Perpendicular
plate of
ethmoid

Sphenoid sinus

Vomer

Horizontal plate
of palatine bone

Fig. 6.10. The nasal septum.

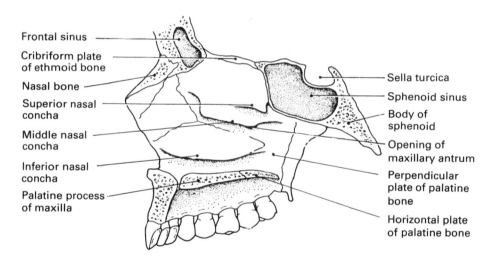

Frontal sinus

Cribriform plate
of ethmoid bone

Nasal bone

Superior nasal
concha

Middle nasal
concha

Inferior nasal
concha

Palatine process
of maxilla

Sella turcica

Sphenoid sinus

Body of
sphenoid

Opening of
maxillary antrum

Perpendicular
plate of palatine
bone

Horizontal plate
of palatine bone

Fig. 6.11. The lateral wall of the nasal cavity.

with mucous membrane. The mucus secreted by this membrane is swept
into the nose by cilia on its surface. The function of the sinuses is not
clearly understood, but they are thought to add resonance to the voice, and
they lighten the head. The sinuses are found in the frontal bone, the
maxillae, the ethmoid bone and the sphenoid bone. They are virtually
absent at birth but gradually enlarge during childhood and puberty. Even
so their size varies considerably in different adult individuals.

The frontal sinuses

There are two frontal sinuses which are found between the inner and outer tables of the frontal bone, behind the superciliary arches. The two sinuses are separated by a septum which is rarely in the midline so that one frontal sinus is frequently larger than the other. Each sinus is roughly triangular in shape, the apex of the triangle being at the nasion. The sinus also extends slightly backwards into the roof of the orbit. The frontal sinuses are not infrequently partly divided by incomplete bony partitions into several inter-communicating portions. They extend upwards for a variable distance into the frontal bone. Seen from the side, as in a lateral view of the skull, each frontal sinus is roughly triangular in shape. Each frontal sinus opens into the middle meatus of the nose, below the middle concha, by the *frontonasal duct* which runs through the anterior part of the ethmoidal labyrinth. Not infrequently both frontal sinuses are absent.

The ethmoidal sinuses

The ethmoidal sinuses are air cells which lie in the ethmoidal labyrinth between the orbit and the nasal cavity. They are separated from the orbit by a thin plate of bone, the orbital plate of the ethmoid bone. The cells vary considerably in number in different individuals, but they are divided into an anterior, a middle and a posterior group. The cells in each group are separated by incomplete bony septa.

The anterior group of air cells opens into the nose, together with the frontal sinuses, through the frontonasal duct.

The middle group of air cells opens into the middle meatus of the nose under the middle concha.

The posterior group of air cells opens into the superior meatus of the nose under the superior concha.

The sphenoid sinuses

The two sphenoidal sinuses, one on each side of the midline, are located in the body of the sphenoid bone behind the upper part of the nasal cavity. They are separated from each other by a thin bony septum which is rarely exactly in the midline so that one sinus is usually larger than the other. They vary in size but frequently extend backwards under the sella turcicia and rarely extend laterally into the base of the greater wing of the sphenoid.

Each sinus has an opening in the upper part of its anterior wall through which it communicates with the upper posterior part of the nasal cavity.

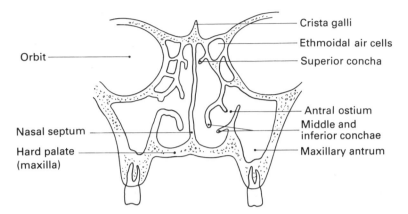

Fig. 6.12. The maxillary antrum in coronal section.

The maxillary sinuses (maxillary antra)

The two maxillary antra are the largest of the paranasal sinuses and are pyramidal-shaped spaces in the body of each maxilla. The *medial wall* of the antrum forms part of the lateral wall of the nasal cavity and bears, on its nasal side, the inferior nasal concha. Above this concha is the *antral ostium* through which the sinus opens into the middle meatus of the nose. The *apex* of the antrum extends into the zygomatic process of the maxilla. The *roof* or upper wall of the antrum is the floor of the orbit. The *floor* of the antrum is formed by the alveolar process of the maxilla. Occasionally the roots of the molar and premolar teeth project into the antrum.

The individual bones of the skull

The mandible

The mandible is the bone of the lower jaw. It is composed of a horseshoe-shaped body, which lies horizontally, and two broad processes which run vertically and are called the rami.

1 **The body of the mandible** has inner and outer surfaces and upper and lower borders.

The outer surface bears a faint line in the midline anteriorly which marks the line of fusion of the two halves of the bone. This is the *symphysis menti*. Anteriorly, on the lower part of the external surface, is a triangular elevation which is called the *mental protuberance* and on either side of this is a small tubercle, the *mental tubercle*. One either side, about 25 cm lateral to the symphysis, is a small foramen, the opening of which faces upwards and

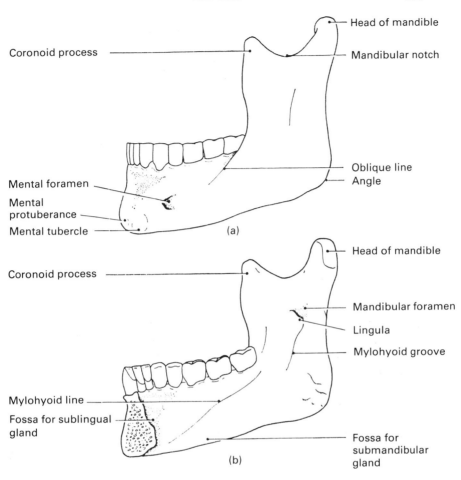

Fig. 6.13. The mandible: (a) external surface, and (b) internal surface.

backwards. This is the *mental foramen* and it transmits vessels and nerves. Running upwards and backwards from the mental foramen is a faint ridge, called the *oblique line*, and this becomes continuous with the anterior border of the ramus.

The **lower border** of the body is rounded and runs backwards and laterally from the mental symphysis and becomes continuous posteriorly with the lower border of the ramus.

The **upper border** of the body is the *alveolar margin* and it bears the lower teeth.

The **inner surface** is divided into an upper and a lower half by a line, called the *mylohyoid line*, which starts near the lower border at the symphysis menti and runs obliquely backwards, laterally and upwards to end just behind and below the third molar tooth. Below this line there is a smooth concave area, called the *submandibular fossa*, which, in the living body, is related to the submandibular gland. Above the line anteriorly is a less well-defined indentation, the *sublingual fossa*, which, in the living body, is related to the sublingual gland.

2   The ramus. The mandibular ramus is roughly quadrilateral in shape and bears two prominent processes, the coronoid process anteriorly and the condylar process posteriorly. The ramus has internal and external surfaces, and anterior, posterior, superior and inferior borders.

The external surface is flat and is marked by one or two oblique ridges.

The internal surface is rough in its lower posterior part where a muscle, the medial pterygoid muscle, is attached. At about the centre of the internal surface there is an opening, called the *mandibular foramen.* This opening leads into the *mandibular canal* which runs downwards and forwards to open at the mental foramen. The canal transmits vessels and nerves. The lower and medial sides of the mandibular foramen are partially covered by a small projection of bone called the *lingula*. A groove, called the *mylohyoid groove*, runs from below the lingula downwards and forwards onto the body of the mandible.

The inferior border of the ramus is continuous with the inferior border of the body. The inferior border and the posterior border of the ramus meet at the *angle of the mandible.*

The posterior border of the ramus is slightly concave and runs from the angle of the mandible to the posterior surfaces of the condylar process.

The superior border of the ramus bears the coronoid process anteriorly and the condylar process posteriorly. The superior border is slightly concave between these two processes and this concavity is referred to as the *mandibular notch.*

The coronoid process is a flattened triangular process which points upwards and slightly forwards. It gives attachment to the temporalis muscle.

The condylar process has a transverse expansion on its upper surface. This expansion is called the *head of the mandible* and it forms the articular surface which articulates with the temporal bone at the temporo-mandibular joint.

The anterior border of the ramus extends from the anterior border of the coronoid process to the upper border of the body.

The hyoid bone

The hyoid is a U-shaped bone which lies below the floor of the mouth at the front of the neck. It consists of a body and four processes: two large called the greater cornua and two small called the lesser cornua.

The body of the bone is composed of a curved bar of bone which is flattened from front to back.

The greater cornua project backwards from the lateral limits of the body and each bears a tubercle at its posterior end.

The lesser cornua project upwards and backwards at the junction of the body with the greater cornua.

The lesser cornua are attached to the styloid process of the temporal bone by the stylo-hyoid ligament.

The frontal bone

The frontal bone is shaped rather like a cockleshell. It forms the anterior portion of the skull vault and on either side it posteriorly bears a horizontal plate which forms the greater part of the roof of each orbit.

The upper border of the frontal bone articulates with the parietal bones at the coronal suture. The lower border forms the upper margin of the orbital opening on either side and in the midline it articulates with the two nasal bones. Projecting downwards in the midline is the *nasal spine*. Above the orbital margins are the two curved *superciliary arches* which meet in the midline to form the *glabella*. On either side of the lower border the *zygomatic process* projects downwards to articulate with the frontal process of the zygomatic bone. The medial third of the supraorbital margin is rounded, the lateral two-thirds is sharp and at the junction of the medial third and lateral two-thirds is the *supraorbital notch* or foramen.

The posterior surface of the frontal bone is grooved in the midline by the *sagittal sinus* and at the lower end of this groove is the *frontal crest* which articulates with the crista galli of the ethmoid bone.

The horizontal orbital plates project backwards from the lower margin of the posterior surface and are separated in the midline by a deep notch which articulates with the cribriform plate of the ethmoid bone.

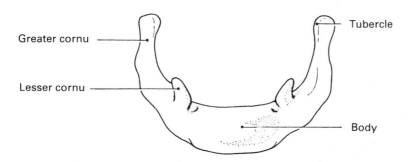

Greater cornu

Lesser cornu

Tubercle

Body

Fig. 6.14. The hyoid bone.

The lower surface of each orbital plate bears, near its lateral margin, a shallow depression, called the *lacrimal fossa*, which lodges the lacrimal gland.

At the junction of the orbital plates with the anterior margin of the bone there are cavities in the bone, in the midline, which are the frontal sinuses.

The parietal bone

Each parietal bone has four margins and internal and external surfaces. They are roughly quadrilateral in shape and form a large part of the roof of the cranium.

The external surface is marked by the *temporal line*.

The internal surface is grooved by the *middle meningeal vessels* and, near the upper border, by the *sagittal sinus*.

The upper margin is serrated and articulates with the parietal bone of the opposite side at the sagittal suture.

The anterior margin is serrated and articulates with the frontal bone at the coronal suture.

The posterior margin is serrated and articulates with the occipital bone at the lambdoid suture.

The lower margin articulates from front to back with the greater wing of the sphenoid bone, the squamous part of the temporal bone and the mastoid part of the temporal bone. The junction of the lower margin of the parietal bone with the greater wing of the sphenoid bone and the squamous part of the temporal bone is bevelled, but the junction with the mastoid part of the temporal bone is serrated.

The temporal bone

The temporal bone is composed of four parts which, in the adult, are fused to form one bone. The four parts are the squamous part, the petromastoid part, the tympanic part and the styloid process.

The squamous part of the temporal bone forms the anterior and upper part of the bone. It forms a large part of the temporal fossa and a small part of the base of the skull. The *zygomatic process* arises by a triangular base from the lower portion of the temporal surface and projects forwards to articulate with the zygomatic bone. At the posterior end of the under surface of the process is a small conical projection called the *postglenoid tubercle*. In front of this tubercle is the *articular fossa* which articulates with the mandible at the temporomandibular joint. The anterior margin of the articular fossa is bounded by the *articular tubercle*. The medial part of the articular fossa is separated from the tympanic part of the bone by the squamotympanic fissure.

The superior margin of the squamous part articulates with the parietal

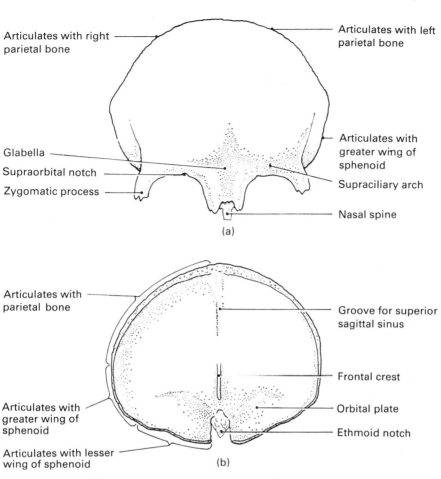

Articulates with right parietal bone

Articulates with left parietal bone

Glabella

Supraorbital notch

Zygomatic process

Articulates with greater wing of sphenoid

Supraciliary arch

Nasal spine

(a)

Articulates with parietal bone

Groove for superior sagittal sinus

Frontal crest

Articulates with greater wing of sphenoid

Orbital plate

Ethmoid notch

Articulates with lesser wing of sphenoid

(b)

Fig. 6.15. The frontal bone; (a) frontal view, and (b) internal view.

bone, and the anteroinferior margin articulates with the greater wing of the sphenoid.

The petromastoid part of the temporal bone is developed as one unit but it is usually described as two parts: the mastoid part and the petrous part.

The mastoid part of the temporal bone lies behind the external acoustic meatus and forms the posterior part of the bone.

The *mastoid process* projects downwards from the lower surface of the mastoid portion. On the medial side of the mastoid process is a deep groove called the *mastoid notch*.

On the inner surface of the mastoid portion is a deep curved groove, the *sigmoid sulcus*, which contains the sigmoid sinus. Close to the posterior portion of the bone there is frequently a small foramen, called the *mastoid foramen*, which transmits a small vein.

The upper and posterior borders of the mastoid part of the temporal bone are serrated. The upper border articulates with the parietal bone and the posterior border with the occipital bone.

The petrous part of the temporal bone is triangular in shape. It projects forwards and medially from the inner surface of the temporal bone to lie like a wedge between the sphenoid bone and the occipital bone. It contains the organs of hearing and of balance.

The petrous part of the temporal bone has a base, an apex, an anterior surface, a posterior surface and an inferior surface.

The apex of the petrous portion forms the posterior boundary of the *foramen lacerum*, and opening into it is the *carotid canal*.

The base is attached to the inner surface of the squamous and mastoid parts of the temporal bone.

The anterior surface of the petrous part forms part of the floor of the middle cranial fossa. At about the middle of this surface there is a slightly raised area called the *arcuate eminence* which is elevated by the superior semicircular canal.

The posterior surface of the petrous portion forms the anterior part of the posterior cranial fossa. At about the centre of this surface there is an opening, the *internal acoustic meatus*, through which the vestibulocochlear nerve passes.

The interior surface of the petrous portion forms part of the skull base. Behind the apex is a rough area where the cartilaginous portion of the auditory tube is attached. Lateral to this is the round opening of the *carotid canal*. Behind the carotid canal is a smooth deep depression, the *jugular fossa*, and posterolaterally to this is the *stylomastoid foramen*.

The anterior border of the petrous part articulates with the greater wing of the sphenoid.

The superior border forms the junction between the middle and posterior cranial fossae. This border is grooved by the superior petrosal sinus and gives attachment to the tentorium cerebelli, a fold of dura mater which almost completely roofs in the posterior cranial fossa.

The posterior border articulates with the occipital bone.

The tympanic part of the temporal bone is a curved plate of bone which lies in front of the mastoid process and forms most of the bony wall of the external acoustic meatus. Its medial end is fused to the petrous portion of the temporal bone. Its anterior surface forms the posterior wall of the mandibular fossa.

The styloid process of the temporal bone is a slender pointed process of bone about 2.5 cm long which projects downwards and forwards from the under surface of the bone, between the tympanic plate and the mastoid process.

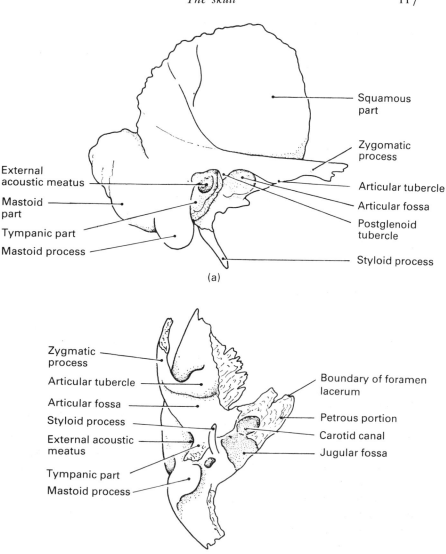

Fig. 6.16. The right temporal bone: (a) viewed from the side, and (b) viewed from below.

The occipital bone

The occipital bone is situated at the posterior and inferior part of the cranium and it encloses the foramen magnum. The occipital bone is divided into four parts: a squamous part, behind the foramen magnum, a basilar part, in front of the foramen magnum, and two condylar parts, one on either side of the foramen magnum.

The squamous part of the occipital bone is situated behind and above the foramen magnum. The external surface of the squamous part is convex from side to side and from above downwards. At about the middle of the external surface there is a protuberance, the *external occipital protuberance*, and running laterally from this on either side is the *superior nuchal line*. The *external occipital crest* runs downwards from the external occipital protuberance to the foramen magnum. Running laterally from this crest at about its midpoint is the *inferior nuchal line.*

The internal surface of the squamous part is concave and bears, at about its midpoint, the *internal occipital protuberance*. A crest, the *internal occipital crest*, runs downwards from this to the foramen magnum. Passing laterally, on either side of the internal occipital protuberance is a groove which contains the transverse sinus. The lips of this groove give attachment to the tentorium cerebelli. Passing upwards from the internal occipital protuberance is the groove for the sagittal sinus.

The upper borders of the squamous part articulate with the two parietal bones at the lambdoid suture. The lateral borders articulate with the mastoid parts of the two temporal bones.

The condylar parts of the occipital bone are situated on either side of the foramen magnum. On their inferior surfaces are oval articular facets, the *occipital condyles*, which articulate with the atlas at the atlanto-occipital joint. The long axes of the condyles run forwards and medially and their surfaces are convex from side to side. Above the anterior part of each

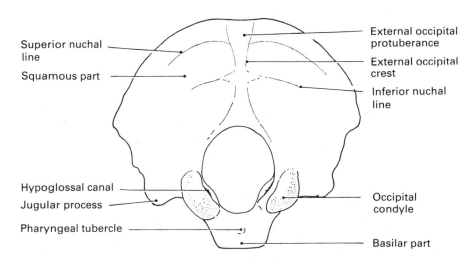

Fig. 6.17. The occipital bone viewed from below.

condyle is a canal, the *hypoglossal canal*. At the posterior end of each condyle is a small canal, the condylar canal, which transmits a vein. Running laterally from each condyle is a process of bone, the *jugular process*. The anterior surface of this process bears the *jugular notch* which forms the posterior margin of the jugular foramen.

The internal surface of the condylar part has opening into it the *hypoglossal canal*, the *condylar canal* and the *jugular foramen*. Running laterally from the jugular foramen is a groove for the sigmoid sinus.

Posteriorly the condylar part of the bone is fused to the squamous part and laterally it is joined to the petrous part of the temporal bone.

The basilar part of the occipital bone runs upwards and forwards from the foramen magnum. Anteriorly, after the age of 25 years, it is fused to the body of the sphenoid. Before this age the two bones are joined by a bar of cartilage. Laterally it articulates with the petrous part of the temporal bone.

On the inferior surface, about 1 cm in front of the foramen magnum, is a small tubercle, called the *pharyngeal tubercle*, to which the upper posterior part of the pharynx is attached.

The sphenoid bone

The sphenoid bone lies at the base of the skull in front of the occipital bone. Its shape is frequently likened to a bat with its wings spread. It is composed of a central body with a greater and a lesser wing on either side. Projecting downwards from the junction of the body and each greater wing is the pterygoid process.

The body of the sphenoid is roughly cuboidal in shape and thus has six surfaces.

The superior surface articulates anteriorly with the cribriform plate of the ethmoid bone and posteriorly it fuses with the basilar part of the occipital bone. The superior surface forms a small part of the floor of the anterior cranial fossa, the whole of the floor of the central raised portion of the middle cranial fossa, and a small portion of the floor of the posterior cranial fossa. The portion which forms the central part of the floor of the middle cranial fossa is hollowed out by a deep depression called the *sella turcica* (hypophyseal fossa). The deep portion of the sella turcica contains the pituitary gland. In front of the sella turcica is the *optic groove*, which contains the optic chiasma. Behind the sella turcica is a square plate of bone, the *dorsum sellae*. The upper angles of the dorsum sellae bear two processes called the *posterior clinoid processes*.

The lateral surface of the body fuses on either side with the greater wing of the sphenoid and the pterygoid process. The upper lateral surface of the body is joined to the lesser wing of the sphenoid on either side.

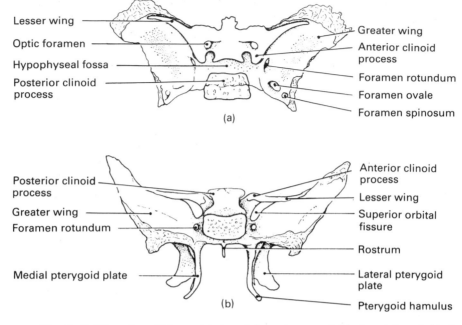

Lesser wing

Optic foramen

Hypophyseal fossa

Posterior clinoid process

Greater wing

Anterior clinoid process

Foramen rotundum

Foramen ovale

Foramen spinosum

(a)

Posterior clinoid process

Greater wing

Foramen rotundum

Medial pterygoid plate

Anterior clinoid process

Lesser wing

Superior orbital fissure

Rostrum

Lateral pterygoid plate

Pterygoid hamulus

(b)

Fig. 6.18. The sphenoid bone: (a) viewed from above; and (b) viewed from behind.

Above the junction with the greater wing the lateral surface of the body is grooved by the *carotid sulcus* which contains the internal carotid artery.

The posterior surface of the body fuses with the basilar part of the occipital bone.

The anterior surface of the body bears, in the midline, a crest which articulates with the perpendicular plate of the ethmoid bone. On either side of this crest is the round opening of the sphenoid sinus. The *sphenoid sinuses* are located, one on either side of the midline, within the body of the sphenoid.

The inferior surface of the body bears, in the midline, a spine, called the rostrum, which articulates with the vomer.

The greater wings of the sphenoid bone curve upwards and laterally from each side of the body. The greater wings are composed of a medial horizontal portion, which forms part of the floor of the middle cranial fossa, a lateral portion which curves upwards and forms part of the floor of the temporal fossa, and an anterior portion which forms the posterior part of the lateral wall of the orbit.

The upper surface of the medial horizontal part is pierced at its anteromedial end by the *foramen rotundum*. Behind and lateral to this is the *foramen ovale*, and behind and lateral to the foramen ovale is the *foramen spinosum*.

The lateral surface of the greater wing forms part of the floor of the temporal fossa. The junction of the lateral surface and the inferior surface of the greater wing is marked by a crest, called the *infratemporal crest*. Posteriorly this crest bears a small blunt projection, the *spine of the sphenoid*.

The anterior, or orbital surface of the greater wing is roughly quadrilateral and it faces medially and forwards. Its medial margin forms the lower border of the *superior orbital fissure*. Its upper serrated margin articulates with the orbital plate of the frontal bone, and laterally it articulates with the zygomatic bone. Its inferior margin forms the posterolateral border of the *inferior orbital fissure*.

The posteromedial part of the greater wing forms the anterior border of the *foramen lacerum*, which separates it from the petrous part of the temporal bone. The posterolateral part articulates with the petrous part of temporal bone, and the lateral part articulates with the squamous part of the temporal bone. The upper surface of the lateral part articulates with the frontal bone and the parietal bone.

The lesser wings of the sphenoid bone project laterally from the upper anterior part of the body of the sphenoid and are roughly triangular in shape. The upper surface of the lesser wing forms part of the floor of the anterior cranial fossa. The lower surface forms the posterior part of the roof of each orbit and the inferior margin forms the upper border of the *superior orbital fissure*. The posterior margin of the lesser wing is smooth and, close to the junction with the body of the sphenoid, it projects posteriorly to form the *anterior clinoid process*. The lesser wing arises from the body by two roots which are separated to enclose the *optic foramen*.

The pterygoid processes of the sphenoid run downwards from the junction of the greater wing and the body of the sphenoid. Each process is composed of a lateral and a medial plate which are joined anteriorly. The two plates are separated by a cleft, the margins of which articulate with the pyramidal process of the palatine bone. The anterior surface of the pterygoid process forms the posterior wall of the pterygopalatine fossa.

The lateral pterygoid plate is broad and thin and curves outwards. Its lateral surface forms part of the medial wall of the infratemporal fossa. Its medial surface forms the lateral wall of the pterygoid fossa.

The medial pterygoid plate is narrower and longer than the lateral, and its lower end curves laterally to form a blunt projection called the *pterygoid hamulus*. The lateral surface of the medial plate forms the medial wall of the pterygoid fossa.

The ethmoid bone

The ethmoid bone is composed of an upper horizontal plate, called the cribriform plate, a perpendicular plate which extends downwards from the midline of the cribriform plate, and two structures, called the ethmoidal labyrinths, which extend downwards from the lateral edges of the cribriform plate.

The cribriform plate fits into the ethmoidal notch of the frontal bone to form the roof of the nose and part of the floor of the anterior cranial fossa. A smooth triangular process, called the *crista galli*, projects upwards from the midline of the cribriform plate. This process gives attachment to the falx cerebri which is a fold of dura mater which lies between the two cerebral hemispheres.

The perpendicular plate of the ethmoid forms the upper posterior part of the nasal septum. It articulates posteroinferiorly with the vomer and anteriorly with the nasal spine of the frontal bone, the nasal bones and the septal cartilage.

The labyrinths, one on each side, are composed of a number of thin walled air cells, the anatomy of which has already been described (p. 109).

The bone on the lateral side of the labyrinth forms part of the medial wall of the orbit. It articulates above with the frontal bone, below with the maxilla, anteriorly with the lacrimal bone and posteriorly with the body of the sphenoid and the orbital process of the palatine bone.

The superior surface of the labyrinth articulates with the orbital plate of the frontal bone.

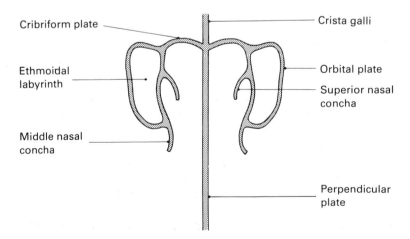

Fig. 6.19. Diagrammatic cross-section of the ethmoid bone to illustrate its component parts.

The posterior surface of the labyrinth articulates with the body of the sphenoid.

The anterior surface articulates with the lacrimal bone and the frontal process of the maxilla.

On the medial wall of the labyrinth are the *superior* and *middle nasal conchae*.

The inferior nasal concha

The inferior nasal concha is a thin plate of bone which projects medially from the lateral wall of the nose. It forms part of the medial wall of the maxillary antrum.

The vomer

The vomer is a thin plate of bone which forms the posteroinferior portion of the nasal septum. Its lower margin articulates with the nasal crest of the palatine bone posteriorly, and the nasal crest of the maxilla anteriorly. Its upper margin articulates with the lower surface of the sphenoid posteriorly and the perpendicular plate of the ethmoid anteriorly. Its posterior margin is free. Its anterior margin articulates with the septal cartilage.

The nasal bones

The nasal bones, one on each side, form the bridge of the nose. They articulate with each other in the midline, with the frontal bone above, and with the frontal processes of the maxillae at the side.

The lacrimal bones

The lacrimal bones lie at the medial side of each orbit and are the smallest of the cranial bones.

The lateral surface of the lacrimal bone forms part of the medial wall of the orbit. The lateral surface is divided into two parts by a vertical ridge, called the *lacrimal crest*. In front of this crest is the *lacrimal fossa* which contains the lacrimal sac.

The medial surface of the lacrimal bone forms part of the lateral wall of the nasal cavity in the region of the middle meatus.

The anterior border of the lacrimal bone articulates with the frontal process of the maxilla.

The posterior border articulates with the orbital plate of the ethmoid bone.

The inferior border articulates with the orbital surface of the maxilla.

The superior border articulates with the frontal bone.

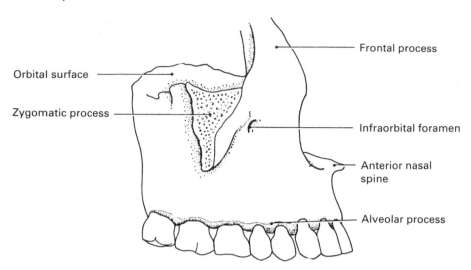

Frontal process

Orbital surface

Zygomatic process

Infraorbital foramen

Anterior nasal spine

Alveolar process

Fig. 6.20. The right maxilla viewed from the lateral side.

The maxilla

The two maxillae form a large part of the facial skeleton and the whole of the upper jaw. Each maxilla is composed of a body and four processes, the frontal, the alveolar, the palatine and the zygomatic processes.

   The body of the maxilla is roughly pyramidal in shape and it contains the *maxillary sinus*. The body has four surfaces, anterior, posterior, orbital and nasal.

   The anterior surface faces forwards and laterally. The lower part is continuous with the alveolar process and is marked by elevations which correspond to the roots of the upper teeth. The medial border of the anterior surface is concave above and forms the lateral and inferior margins of the nasal aperture. Immediately below the nasal aperture is a pointed process which fuses with the process of the maxilla of the opposite side to form the *anterior nasal spine*. The middle portion of the upper border of the anterior surface forms the inferior margin of the orbit. Opening onto the anterior surface is a foramen, the *infraorbital foramen.*

   The infratemporal (posterior) surface is convex and faces backwards and laterally. It forms the anterior wall of the infratemporal fossa laterally and the anterior wall of the pterygopalatine fossa medially. It is separated from the anterior surface by the zygomatic process and by a ridge which runs downwards from this process to the socket of the first molar tooth.

   The superior, orbital surface is smooth and triangular in shape and forms part of the floor of the orbit. It articulates with the zygomatic bone laterally. The medial border articulates posteriorly with the palatine bone and the ethmoid bone, and anteriorly with the lacrimal bone. The posterior border forms most of the anterior margin of the *inferior orbital fissure*. Running forwards from this fissure is a groove which ends in the *infraorbital canal*. The infraorbital canal opens on the anterior surface at the infraorbital foramen.

The medial, or nasal, surface forms part of the lateral wall of the nasal cavity. In its upper posterior portion there is a large opening, the *maxillary hiatus*, through which the maxillary sinus communicates with the nasal cavity. In front of this opening is the *nasolacrimal groove*. The anterior part of this surface bears a rough crest which articulates with the inferior nasal concha. The size of the maxillary hiatus is reduced by the ethmoid bone, the lacrimal bone, the inferior nasal concha and the perpendicular plate of the palatine bone all of which overhang its margins.

The zygomatic process is a pyramidal process which arises from the maxilla at the junction of the anterior, infratemporal and orbital surfaces. Its superolateral border articulates with the zygomatic bone.

The frontal process projects upwards and backwards from the upper medial angle of the body of the maxilla. The medial surface of this process forms part of the lateral wall of the nasal cavity. Superiorly it articulates with the frontal bone. Anteriorly it articulates with the nasal bone. Posteriorly it articulates with the lacrimal bone.

The alveolar process fuses with the alveolar process of the opposite maxilla to form the horseshoe-shaped arch of the upper jaw. The inferior surface of the alveolar process is marked by the sockets for the upper teeth.

The palatine process is a horizontal plate of bone which projects medially from the lower part of the body above the alveolar process. Its inferior surface, together with the inferior surface of the palatine process of the opposite side, forms the anterior three-quarters of the bony palate. The upper surface forms the floor of the nasal cavity, and at the site of fusion of the two processes in the midline, there is an upwardly directed crest which forms part of the nasal septum.

The posterior border of the palatine process articulates with the horizontal plate of the palatine bone of that side.

The palatine bone

The palatine bone is composed mainly of a horizontal and a vertical plate and is roughly L-shaped.

The horizontal plate of the palatine bone articulates medially with the horizontal plate of the opposite palatine bone to form the posterior part of the bony palate. The upper surface of the horizontal plate thus forms part of the floor of the nasal cavity. At the junction of the two horizontal plates there is an upwardly directed crest which forms part of the nasal septum.

The anterior border articulates with the posterior border of the palatine process of the maxilla. The posterior border is free. The lateral border is fused to the perpendicular plate.

The perpendicular plate of the palatine bone bears on its upper border two processes, the orbital process and the sphenoidal process. The *orbital process* runs laterally to form a small part of the floor of the orbit. The *sphenoidal process* run medially to articulate with the body of the sphenoid.

The medial surface of the perpendicular plate forms the posterior part of the lateral wall of the nose.

The lateral surface of the perpendicular plate articulates anteriorly with the nasal surface of the body of the maxilla. The upper posterior part of the lateral surface is smooth and forms the medial wall of the pterygopalatine fossa. The anterior border articulates with the inferior nasal concha and the posterior border articulates with the medial pterygoid plate of the sphenoid bone.

A process, the *pyramidal process*, projects downwards, backwards and laterally from the posterior end of the junction of the horizontal and perpendicular plates. This process fits into the notch between the lateral and medial pterygoid plates of the sphenoid bone.

The zygomatic bone

The zygomatic bone forms the prominence of the cheek. Each bone has lateral, medial and posterior surfaces and two processes, the frontal and temporal processes.

The lateral surface is subcutaneous.

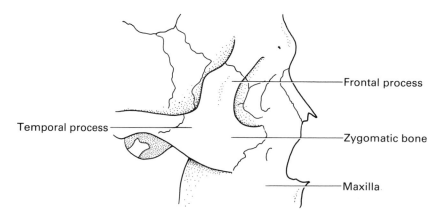

Fig. 6.21. The right zygoma *in situ*.

The medial surface forms the anterolateral part of the floor of the orbit and the adjacent part of the lateral wall. It is continued upwards onto the *frontal process*, which runs upwards to articulate with the zygomatic process of the frontal bone.

The posterior surface is concave and forms the anterior wall of the temporal fossa. Projecting backwards is the *temporal process*, which articulates with the zygomatic process of the temporal bone to form the zygomatic arch.

The anterosuperior border of the zygomatic bone forms part of the inferior and lateral margins of the orbital opening.

The anteroinferior border articulates with the zygomatic process of the maxilla.

The posteroinferior border is rough and gives attachment to the masseter muscle.

The posteromedial border articulates with the greater wing of the sphenoid bone.

The posterosuperior border is concave and runs from the posterior border of the frontal process to the upper border of the temporal process.

CHAPTER 7

The vertebral column, ribs and sternum

The vertebral column

The vertebral column forms the central axis of the trunk. It is composed of 33 irregular bones called the vertebrae, most of which are connected to each other by joints at which a small range of movement is possible. The large number of joints, however, results in a considerable range of movement of the column as a whole.

The vertebral column has two main functions:

Support. The weight of the head, the upper limbs and the trunk is transferred to the vertebral column and from the vertebral column to the hip bones and thus to the lower limbs.

Protection. The vertebral column provides protection for the spinal cord.

The vertebral column is divided into five regions:

1 The *cervical* region consisting of seven vertebrae.
2 The *thoracic* region consisting of twelve vertebrae.
3 The *lumbar* region consisting of five vertebrae.
4 The *sacral* region consisting of five vertebrae, which in the adult are fused to form a single unit, the sacrum.
5 The *coccygeal* region consisting usually of four vertebrae which in the adult are fused to form a single unit, the coccyx.

Although the vertebrae all have certain structural characteristics in common each region is different and the vertebrae within each region differ in detail.

The structure of a typical vertebra

A typical vertebra is composed of two parts, an anterior part called the *body* and a posterior part called the *vertebral arch*. The posterior surface of the body and the vertebral arch enclose a foramen called the *vertebral foramen* through which the spinal cord runs.

The body of a vertebra is roughly cylindrical, but its shape varies from one region of the spinal cord to another. The body has flattened, rough upper and lower surfaces and is connected to the bodies of the vertebrae above and below by a flat disc of fibrocartilage called the *intervertebral disc*. The body transmits the weight from the vertebra above to the vertebra below. For this reason the bodies of the vertebrae gradually increase in size from the cervical region to the sacrum, where the weight is transmitted to the hip bones.

The anterior surface of the body is convex from side to side, but the posterior surface is generally concave from side to side where it borders on the vertebral foramen.

The vertebral arch is made up of two *pedicles* and two *laminae*, and

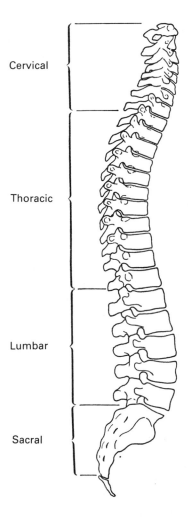

Fig. 7.1. The vertebral column.

from the arch seven processes arise. There are two *transverse processes*, two *superior* and two *inferior articular processes* and one *spinous process*.

The pedicles are two bars of bone which project backwards from either side of the body at the junction of the posterior and lateral surfaces. Each pedicle has a notch on its upper surface and a similar notch on its lower surface. When the vertebrae are articulated together these notches form a foramen which is called the *intervertebral foramen*. Through these foramina nerves pass outwards from the spinal cord.

The laminae are two flat plates of bone which pass medially, from the posterior ends of the pedicles, to fuse in the midline thus enclosing the vertebral foramen. The shape of the foramen varies from region to region.

The spinous process projects backwards in the midline from the site of fusion of the two laminae.

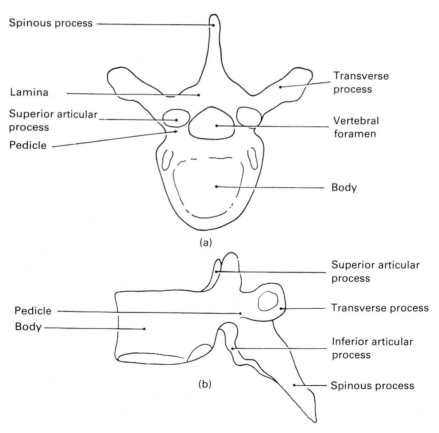

Fig. 7.2. A typical vertebra: (a) superior view; and (b) lateral view.

The transverse processes project laterally, one on either side, from the site of fusion of the pedicles and laminae.

The articular processes project upwards and downwards, an upper and a lower on each side, from the junction of the pedicles and laminae. These have articular facets which articulate with the corresponding facets on the articular processes of the vertebrae above and below.

The vertebrae in each region are in many respects similar, and so a typical vertebra of each region will first be described and then the atypical features of individual vertebrae will be listed.

The cervical vertebrae

There are seven cervical vertebrae of which the 3rd, 4th, 5th and 6th are typical.

A typical cervical vertebra

The body of a typical cervical vertebra is oval, with the long axis lying transversely. The height of the body is small in comparison to its width. The upper surface of the body is concave from side to side and the lower surface is convex from side to side. The anterior border of the lower surface projects slightly downwards to form a lip. The posterior surface of the body is flattened where it forms part of the vertebral foramen.

The vertebral arch. The **pedicles** are small and short and project backwards and laterally. The **laminae** are long and narrow and project backwards and medially. As a result the vertebral foramen is triangular in shape, and is large in comparison with the size of the body. The superior and inferior surfaces of the pedicles are grooved to form the *intervertebral notches*, the notch on the superior surface being deeper than that on the inferior surface.

The transverse processes are short and are formed from an anterior root which arises from the side of the body of the vertebra and a posterior root which arises from the junction of the pedicles and laminae. The two roots are joined around a foramen, called the *transverse foramen*, which transmits the vertebral artery. The transverse processes end in two blunt projections called the *anterior* and *posterior tubercles*.

The spinous process, which is short and bifid, projects backwards and slightly downwards.

The articular processes form a pillar of bone at the junction of the pedicles and laminae. The inferior articular facets face downwards and forwards and the superior articular facets face upwards and backwards.

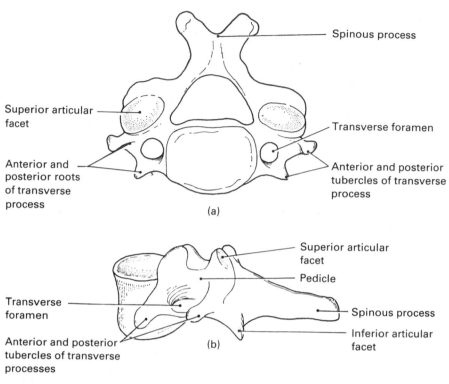

Spinous process

Superior articular facet

Transverse foramen

Anterior and posterior roots of transverse process

Anterior and posterior tubercles of transverse process

(a)

Superior articular facet

Pedicle

Transverse foramen

Spinous process

Inferior articular facet

Anterior and posterior tubercles of transverse processes

(b)

Fig. 7.3. A typical cervical vertebra: (a) superior view, and (b) lateral view.

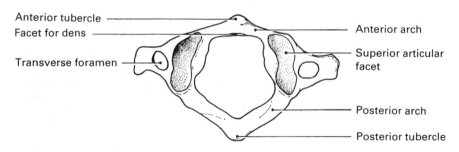

Anterior tubercle

Facet for dens

Transverse foramen

Anterior arch

Superior articular facet

Posterior arch

Posterior tubercle

Fig. 7.4. The atlas, superior aspect.

The atypical cervical vertebrae

1   The 1st cervical vertebra (the atlas) differs from the other vertebrae in that it has no body. The atlas is composed of two *lateral masses* which carry the articular facets and which are joined by a short *anterior arch* and a longer *posterior arch*.

The anterior arch bears, on its posterior surface, a round articular facet which articulates with the dens (odontoid process), which projects upwards from the 2nd cervical vertebra.

The anterior surface of the arch bears, in the midline, a blunt projection called the *anterior tubercle.*

The posterior arch is longer and bears on its posterior surface a blunt *posterior tubercle* which represents the spinous process.

The lateral masses lie between the ends of the arches. Each possesses a kidney-shaped concave superior articular facet which articulates with the condyle of the occipital bone of the skull, and a rounded inferior articular facet which articulates with the superior articular facet of the 2nd cervical vertebra.

The transverse processes are long and arise by two roots, an anterior root which arises from the side of the lateral mass and a posterior root which arises from the posterior arch. The tips of the transverse processes are square and possess no tubercles but the roots enclose a transverse foramen as in the other cervical vertebrae. Because of the length of the transverse processes the atlas is wider than the other cervical vertebrae.

2   **The 2nd cervical vertebra (the axis)** possesses on the upper surface of its body a blunt process, called the *dens (odontoid process).* The dens bears, on its anterior surface, a smooth round articular facet which articulates with the facet on the posterior surface of the anterior arch of the atlas. The posterior surface of the dens is grooved by the transverse ligament of the atlas which helps to keep it in position.

The superior articular facets lie on the lateral side of the upper surface of the body and the adjoining part of the pedicle, and face upwards and laterally. **The inferior articular facets** face downwards and forwards, as in a typical cervical vertebra. They lie therefore more posteriorly than the superior articular facets. The spinal process of the 2nd cervical vertebra is short and bifid.

3   **The 7th cervical vertebra** differs from a typical cervical vertebra in three ways:

a)   **the spinous process** is longer and is not bifid, but ends in a roughened tubercle;

b)   **the transverse processes** have a large posterior root and have no anterior tubercle; and

c)   **the transverse foramen** is small and the vertebral artery does not run through it.

The thoracic vertebrae

The thoracic vertebrae are larger than the cervical vertebrae and they form the posterior part of the bony cage of the thorax. There are twelve thoracic

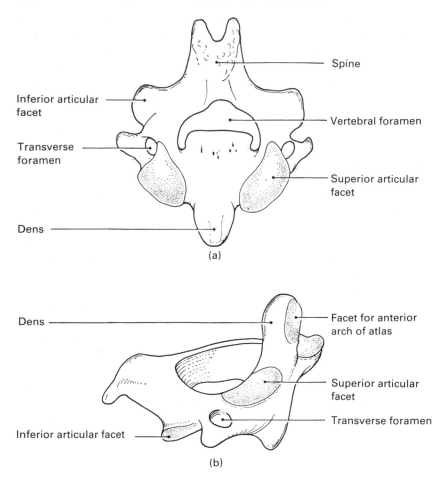

Spine

Inferior articular facet

Vertebral foramen

Transverse foramen

Superior articular facet

Dens

(a)

Dens

Facet for anterior arch of atlas

Superior articular facet

Transverse foramen

Inferior articular facet

(b)

Fig. 7.5. The axis: (a) superior aspect, and (b) oblique aspect.

vertebrae and they increase in size from the 1st to the 12th. They can easily be recognized as they have at least one articular facet on each side of the body for articulation with the head of a rib. The 2nd to 8th vertebrae are typical, the remainder differing in minor respects.

A typical thoracic vertebrae

The body of a typical thoracic vertebra is roughly heart-shaped and its measurements from side to side and front to back are roughly equal. The anterior and lateral sides of the body are convex, but the posterior surface where it borders on the vertebral foramen is concave. The upper few vertebrae are rather more oval than those in the middle of the series and the bodies of the lower vertebrae are more elliptical in shape, resembling the bodies of the lumbar vertebrae. On the upper and lower margins of the lateral side of the body, and close to the roots of the pedicles, are *demi-facets* which, together with the demi-facets on the bodies of the vertebrae above and below, and an indentation of the intervening intervertebral disc, form an articular facet for articulation with the head of a rib.

The vertebral arch. The *pedicles*, which project directly backwards, arise from the upper part of the posterior surface of the body, and consequently the inferior vertebral notch is deep. The *laminae* are wide, flat bars of bone and the inferior margins overlap the superior margins of the laminae of the vertebra below so that the posterior margin of the vertebral canal is closed. The vertebral foramen in the thoracic region is round, and is smaller than that of the cervical or lumbar region.

The transverse processes project backwards and laterally from the junction of the body and pedicles. On the anterior surface, close to the tip of each transverse process, is an oval articular facet which articulates with the facet on the tubercle of the appropriate rib.

The spinous processes are long and tapering and project backwards and downwards to end in a rough tubercle. The spinous processes of the middle vertebrae of the series run more obliquely downwards than those of the upper or lower parts of the series, and indeed are so oblique that their tips lie opposite the upper surface of the vertebra next but one below.

The superior articular processes project upwards from the junction of the pedicles and laminae and bear a flat articular facet which faces backwards and slightly laterally. **The inferior articular processes** are joined to the lateral ends of the laminae and their articular facets face forwards and slightly medially.

The atypical thoracic vertebrae

1 **The 1st thoracic vertebra**. The upper surface of the 1st thoracic vertebra is concave from side to side and resembles that of a cervical vertebra. The lower surface of the body is flat. On the upper part of the lateral side of the body is a complete facet for articulation with the head of the 1st rib. The lower surface of the lateral side of the body bears a demi-facet which is similar to those of the typical vertebrae. The *pedicles* arise from the lateral side of the posterior surface of the body, midway between the upper and lower borders, and run backwards and slightly laterally. The *spinous process* projects only slightly downwards.

2 **The 9th thoracic vertebra** usually has no demi-facet on the lower lateral surface of its body as the 10th thoracic vertebra has a complete facet for articulation with the head of the 10th rib.

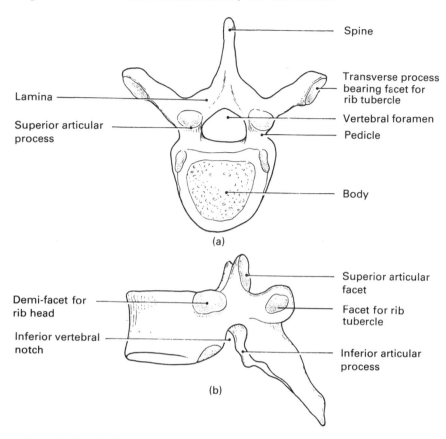

Spine

Transverse process bearing facet for rib tubercle

Lamina

Vertebral foramen

Superior articular process

Pedicle

Body

(a)

Superior articular facet

Demi-facet for rib head

Facet for rib tubercle

Inferior vertebral notch

Inferior articular process

(b)

Fig. 7.6. A typical thoracic vertebra: (a) superior aspect, and (b) lateral aspect.

3 **The 10th thoracic vertebra** has a large oval facet on each side of the body and this facet frequently extends onto the root of the pedicle.

4 **The 11th thoracic vertebra** has a single complete facet on each side which extends onto the pedicle. There are no articular facets on the transverse processes.

5 **The 12th thoracic vertebra** has a complete facet on either side and this lies mainly on the pedicle. The shape of the body and the transverse processes closely resembles that of the 1st lumbar vertebra. There are no articular facets on the transverse processes. The superior articular facets face downwards as in a typical thoracic vertebra, but the inferior articular facets face laterally as in a typical lumbar vertebra.

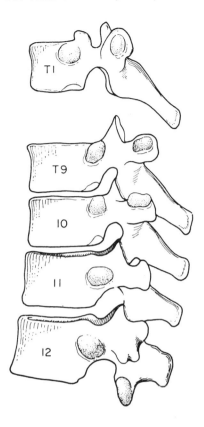

Fig. 7.7. The atypical vertebra (T_1, T_9-T_{12}).

The lumbar vertebrae

The lumbar vertebrae are larger than the thoracic vertebrae and can be distinguished by the absence of facets for articulation with the ribs.

The body is roughly elliptical in shape with the long axis lying transversely. The bodies are larger than those of the cervical or thoracic region, and are slightly higher anteriorly than posteriorly. This difference in height is most marked in the 5th lumbar vertebra.

The vertebral arch. The *pedicles* are thick but rather short. The *laminae* are broad bars of bone which run backwards and medially. The vertebral foramen is triangular in shape. The pedicles arise from the upper part of the body so that the inferior vertebral notches are deeper than the superior vertebral notches.

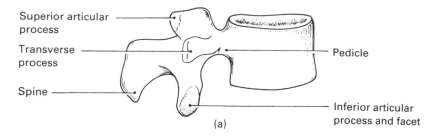

Superior articular process

Transverse process

Spine

Pedicle

Inferior articular process and facet

(a)

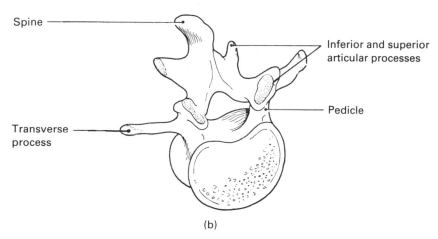

Spine

Inferior and superior articular processes

Pedicle

Transverse process

(b)

Fig. 7.8. A typical lumbar vertebra: (a) lateral aspect, and (b) anterosuperior aspect.

The transverse processes are thin and are directed backwards and laterally. The transverse processes of the 3rd lumbar vertebra are the longest, while the transverse processes of the 5th lumbar vertebra are shorter and thicker than those of the other lumbar vertebrae. At the posterior part of the lower border of each transverse process, and close to the body of the vertebra, there is bony tubercle called the *accessory process*.

The spinous processes of the lumbar vertebrae project directly backwards and are square, or hatchet-shaped. The posterior and inferior borders of the spinous processes are somewhat thickened.

The articular processes arise from the junction of the pedicles and laminae. The **superior articular facets** are concave and face backwards and medially. The **inferior articular facets** are convex and face forwards and laterally. When the vertebrae are articulated the inferior articular facets lie between the superior articular facets of the vertebra below. The superior articular processes bear a bony tubercle on their posterior surfaces. This tubercle is called the *mamillary process*.

The sacrum

The sacrum is composed of five vertebrae which, in the adult, are completely fused to form a single unit. The sacrum together with the coccyx forms the posterior wall of the bony pelvis. The sacrum is roughly triangular in shape, and has *anterior, posterior* and *lateral* surfaces, a *base* which lies superiorly and an *apex* which lies inferiorly. The sacrum lies like a wedge between the two hip bones and is situated obliquely so that the posterior surface faces partly upwards.

The anterior surface of the sacrum is markedly concave from above downwards. The central part of the anterior surface is marked by four transverse ridges, which represent the site of fusion of the five vertebrae which make up the bone. Lateral to these ridges are four foramina on each side which communicate with the sacral canal. These are the anterior sacral foramina, and they represent the intervertebral foramina. The sacrum is expanded laterally to these foramina to form the lateral masses of the sacrum.

The posterior surface of the sacrum is irregular. The spines of the sacral vertebrae are represented by three or four tubercles which are joined by a central longitudinal ridge. The whole structure is called the *sacral crest*. On either side of the sacral crest is a flat area of bone which encloses the posterior aspect of the sacral canal. This flat area represents the fused

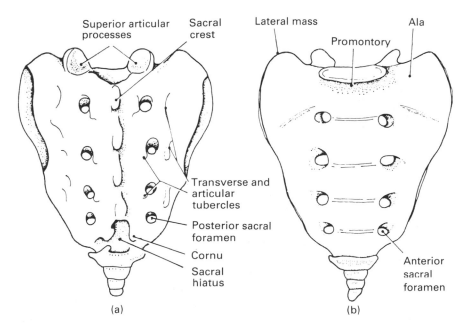

Fig. 7.9. The sacrum and coccyx: (a) posterior aspect, and (b) anterior aspect.

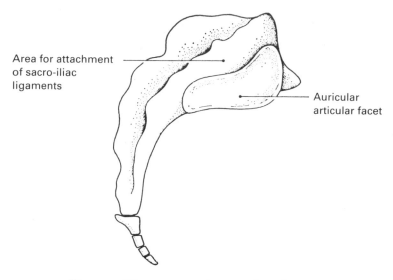

Area for attachment
of sacro-iliac
ligaments

Auricular
articular facet

Fig. 7.10. The sacrum and coccyx, lateral aspect.

laminae. Below the sacral crest there is a deficiency, shaped like an inverted U, due to failure of the laminae of the 5th sacral vertebra to meet in the midline. This deficiency is called the *sacral hiatus*. Lateral to the flat area there are four posterior sacral foramina on each side which correspond to the anterior sacral foramina and which also communicate with the sacral canal. Just medial to the sacral foramina are three or four bony tubercles which represent the articular processes. At the upper end of the posterior surface, in line with these tubercles, two articular processes project upwards. These processes resemble the superior articular processes of a lumbar vertebra and articulate with the inferior articular processes of the 5th lumbar vertebra. The 5th sacral articular processes are called the *sacral cornua* and are two blunt processes which project downwards on either side of the sacral hiatus. Laterally to the posterior sacral foramina, on either side of the sacrum, are three or four more tubercles which represent the sacral transverse processes. They are called the transverse tubercles.

The lateral surface of the sacrum is roughly triangular in shape; wide above, it gradually narrows towards its lower end. On the upper wide portion is the *auricular* (ear-shaped) *articular surface* which articulates with a similar articular surface on the iliac bone to form the sacroiliac joint. Behind the auricular area the bone is roughened by the attachment of the sacroiliac ligaments.

The base of the sacrum faces upwards and forwards. Anteriorly, in the midline, can be seen the body of the 1st sacral vertebra which, in shape, resembles the body of the 5th lumbar vertebra. The anterior border of the body of the 1st sacral vertebra is called the *sacral promontory*. Behind the body is the triangular-shaped opening of the sacral canal. On either side of the body are the smooth upper surfaces of the lateral masses of the sacrum.

The apex of the sacrum bears a smooth oval articular facet which articulates with the coccyx.

The coccyx

The coccyx is a small triangular bone which is composed of four (occasionally three or five) fused vertebrae. On occasions the first coccygeal vertebra is a separate bone. The anterior surface of the coccyx is concave and the posterior surface is convex. The upper surface of the first coccygeal vertebra has an oval articular facet which articulates with the facet on the apex of the sacrum. Projecting upwards, from the posterolateral corners of this articular surface, are two cornua which articulate with the cornua of the lower end of the sacrum. On either side of the upper end of the coccyx a rudimentary transverse process projects laterally.

The vertebral column as a whole

When the articulated vertebral column is examined from the side it is noted to describe a series of curves. During intrauterine life the vertebral column has a single continuous curve which is convex posteriorly. This curvature is called the *primary curvature*. The *secondary curvatures* develop later:
1   At about the age of 3−4 months, at the time the child starts to lift its head, the cervical region develops a curve which is convex forward.
2   At about the age of 18 months, when the child begins to walk, the lumbar region develops a curve which is convex forwards.

Thus, in the adult, there are four curves:
1   *the cervical curvature*, convex forwards,
2   *the thoracic curvature*, convex backwards,
3   *the lumbar curvature*, convex forwards,
4   *the sacral curvature*, convex backwards.

The ribs

The ribs, twelve on each side, constitute the greater part of the bony wall of the thorax. They articulate posteriorly with the thoracic vertebrae. The first seven ribs on each side are attached anteriorly by a cartilaginous bar, called the *costal cartilage*, to the sides of the sternum. The remaining five

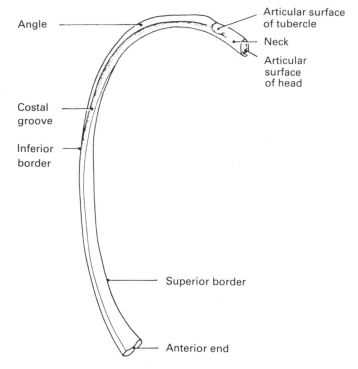

Fig. 7.11. A typical rib, inferior aspect.

ribs are called false ribs. The 8th, 9th and 10th ribs are attached by a bar of cartilage to the costal cartilage of the rib above. The 11th and 12th ribs have no anterior attachment and are called floating ribs.

The ribs slope obliquely downwards from front to back; the degree of slope increases gently from the 1st to the 9th rib and then decreases. The length of the ribs increases from the 1st to the 7th rib and below this the length decreases. The first two and the last three ribs differ somewhat from the 3rd to the 9th ribs which can be described as typical.

A typical rib

A typical rib has a *posterior* (vertebral) end, an *anterior* end and a *shaft*.

The posterior end is composed of a *head*, a *neck* and a *tubercle*.

The head has two articular facets separated by a ridge called the *crest*. The lower articular facet articulates with the demi-facet on the upper border of the body of the vertebra from which the rib takes its number. The upper facet articulates with the facet on the lower part of the body of the vertebra above. The crest lies opposite the intervertebral disc.

The neck of the rib is the narrow portion of bone immediately lateral to the head. The neck is flattened from front to back; the upper border is sharp and the lower border is rounded. The anterior surface of the neck is smooth and is in contact with the pleura. The posterior surface is rough where ligaments are attached.

The tubercle of the rib lies posteriorly between the neck and the shaft. On the medial part of the tubercle there is a smooth oval facet which articulates with the facet on the transverse process of the thoracic vertebra of the same number. The lateral part of the tubercle is roughened by the attachment of ligaments.

The shaft is thin and flattened from front to back so that it possesses *upper* and *lower borders*, and *inner* and *outer surfaces*. The shaft curves in a gentle arc, but 5 to 6 cm from the tubercle it is bent at a point called the *angle of the rib*. At this point also the shaft is slightly twisted; behind the angle the inner surface faces slightly upwards, but beyond the angle the inner surface faces slightly downwards. The angle is marked on the outer surface by a ridge which runs downwards and laterally; to this ridge is attached part of the sacrospinalis muscle.

The *outer surface* of the shaft is smooth and convex. The *inner surface* is also smooth but at the inferior border there is a groove, the *costal groove*, which lodges the intercostal vessels and nerve. The costal groove becomes increasingly shallow as it is traced forwards and it disappears at the junction of the middle and anterior thirds of the shaft. The *upper border* of the shaft is rounded but the *lower border* is sharp.

The anterior end of the rib has an oval concave surface which is attached to the lateral end of the costal cartilage.

The atypical ribs

1   **The 1st rib** is usually the shortest rib. Unlike the other ribs it has an *upper* and a *lower surface*, and *internal* and *external borders*. At the posterior end is a single facet for articulation with the single facet on the upper border of the 1st thoracic vertebra. The tubercle and the angle are merged, but the angle bears a facet for articulation with the facet on the transverse process of the 1st thoracic vertebra. On the superior surface of the anterior part of the rib is a ridge, the *scalene tubercle*, to which the scalenus anterior muscle is attached. On either side of the ridge is a groove, the subclavian vein lying in the anterior groove and the subclavian artery in the posterior groove. The inferior surface of the 1st rib is smooth and has no costal groove.

2   **The 2nd rib** is of the same shape as the 1st rib, but is longer. The head bears two small facets for articulation with the adjacent facets on the upper border of the 2nd thoracic vertebra and the lower border of the 1st

thoracic vertebra. The angle of the rib is close to the tubercle, which is small. The lower surface of the 2nd rib has a poorly marked costal groove.

3 The 10th rib resembles a typical rib, but it has only one facet on its head for articulation with the facet on the upper border of the body of the 10th thoracic vertebra.

4 The 11th rib is rather short and, like the 10th rib, has only a single facet on its head. It has no tubercle and its anterior end is pointed. There is a poorly marked costal groove on the lower part of the inner surface.

5 The 12th rib is shorter than the 11th, and has no tubercle, no angle and no subcostal groove. There is a single facet on its head for articulation with the facet on the body and pedicle of the 12th thoracic vertebra.

The costal cartilages

The costal cartilages are bars of hyaline cartilage which are attached to the anterior ends of the 1st to the 10th ribs.

The costal cartilages of the first seven ribs articulate with the sides of the sternum. The costal cartilages of the 8th, 9th and 10th ribs articulate with the costal cartilage of the rib immediately above. The 11th and 12th ribs have small cartilages which end in the muscles of the abdominal wall.

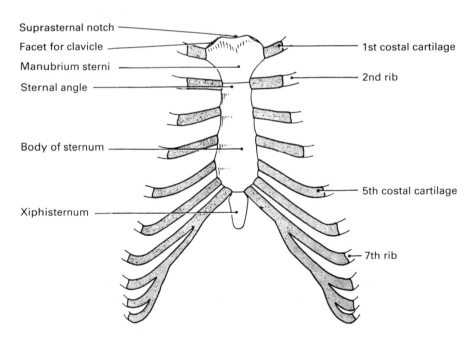

Suprasternal notch
Facet for clavicle
Manubrium sterni
Sternal angle
Body of sternum
Xiphisternum
1st costal cartilage
2nd rib
5th costal cartilage
7th rib

Fig. 7.12. The sternum and the costal cartilages.

The costal cartilages increase in length from the 1st to the 7th and below this they decrease in length. The 1st costal cartilage is the widest and they gradually decrease in width down the series.

The sternum

The sternum is a long flat bone which lies centrally in the anterior wall of the thorax. Its upper end articulates with the clavicles, and the first seven costal cartilages are attached to its lateral margins. It is composed of three parts which are from above downward: the *manubrium*, the *body of the sternum*, and the *xiphisternum*.

The manubrium

The manubrium is quadrilateral in shape and is wider above than below. The *anterior surface* is smooth and is concave from above downwards, and convex from side to side. Its *posterior surface* is smooth and concave. The *superior border* is thick and concave. The concavity is called the *suprasternal notch*.

On each side of the manubrium, at the junction of the superior and lateral borders, is an oval articular facet which faces upwards and laterally and articulates with the inner end of the clavicle at the sternoclavicular joint.

Below this facet, on the lateral border, is a rough facet to which the 1st costal cartilage is attached.

On the junction of the lateral and lower borders there is a notch, which, together with a similar notch on the upper lateral border of the body of the sternum, forms the articular area for attachment of the 2nd costal cartilage.

The *inferior border* of the manubrium is roughened and oval in shape. It articulates with the upper border of the body of the sternum. The manubrium makes a slight angle, convex forwards, with the body of the sternum and this is called the *sternal angle*. The sternal angle can be felt subcutaneously as a ridge and it is an important landmark as the 2nd rib lies at its lateral limit, and thus the ribs can be numbered accurately by first identifying the sternal angle.

The body of the sternum

The body of the sternum is thinner and longer than the manubrium. It is also narrower than the manubrium and its width decreases slightly from above downwards.

The *anterior surface* of the body of the sternum faces forwards and slightly upwards, and is marked by three transverse ridges, which indicate the lines of fusion of the four segments which, in early life, make up the sternum. At each end of these ridges on the lateral surface are rough depressions for attachment of the 3rd, 4th and 5th costal cartilages.

The *posterior surface* of the body of the sternum is slightly concave and also has three transverse ridges which are less well marked than those on the anterior surface.

The *upper border* is oval and articulates with the lower border of the manubrium. The lateral angles of the upper border are notched, and form part of the articular area for the 2nd costal cartilage.

The *lateral borders* of the body bear notches opposite each of the transverse ridges. Below the lowest ridge, the sternum narrows more rapidly and the lateral borders on each side bear two rough depressions which are close together and which give attachment to the 6th and 7th costal cartilages.

The *lower border* of the body of the sternum is narrow and articulates with the xiphisternum.

The xiphisternum

The xiphisternum, or xiphoid process, is a thin triangular-shaped piece of bone which is attached above to the lower border of the body of the sternum. The upper lateral angles have small notches which complete the notch on the lower part of the body of the sternum to which the 7th costal cartilage is attached.

The bones of the upper limb

The upper limb is attached to the trunk by the shoulder girdle, which is made up of two bones: the *clavicle* anteriorly and the *scapula* posteriorly. The shoulder girdle is attached to the trunk by one joint, the sternoclavicular joint, and by muscles. The shoulder girdle is able to move in several directions, and its movements are important in the range of movements that are possible with the upper limb.

The upper limb consists of three segments:

1 the upper arm, which has one bone, the *humerus*;

2 the forearm, which has two bones, the *radius* and the *ulna*;

3 the wrist and hand, which is made up of eight *carpal bones*, five *metacarpal bones*, and 14 *phalanges*.

The clavicle

The clavicle, or collar bone, is a long bone. It differs from other long bones in two ways:

1 it has no medullary cavity;

2 it is ossified in membrane.

The shaft

The shaft of the clavicle is S-shaped; the medial two-thirds of the shaft is convex anteriorly, and the lateral third is concave anteriorly.

The lateral third of the shaft is flattened having superior and inferior surfaces and anterior and posterior borders. In the middle third of the shaft the anterior and posterior borders become progressively less prominent and the medial third of the shaft is cylindrical.

The superior surface of the bone is subcutaneous throughout its length. The medial two-thirds of this surface is smooth but the lateral third is roughened where the deltoid and trapezius muscles are attached to it.

The inferior surface has a roughened area at its medial end where a ligament, the costoclavicular ligament, is attached. This ligament binds the medial end of the clavicle to the 1st rib. At the outer end of the inferior surface, and close to the posterior border, there is a projection, the *conoid tubercle*, and running laterally and forwards from this tubercle is a ridge— the *trapezoid line.* The conoid tubercle and the trapezoid line give attachment to the coracoclavicular ligament which binds the lateral part of the clavicle to the coracoid process of the scapula.

The medial end

The medial end of the bone widens out to a large quadrangular articular surface, which entends for a short distance onto the inferior surface of the bone. This articular surface articulates with the sternum at the sternoclavicular joint.

The lateral end

The lateral end of the clavicle has a small oval articular facet which faces laterally and articulates with the acromion process of the scapula at the acromioclavicular joint.

The scapula

The scapula, or shoulder blade, is the posterior bone of the shoulder girdle. It is a flattened triangular bone consisting of a triangular body at the lateral angle of which is an expanded portion, the head of the scapula. This bears the glenoid cavity which articulates with the head of the humerus at the shoulder joint. A process, called the coracoid process, projects forwards from the anterior part of the head of the scapula. The posterior surface of the body is divided into two parts by the spine of the scapula, which widens out at its lateral end to form the acromion process.

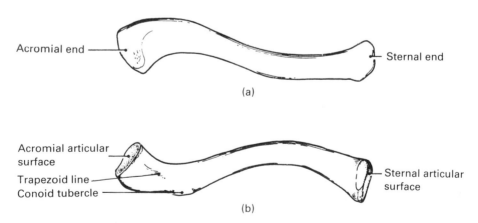

Fig. 8.1. The right clavicle: (a) superior aspect, and (b) inferior aspect.

The body

The body of the scapula is a flat triangle having anterior and posterior surfaces and lateral, medial and superior borders.

The *anterior*, or costal *surface*, is concave and is called the *subscapular fossa*. Attached to the medial two-thirds of the floor of this fossa is a muscle, the subscapular muscle, and the floor is, therefore, slightly ridged.

The *posterior surface* of the body is slightly convex. The spine of the scapula divides the posterior surface into an upper third called the *supraspinous fossa* and a lower two-thirds called the *infraspinous fossa*.

The *superior border* of the body runs slightly upwards from the root of the coracoid process to the superior angle. Close to the root of the coracoid process there is a notch in the superior border called the *suprascapular notch*.

The *medial border* of the scapula is the longest border and it extends from the superior to the inferior angles. It is slightly convex, reaching a prominence at the root of the spine of the scapula.

The *lateral border* runs from the inferior angle to the head of the scapula. The head of the scapula is the thickest part of the bone and forms the lateral angle. The lateral surface of the head bears the *glenoid cavity*, which is an oval, slightly concave articular surface. Between the head of the scapula and the body there is a slight constriction called the *neck of the scapula*.

The coracoid process

The coracoid process arises from the upper part of the anterior side of the head of the scapula and bends sharply to run forwards and laterally, so that it has the appearance of a bent finger. The outer part of the coracoid process is slightly flattened and overhangs the glenoid cavity.

The spine of the scapula

The spine of the scapula projects from the posterior surface of the body dividing it into the supraspinous and infraspinous fossa. The spine projects increasingly from the body as it runs laterally. The crest of the spine has a flat subcutaneous surface, the upper and lower borders of which give attachments to muscles.

The acromion process

The acromion process is the lateral continuation of the spine of the scapula. It runs forwards and laterally from the lateral end of the spine and

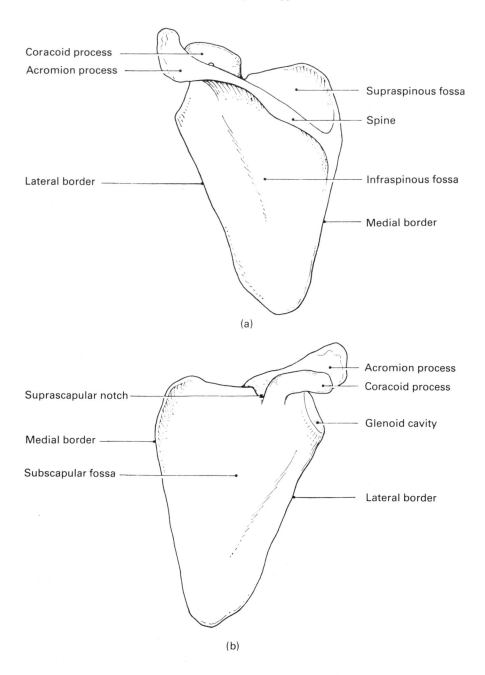

Fig. 8.2. The left scapula: (a) posterior aspect, and (b) anterior aspect.

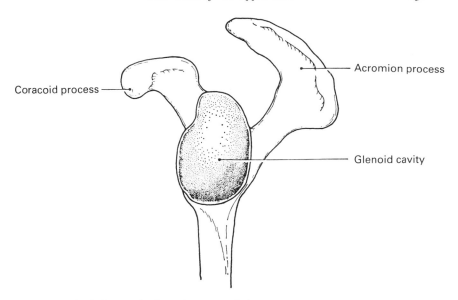

Coracoid process

Acromion process

Glenoid cavity

Fig. 8.3. The left scapula, lateral aspect of the upper part.

overhangs the glenoid cavity. A strong ligament, the coracoacromial ligament, runs between the coracoid process and acromion process forming an arch above the shoulder joint. The outer end of the acromion process bears an oval articular surface which articulates with the outer end of the acromioclavicular joint.

The humerus

The humerus is a long bone. It is the largest bone of the upper limb and it possesses a shaft and upper and lower ends.

The upper end of the humerus

The head of the humerus is the most prominent feature of the upper end. It is a smooth convex articular surface which forms slightly less than half a sphere. It faces upwards, medially and slightly backwards. The central axis of the head makes an angle of about 120° with the long axis of the shaft. The head is separated from the rest of the upper end of the bone by a shallow groove called the *anatomical neck* of the humerus. On the lateral side of the upper end is a prominent blunt projection called the *greater tubercle* (tuberosity). On the anterior aspect of the upper end is a smaller projection called the *lesser tubercle* (tuberosity). The two tubercles are important sites for the attachment of muscles. Separating the greater and lesser tubercles is a shallow groove, the *intertubercular sulcus* (bicipital groove). In the living body this groove contains the tendon of the long head of the biceps brachii muscle. The upper end of the humerus is separated from the shaft of the humerus by a poorly defined constriction called the *surgical neck* of the humerus.

The lower end of the humerus

The lower end of the humerus is triangular and is flattened from front to back. It is slightly concave forwards. The two angles of the triangle are formed by the *medial* and *lateral epicondyles*. The medial epicondyle is thicker and more prominent than the lateral, and can be felt through the skin. Running upwards from the medial epicondyle is a sharp border called the *medial supracondylar ridge*. The lateral epicondyle, although less prominent than the medial epicondyle, can also be felt through the skin. A sharp border, the *lateral supracondylar ridge*, runs upwards from the lateral epicondyle.

The lower end of the humerus articulates with the radius and ulna at the elbow joint. The articular surface of the lower end of the humerus is divided into two parts.

The lateral part is called the *capitulum*. The capitulum is a smooth rounded articular surface on the anterior and inferior aspects of the lower end of the humerus. The capitulum articulates with the upper surface of the head of the radius. Above the capitulum, on the anterior surface, is a shallow depression. This depression is called the radial fossa because the head of the radius is accommodated in this fossa when the elbow joint is fully flexed.

The medial part of the articular surface of the lower end of the humerus is pulley-shaped and is called the *trochlea*. The trochlea articulates with the trochlear notch of the ulna at the elbow joint. The trochlea lies on the anterior, inferior and posterior surfaces of the lower end of the humerus and thus almost forms a complete cylinder. The lateral side of the trochlea is separated from the capitulum by a lateral lip. The medial side of the capitulum projects downwards below the lower surface of the bone as the prominent medial lip. The surface of the trochlea is concave from side to side between the two lips. When the forearm is fully extended and supinated, as in the '*anatomical position*', the downward projection of the medial lip of the trochlea causes the long axis of the forearm to lie at an angle of about 170° to the long axis of the humerus. This angle is called the '*carrying angle*', because the ulnar surface of the extended and supinated forearm cannot touch the lateral surface of the thigh when the arm is by the side. This angle is more acute in women than in men.

Above the anterior surface of the trochlea there is a depression in the

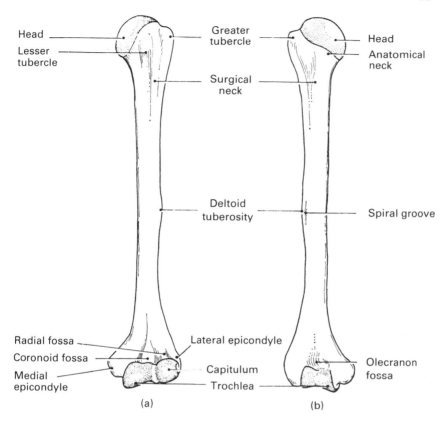

Fig. 8.4. The left humerus: (a) anterior aspect, and (b) posterior aspect.

bone, called the *coronoid fossa*, which accommodates the coronoid process of the ulna when the elbow joint is fully flexed. Above the posterior surface of the trochlea is another depression, called the *olecranon fossa*, which accommodates the olecranon process of the ulna when the elbow joint is fully extended.

The shaft of the humerus

The shaft of the humerus is cylindrical in its upper two-thirds, but the lower third is flattened from front to back. On the anterolateral surface of the shaft is a roughened projection which extends halfway down the shaft. This projection is called the *deltoid tuberosity* and it gives attachment to the deltoid muscle.

The shaft of the humerus has three borders.

The medial border starts below as the upward continuation of the medial supracondylar ridge, and can be traced upwards to join the medial lip of the intertubercular sulcus.

The lateral border starts below as the upward continuation of the lateral supracondylar ridge, but it gradually fades out as it is traced upwards. At about the middle of the shaft, however, the lateral border is crossed by a shallow groove, the *spiral groove*, which runs obliquely downwards from the posterior to the anterior aspect of the shaft, and contains, in the living body, the radial nerve.

The anterior border starts at the upper end of the shaft as the downward continuation of the lateral lip of the intertubercular sulcus. It runs along the anterior border of the deltoid tuberosity but gradually fades out in the lower third of the shaft.

The humerus thus has three surfaces.

The posterior surface lies between the lateral and medial borders.

The anterolateral surface lies between the anterior and lateral borders, and bears, in its upper half, the deltoid tuberosity.

The anteromedial surface lies between the anterior and medial borders.

The radius

The radius is the lateral bone of the forearm and it is a long bone. It has a shaft, a small circular upper end and a wider lower end.

The upper end of the radius

The upper end of the radius is composed of the head, the neck and the radial tuberosity.

The head of the radius is disc-shaped and its upper surface is a concave articular surface which articulates with the capitulum of the humerus at the elbow joint. The circumference of the head is a narrow, smooth articular surface, which articulates with the radial notch of the ulna at the superior radioulnar joint.

Below the head is a constricted portion, the *neck of the radius*.

On the medial side of the bone, just below the neck, is a blunt projection called the *radial tuberosity*, which gives attachment to the biceps brachii muscle.

The lower end of the radius

The lower end of the radius is the widest portion of the bone, and is quadrilateral in cross-section.

The lateral surface projects downwards below the rest of the bone as the *styloid process* of the radius.

The medial surface has a narrow, concave articular surface, the *ulnar notch*, which articulates with the lower end of the ulna at the inferior radioulnar joint.

The anterior surface is wide and is roughened by the attachment of the anterior ligament of the wrist joint.

The posterior surface is also wide and is grooved by the extensor tendons which pass from the forearm to the hand.

The inferior surface of the lower end of the radius is a concave articular surface which is divided, by a ridge, into a lateral and a medial portion. The lateral portion of the articular surface is triangular and articulates with the scaphoid. The medial portion is quadrilateral and articulates with the lunate.

The shaft of the radius

The shaft of the radius is slightly convex laterally and has three borders and three surfaces.

The most prominent border is the medial, or *interosseous*, border which is noted as a sharp edge in the middle third of the shaft. The interosseous border can be traced upwards to the radial tuberosity and downwards to the ulnar notch, above which it divides into two ridges which pass to the anterior and posterior borders of the notch.

The anterior and posterior borders are rounded and indistinct.

The ulna

The ulna is the medial bone of the forearm and it is a long bone. It has a shaft, a wide upper end, and a small rounded lower end.

The upper end of the ulna

The upper end of the ulna is composed of two large processes, the *olecranon* and *coronoid* processes, and two articular notches, the *trochlear* and *radial notches*, which articulate with the humerus and radius respectively.

The olecranon process forms the uppermost part of the bone. It curves forwards at its summit to overhang the trochlear notch. This forward projection is accommodated by the olecranon fossa of the lower end of the humerus when the elbow joint is fully extended. The posterior surface of the olecranon is smooth and triangular and can be felt through the skin as the point of the elbow. The upper surface of the olecranon is roughened by

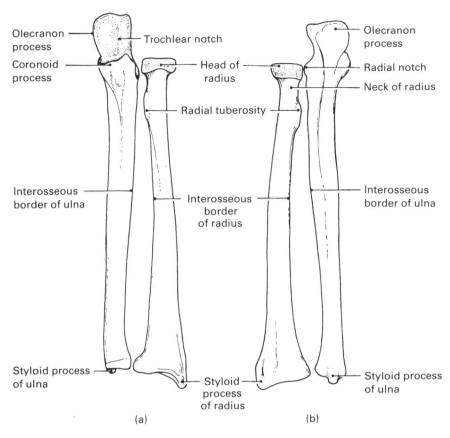

Fig. 8.5. The left radius and ulna: (a) anterior aspect, and (b) posterior aspect.

attachment of the triceps muscle. The anterior surface of the olecranon process is smooth and forms the upper part of the trochlear notch.

The coronoid process projects forwards from the bone below the olecranon process. Its upper surface is smooth and forms the lower part of the trochlear notch. The upper part of the lateral surface of the coronoid process bears the *radial notch*, and below this the bone is hollowed out to accommodate the radial tuberosity when the forearm is pronated. The anterior surface of the coronoid process is triangular and gives attachment to the brachialis muscle. Below this is a roughened area which is called the *tuberosity of the ulna*.

The radial notch is a smooth oblong concave articular surface on the lateral side of the coronoid process. It articulates with the articular surface on the circumference of the head of the radius at the superior radioulnar joint. It is divided from the trochlear notch by a ridge at its upper margin.

The trochlear notch articulates with the trochlea of the lower end of the humerus at the elbow joint. Its upper part is formed by the anterior surface of the olecranon process, and its lower part by the upper surface of the coronoid process. The junction of the two parts is marked by a shallow groove. A smooth ridge, which matches the groove of the trochlea, divides the trochlear notch into a larger medial and a smaller lateral portion.

The lower end of the ulna

The lower end of the ulna is composed of the small rounded head and the styloid process.

The head of the ulna has, on its lateral side, a smooth convex articular surface which articulates with the ulnar notch of the radius. The inferior surface of the head bears a smooth crescentic articular surface which articulates with the articular disc of the wrist joint. The apex of this articular disc is attached to a small rough area between the articular surface and the styloid process.

The styloid process is a smooth rounded projection which arises from the posteromedial side of the lower end of the ulna.

The shaft of the ulna

The shaft of the ulna is roughly triangular in cross-section and has three borders and three surfaces.

The lateral, or *interosseous border,* forms a sharp edge in the middle portion of the shaft. It can be traced downwards to the lateral side of the head and upwards to a prominent ridge called the *supinator crest.*

The anterior border is thick and rounded. It starts above, at the lower surface of the tuberosity of the ulna, and can be traced downwards, curving slightly backwards in its lower part, to the base of the styloid process.

The posterior border can be felt beneath the skin throughout its length. It begins above at the apex of the smooth triangular area on the posterior surface of the olecranon process and runs downwards to the back of the styloid process.

The anterior surface of the shaft of the ulna lies between the anterior and the interosseous borders.

The medial surface lies between the anterior and the posterior borders.

The posterior surface lies between the interosseous and posterior borders.

The carpal bones

The carpal bones are eight small bones which are arranged in a proximal and a distal row. Those of the proximal row are named, from the lateral to the medial side, the *scaphoid*, the *lunate*, the *triquetral* and the *pisiform*. The distal row are named, from the lateral to the medial side, the *trapezium*, the *trapezoid*, the *capitate* and the *hamate*. The pisiform lies on the palmar surface of the triquetral and articulates with that bone alone, but all the other bones of the carpus articulate with their neighbours.

The proximal row of the carpus forms an arch which is convex towards the arm, and which articulates with the distal surface of the radius and the articular disc of the wrist joint. The capitate and the hamate project into the concavity formed by the proximal row of bones and the trapezium and the trapezoid articulate with the distal surface of the scaphoid. The distal surfaces of the distal row of the carpal bones articulate with the bases of the metacarpal bones.

The carpus is deeply concave from side to side on its palmar surface, but its dorsal surface is only gently convex from side to side.

The scaphoid bone

The scaphoid bone is the largest bone of the proximal row of the carpus. Its medial surface is hollowed out and it is therefore described as boat-shaped. Its convex proximal surface articulates with the radius. The concavity of its medial surface articulates with the capitate and a flattened area on the proximal part of the medial surface articulates with the lunate. The distal convex surface articulates with the trapezium and trapezoid. On the distal part of the palmar surface there is a laterally directed rounded projection called the *tubercle* of the scaphoid.

The lunate bone

The lunate bone is crescentic in shape, and is convex proximally and concave distally. It articulates proximally with the radius and distally with the capitate. The lateral surface articulates with the scaphoid and the medial surface with the triquetral. The dorsal and palmar surfaces are roughened.

The triquetral bone

The triquetral bone is roughly wedged-shaped. Its palmar surface bears a smooth oval articular facet which articulates with the pisiform. The proximal surface slopes medially and downwards and articulates with the articular disc of the wrist joint. The lateral surface articulates with the lunate. The distal surface articulates with the hamate.

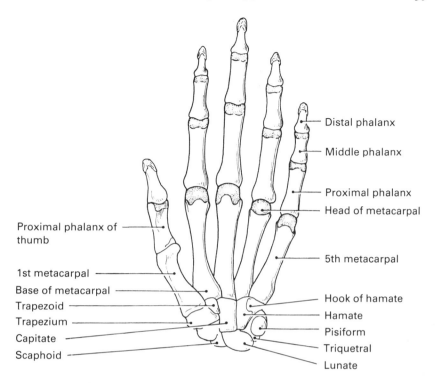

Distal phalanx

Middle phalanx

Proximal phalanx
Head of metacarpal

Proximal phalanx of thumb

5th metacarpal

1st metacarpal

Base of metacarpal

Trapezoid

Trapezium

Capitate

Scaphoid

Hook of hamate

Hamate

Pisiform

Triquetral

Lunate

Fig. 8.6. The left hand, palmar aspect.

The pisiform bone

The pisiform bone is a sesamoid bone. It is pea-shaped and bears, on its dorsal surface, an oval articular facet which articulates with the triquetral.

The trapezium bone

The trapezium bone is roughly cuboidal in shape, but it has, on its palmar surface, a groove through which one of the flexor tendons runs, and lateral to this is a projection called the *tubercle*. The lateral surface is rough where ligaments are attached. The distal surface has a large saddle-shaped articular surface, which faces partly laterally and articulates with the 1st metacarpal bone, and a smaller flat articular surface, which faces partly medially and articulates with the 2nd metacarpal bone. The medial surface is a gently concave articular surface which articulates with the trapezoid. The proximal surface is a small concave articular surface which articulates with the scaphoid.

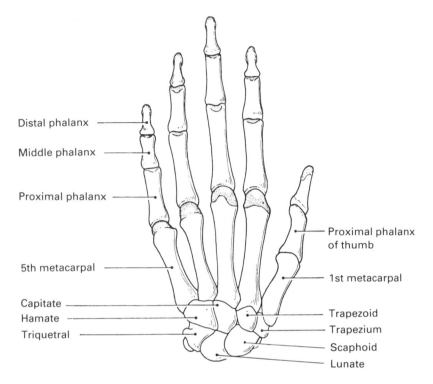

Distal phalanx

Middle phalanx

Proximal phalanx

5th metacarpal

Capitate

Hamate

Triquetral

Proximal phalanx of thumb

1st metacarpal

Trapezoid

Trapezium

Scaphoid

Lunate

Fig. 8.7. The left hand, dorsal aspect.

The trapezoid bone

The trapezoid bone is a small irregular bone. The distal surface articulates with the base of the 2nd metacarpal bone. The medial surface articulates with the capitate. The lateral surface articulates with the trapezium. The proximal surface articulates with the scaphoid.

The capitate bone

The capitate bone is the largest of the carpal bones and it lies opposite the base of the 3rd metacarpal bone. Its distal surface articulates centrally with the base of the 3rd metacarpal bone, but on the lateral side there is a narrow concave articular surface, which articulates with the base of the 2nd metacarpal bone, and on the medial side is a small articular facet, which articulates with the base of the 4th metacarpal bone. The proximal surface is a convex surface, which is called the *head of the capitate*, which articulates with the concavity formed by the scaphoid and lunate bones. The lateral surface of the bone articulates with the trapezoid and the medial surface articulates with the hamate.

The hamate bone

The hamate bone is roughly wedge-shaped but it has, on its palmar surface, a hook-like process. The distal surface is divided into two articular surfaces by a ridge. The lateral facet articulates with the base of the 4th metacarpal bone and the medial facet with the base of the 5th metacarpal bone. The proximal surface is the apex of the wedge and has a narrow facet which articulates with the lunate. The medial surface articulates with the triquetral bone and the lateral surface articulates with the capitate.

The metacarpal bones

The metacarpal bones, five in number, are miniature long bones. They are composed of a rounded head which lies distally, a shaft and an expanded proximal end, called the base.

The heads of the metacarpal bones

The heads of the metacarpal bones form the prominence of the knuckles. On the distal surface of each is a rounded articular facet which extends onto the anterior surface of the head. This articular surface articulates with the base of the proximal phalanx.

Bases of the metacarpal bones

The bases of the metacarpal bones are the expanded proximal portions of each bone. The base of each metacarpal bone articulates with the distal row of the carpus and the 2nd to 5th metacarpal bones articulate with each other.

The shafts of the metacarpal bones

The shafts of the metacarpal bones are roughly triangular in cross-section. The dorsal surface is flat, and the medial and lateral surfaces meet at an anterior rounded border. The anterior rounded border is concave longitudinally.

The 1st metacarpal bone

The 1st metacarpal bone is shorter and broader than the other metacarpal bones. Its base articulates with the trapezium by a saddle-shaped articular surface.

The 2nd metacarpal bone

The 2nd metacarpal bone is the longest of the metacarpal bones. Its base articulates mainly with the trapezoid, but a small area laterally articulates with the trapezium. On the medial side of the base is a large facet which articulates with the base of the 3rd metacarpal bone.

The 3rd metacarpal bone

The 3rd metacarpal bone has, on the dorsal surface of its base, a projection called the styloid process. The base of the 3rd metacarpal bone articulates with the capitate and, on either side of the base, with the 2nd and 4th metacarpal bones respectively.

The 4th metacarpal bone

The 4th metacarpal bone has, on its base, a large articular facet for the hamate and a smaller articular facet for the capitate. On either side of the base are articular facets for the 3rd and 5th metacarpal bones respectively.

The 5th metacarpal bone

The 5th metacarpal bone is the smallest of the metacarpal bones. Its base articulates with the hamate. On the lateral side of the base is an articular facet for the 4th metacarpal bone.

The phalanges

There are 14 phalanges, three for each of the four fingers and two for the thumb. The phalanges of each finger are referred to as the proximal, middle and distal phalanges respectively. The two phalanges of the thumb resemble the proximal and distal phalanges of one of the fingers.

The proximal phalanges

The proximal phalanges articulate proximally with the head of the appropriate metacarpal bone and distally with the middle phalanges. They have a *shaft* which is slightly concave forwards. The anterior surface of the shaft is flat and the posterior surface is rounded. The two surfaces meet at sharp lateral and medial borders.

The proximal end, or *base*, of the proximal phalanx has an oval concave articular facet which articulates with the head of the appropriate metacarpal bone.

The distal end, or *head*, of the proximal phalanx is narrower than the base and bears lateral and medial condyles which carry an articular facet for articulation with the base of the middle phalanx.

The middle phalanges

The middle phalanges are similar in shape to the proximal phalanges, but are smaller. The *base*, or proximal end, has a concave articular surface divided by a ridge into two parts which articulate with the head of the proximal phalanx. The *head* of the middle phalanx is similar in shape to the head of the proximal phalanx.

The distal phalanges

The distal phalanges are the smallest of the phalanges and have a flat posterior surface and a concave anterior surface. The *base* is similar in shape to the base of the middle phalanges, but the distal end widens out into a horseshoe-shaped expansion called the *terminal tuft* which supports the pulp of the fingers.

Arteries and nerves related to the bones of the upper limb

Certain arteries and nerves are related to the bones of the upper limb and are liable to damage if the bone is fractured (see Fig 11.20, p. 296; and Fig 16.20, p. 441).

1 The subclavian artery passes behind the middle of the shaft of the clavicle as it curves over the first rib to enter the axilla. On rare occasions a fracture of the clavicle can damage the subclavian artery.

2 The axillary artery is closely related to the medial side of the surgical neck and upper end of the shaft of the humerus as it runs down the lateral wall of the axilla. Fractures of the surgical neck of the humerus may, on occasions, damage the axillary artery.

3 The brachial artery lies in front of the lower end of the humerus separated from the bone only by the brachialis muscle. The brachial artery is particularly liable to damage in fractures of the lower end of the humerus (supracondylar fractures).

4 The radial artery lies in front of the radius throughout its whole length, separated from it only by muscles. The ulnar artery, which lies in front of the ulna in the lower two-thirds of the forearm is separated from it only by muscles. Both these arteries may be damaged in fractures of the radius and ulna.

5 The circumflex nerve winds round the lateral and posterior aspects of the surgical neck of the humerus. It is liable to damage both in fractures of the surgical neck of the humerus and in dislocation of the shoulder joint.

6 The radial nerve passes obliquely across the posterior aspect of the middle third of the shaft of the humerus in the spiral groove. It is occasionally damaged in fractures of the shaft of the humerus.

7 The ulnar nerve lies in the groove on the back of the medial epicondyle of the humerus. It is liable to damage in fractures of the medial epicondyle and may be damaged by stretching in supracondylar fractures of the humerus.

8 The median nerve lies on the medial side of the brachial artery in front of the lower end of the humerus. It may be damaged in supracondylar fractures of the humerus.

Ossification of the bones of the upper limb

The clavicle

The clavicle is the first bone in the body to ossify. It is ossified directly from a mesodermal membrane and does not have a cartilaginous stage. Two primary centres of ossification appear during the 5th to 6th weeks of intrauterine life and unite to form one centre at the 7th week.

A secondary centre of ossification appears at the *sternal end* of the clavicle during the 15th year and this fuses with the shaft during the 22nd to the 24th years of life.

The scapula

The scapula is ossified from eight centres.

1 *The body of the scapula* is ossified from one centre which appears at the 8th week of intrauterine life.

2 *The coracoid process* is ossified from one centre of ossification which appears during the 1st year of life. It fuses with the rest of the bones during the 15th year.

3 The remaining six centres appear at the time of puberty and all have fused with the rest of the bone by the 20th year. The six centres are distributed as follows: one for the *subcoracoid region*, two for the *acromion process*, one for the *medial border* of the scapula, one for the *inferior angle* of the scapula, and one for the lower part of the *glenoid cavity*.

The humerus

The humerus is ossified from eight centres.

The shaft of the humerus is ossified from one centre which appears during the 8th or 9th week of intrauterine life.

The upper end of the humerus is ossified from three centres:

1 A secondary centre of ossification appears in the *head of the humerus* during the first six months of life.

2 A secondary centre appears in the *greater tubercle* during the 1st or 2nd year of life.

3 A secondary centre of ossification appears in the *lesser tubercle* during the 5th year of life.

These three centres unite to form one large epiphysis during the 6th year of life. The whole epiphysis fuses with the shaft during the 20th year of life.

The lower end of the humerus is ossified from four centres:

1 A secondary centre appears in the *capitulum* during the 1st year of life. This centre forms the capitulum and the lateral part of the trochlea.

2 A secondary centre of ossification appears in the *medial epicondyle* in the 4th to 6th years.

3 A secondary centre of ossification appears in the medial part of the *trochlea* during the 9th to 10th years.

4 A secondary centre of ossification appears in the *lateral epicondyle* during the 12th year.

The centres for the lateral epicondyle, the capitulum, and the trochlea fuse to form one large epiphysis at about the time of puberty. This epiphysis fuses with the shaft during the 15th or 16th year. The medial epicondyle remains as a separate epiphysis which fuses with the rest of the bone during the 20th year.

The radius

The radius is ossified from three centres.

1 *The shaft of the radius* is ossified from one centre which appears during the 8th week of intrauterine life.

2 *The upper end of the radius* is ossified from a secondary centre of ossification which appears in the 4th year of life and fuses with the shaft during the 15th or 16th year of life.

3 *The lower end of the radius* is ossified from a secondary centre of ossification which appears at the end of the 1st year of life and fuses with the shaft during the 17th or 18th year of life.

The ulna

The ulna is ossified from three centres.

The shaft of the ulna is ossified from one centre which appears during the 8th week of intrauterine life.

The upper part of the *olecranon* is ossified from a secondary centre of ossification which appears during the 5th year of life, and fuses with the shaft during the 17th year of life.

The bones of the hand

The carpal bones are each ossified from one centre of ossification. The dates of appearance of these centres are:

The capitate	1st year
The hamate	2nd year
The triquetral	3rd year
The lunate	4th year
The scaphoid	5th year
The trapezium	6th year
The trapezoid	7th year
The pisiform	8th year

The metacarpal bones each ossify from two centres. The 2nd to 5th metacarpal bones ossify from one primary centre of ossification for the shaft and one secondary centre of ossification for the head. The first metacarpal bone has one primary centre of ossification for the shaft and a secondary centre of ossification for the base.

The primary centres for the *shaft* appear during the 9th week of intrauterine life.

The secondary centres appear during the 2nd and 3rd years of life and unite with the shaft between the 16th and 18th years.

The phalanges are each ossified from two centres of ossification.

The shafts of the phalanges are ossified from a primary centre which appears between the 9th and 12th weeks of intrauterine life.

The bases of the phalanges are each ossified from a secondary centre of ossification which appears between the 2nd and 4th years and fuses with the shaft between the 15th and 18th years.

CHAPTER 9

The bones of the lower limb

The lower limb is attached to the trunk by the pelvic girdle, which is formed by the two *hip bones* which articulate with each other in the midline anteriorly and with the sacrum posteriorly.

The lower limb is composed of three segments:

1   the thigh, which has one bone, the *femur*;
2   the leg, which has two bones, the *tibia* and the *fibula*;
3   the foot, which is composed of seven *tarsal bones*, five *metatarsal bones*, and 14 *phalanges*.

The hip bone

The hip bone is a large irregular-shaped bone which is constricted about its middle and is expanded above and below. The lateral surface of the hip bone bears, near its middle, a large cup-shaped structure called the *acetabulum*, which forms a socket for articulation with the head of the femur at the hip joint. The expanded portion of the bone, below the acetabulum, is pierced by a large foramen called the *obturator foramen*. Above the acetabulum is a wide, flat plate of bone which has an upper curved border called the *iliac crest*.

Each hip bone is composed of three portions called the ilium, the ischium and the pubis. In the child these portions are joined by cartilage, but, by adult life, bony fusion has occurred. Each of these parts forms part of the acetabulum. (The lines of fusion of the three parts are shown by dotted lines in Fig. 9.1.)

The ilium

The ilium is a roughly triangular plate of bone of which the lower part forms the upper two-fifths of the acetabulum. The upper border of the ilium is curved and thickened throughout its length, forming the *iliac crest*. The iliac crest ends anteriorly in a blunt process, called the *anterior superior iliac spine* and posteriorly at a blunt process called the *posterior superior iliac spine*. The iliac crest, when viewed from above, is seen to describe an S-shaped curve. It curves outwards anteriorly and inwards posteriorly. The iliac crest is an important site for the attachment of the muscles of the abdominal wall.

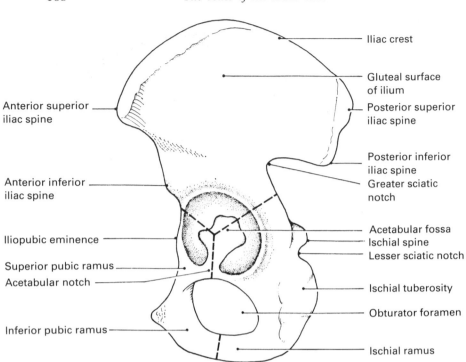

Iliac crest

Gluteal surface of ilium

Anterior superior iliac spine

Posterior superior iliac spine

Posterior inferior iliac spine

Greater sciatic notch

Anterior inferior iliac spine

Iliopubic eminence

Acetabular fossa

Ischial spine

Lesser sciatic notch

Superior pubic ramus

Acetabular notch

Ischial tuberosity

Obturator foramen

Inferior pubic ramus

Ischial ramus

Fig. 9.1. The external aspect of the left hip bone.

The *anterior border* of the ilium begins above at the anterior superior iliac spine. Immediately below this spine is a notch, and the lower margin of this notch is formed by another blunt projection, called the *anterior inferior iliac spine*. Below the anterior inferior iliac spine the anterior border ends at the acetabulum.

The *posterior border* of the ilium runs downwards and forwards from the *posterior superior iliac* spine to a sharp projection called the *posterior inferior iliac spine*. Below the posterior inferior iliac spine the posterior border presents a deep notch, called the *greater sciatic notch*. In the lower part of this notch the ilium is continuous with the ischium.

The *outer surface* of the ilium is called the *gluteal surface*. It is concave posteriorly and convex anteriorly and is marked by three lines called the gluteal lines. The gluteal surface of the ilium gives attachment to the gluteal muscles, which are the muscles of the buttock.

The *inner surface* of the ilium is divided into an upper and a lower portion by a ridge which forms the medial border of the bone. This ridge starts behind the iliac crest, 50−75 mm above the posterior superior iliac spine, it then runs downwards and forwards, as the sharp upper margin of the articular area of the ilium, and then becomes smooth and runs across the ilium as the iliac part of the *arcuate line*. It ends at a prominence, called the *iliopubic eminence*, which marks the site of junction of the ilium and pubis. The arcuate line marks the site of division between the false and the true pelvis (p. 171). The area of the inner surface of the ilium above the arcuate line is called the *iliac fossa* and is gently concave. The iliacus muscle arises from the floor of the iliac fossa. The portion of the ilium below the medial border is called the sacropelvic surface. The anterior and lower part of this surface forms part of the lateral wall of the true pelvis and is smooth and concave. The posterior and upper part of the sacropelvic surface is the articular area of the ilium. On the lower part of the articular area there is the roughened auricular area which articulates with the sacrum at the sacroiliac joint. The portion of the bone behind this is roughened for the attachment of the sacroiliac ligaments.

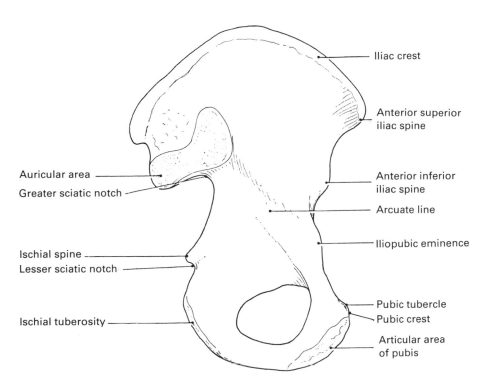

Fig. 9.2. The internal aspect of the left hip bone.

The ischium

The ischium is L-shaped and it forms the posterior and inferior parts of the hip bone. It is composed of a body, a tuberosity and a ramus. The body is the upper part of the bone which unites with the ilium and the pubis to form the acetabulum. From the acetabulum the body runs downwards to the expanded ischial tuberosity on which the weight of the body is supported in the sitting position. The ischial ramus runs upwards and forwards from the tuberosity to join the inferior pubic ramus, thus enclosing the obturator foramen.

The body of the ischium forms the posterior and lower parts of the acetabulum and extends downwards as a flat bar to end at the ischial tuberosity. The posterior border of the body starts as a continuation of the ilium in the lower part of the greater sciatic notch. The lower margin of the greater sciatic notch is marked by a sharp spine, the *ischial spine*. Below the ischial spine is another notch, the *lesser sciatic notch*, at the lower border of which the posterior border of the body is continuous with the ischial tuberosity.

The *anterior border* of the body is sharply defined and forms the lower margin of the acetabulum above, and the posterior margin of the obturator foramen below.

The *lateral*, or femoral, *surface* of the body forms the lower two-fifths of the acetabulum above and is continuous with the ischial tuberosity below. The *medial*, or pelvic, *surface* of the body is smooth and flat and is continuous with the pelvic surface of the ilium.

The ischial tuberosity is a large roughened expansion which lies at the inferior end of the body of the ischium. It can be felt easily through the skin when the thigh is flexed.

The ramus of the ischium projects forwards and upwards from the ischial tuberosity, and fuses with the inferior ramus of the pubis, thus enclosing the obturator foramen. The ramus has a sharp upper border which forms part of the lower margin of the obturator foramen. The upper border of the ischial ramus is continuous above with the anterior border of the body. The anterior surface of the ramus is roughened by the attachment of muscles but the posterior surface is smooth and forms part of the wall of the true pelvis.

The pubis

The pubis is composed of a quadrilateral shaped body, a superior ramus, which forms part of the acetabulum, and an inferior ramus which fuses with the ischial ramus.

The body of the pubis is a flat quadrilateral plate of bone which forms part of the anterior wall of the true pelvis. The lateral margin of the body is sharp where it forms the anterior boundary of the obturator foramen. The medial surface of the body has an oval articular area which articulates with the body of the pubis of the opposite hip bone at the pubic symphysis. The upper surface of the body is called the *pubic crest* and, at its lateral end, there is a prominence called the *pubic tubercle*.

The superior pubic ramus runs laterally, backwards and upwards from the upper part of the body. Its lateral end forms the anterior one-fifth of the acetabulum. The upper border of the superior ramus forms the anterior part of the *arcuate line* which divides the true and false pelvis. This border is called the *pectineal line* and it extends to the iliopubic eminence. The lower border of the superior ramus forms the upper boundary of the obturator foramen. The anterior surface of the superior ramus is marked by the smooth obturator crest which runs from the pubic tubercle to the front of the acetabulum.

The inferior pubic ramus runs downwards and backwards, from the lower part of the body of the pubis, to fuse with the ischial ramus, thus enclosing the obturator foramen.

The acetabulum

The acetabulum is a hemispherical depression on the lateral aspect of the hip bone. Its margin is defined by the prominent acetabular rim which is deficient below at the *acetabular notch*. The acetabulum forms a cup-shaped socket for articulation with the head of the femur. The articular surface of the acetabulum is deficient centrally and below, where it forms the acetabular fossa. The articular surface is thus horsehoe-shaped. The *anterior one-fifth* of the acetabulum is formed by the pubis, the *upper two-fifths* are formed by the ilium and the *lower* and *posterior two-fifths* are formed by the ischium.

The pelvis

The pelvis is a ring of bone which is placed between the lower part of the vertebral column and the lower limbs. It is formed posteriorly by the sacrum and anteriorly and laterally by the two hip bones. It transfers the weight of the trunk and upper limbs from the vertebral column to the lower limbs.

The pelvis is divided into an upper portion called the *false pelvis* and a lower portion called the *true pelvis*. The boundary between the two is formed posteriorly by the *promontory of the sacrum* and laterally and anteriorly by the *arcuate line* and the *pubic crest*. This boundary is called the *pelvic inlet* and in the erect position the plane of the inlet lies at 60° to the horizontal plane. In the erect position the anterior superior iliac spine and the pubic tubercle lie in the same vertical plane.

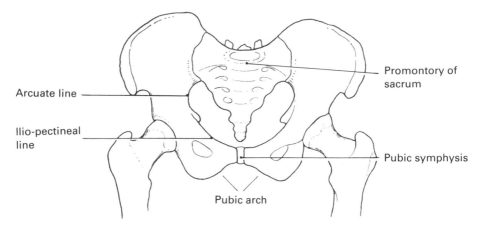

Fig. 9.3. The pelvis, anterior aspect.

The false pelvis is bounded laterally and posteriorly by the iliac fossae which form part of the walls of the abdominal cavity and help to support the abdominal viscera.

The true pelvis lies below the pelvis inlet and has anterior, lateral and posterior walls. The posterior wall is formed by the sacrum and the coccyx. The lateral and anterior walls are formed by the ilium, the ischium and the pubis.

The *inlet* of the pelvis is slightly heart-shaped in the male but is more circular in the female. The *outlet* of the pelvis is the lower margin and is formed anteriorly by the lower border of the symphysis pubis, laterally by the ischial tuberosities and posteriorly by the tip of the coccyx. The anterolateral margins of the outlet are formed by the inferior pubic rami which meet at an angle called the *pubic arch*.

The *cavity* of the true pelvis is a curved canal which is deeper behind than in front. The curvature of the canal follows the curvature of the sacrum and coccyx.

The size of the inlet and outlet of the pelvis are of importance in the female as the fetal head passes through the pelvis during delivery. Certain measurements of the inlet and outlet are frequently measured by radiographic methods in order to assess the adequacy of the pelvis for the passage of the fetal head.

The measurements of importance for the inlet are:

1   *The internal conjugate*, which is the distance between the back of the body of the pubis and the sacral promontory. The average measurement in the female is 110 mm.

2   *The maximum transverse diameter*, which is the measurement of the inlet at its widest point. The average measurement in the female is 135 mm.

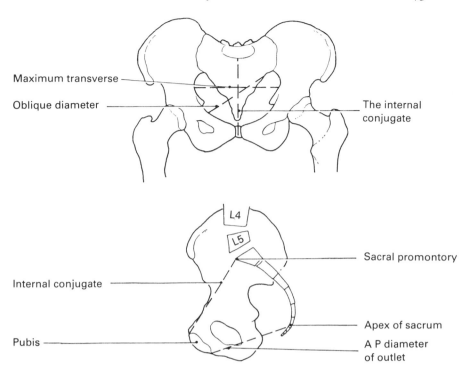

Maximum transverse

Oblique diameter

The internal conjugate

L4

L5

Sacral promontory

Internal conjugate

Apex of sacrum

Pubis

A P diameter of outlet

Fig. 9.4. The measurements of the female pelvis.

3 *The oblique diameter of the inlet,* which is the distance between the iliopectineal eminence and the lower margin of the sacroiliac joint. The average measurement in the female is 125 mm.

The important measurements of the *outlet* are:

1 *The anteroposterior diameter,* which is the distance between the back of the lower part of the symphysis pubis and the tip of the sacrum. The average measurement in the female is about 115 mm.

2 *The intertuberous diameter,* which is the transverse measurement between the ischial tuberosities. The average measurement in the female is 115 mm.

The differences between the female and the male pelvis

The main differences between the male and the female pelvis are concerned with the function of childbearing, the female pelvis being adapted to allow the passage of the fetal head.

1 The inlet of the pelvis is heart-shaped in the male, but is more circular in the female.

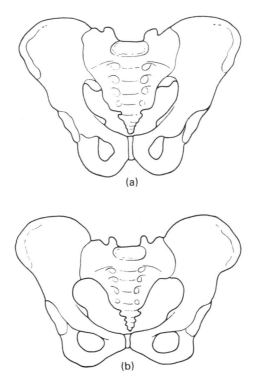

(a)

(b)

Fig. 9.5. (a) Male, and (b) female pelves compared.

2 The female sacrum is wider, shorter and less curved than the male sacrum.
3 The ischial tuberosities are wider apart in the female.
4 The pubic arch is generally more than 90° in the female and less than 90° in the male.
5 The female pelvis is generally less massive than the male and the muscular attachments are generally less well marked.

The femur

The femur is a long bone and is the longest and strongest bone in the body. It consists of an upper end, a lower end and a shaft. In the articulated skeleton the femora lie obliquely; their upper ends are separated by the width of the pelvis but their lower ends are in close proximity at the knees.

The upper end of the femur

The upper end of the femur is composed of the head, the neck, and the lesser and greater trochanters.

The head of the femur is slightly more than a hemisphere in extent. It faces upwards, medially and slightly forwards, and it articulates with the acetabulum at the hip joint. Just below the centre of the head there is a small roughened pit, called the *fovea*. The ligament of the head of the femur is attached at the fovea.

The neck of the femur is about 5 cm long. It connects the head of the femur with the shaft, which it meets at an angle of about 125°. The neck is widest at its junction with the shaft, narrows towards its midpoint and then widens slightly towards its junction with the head. The medial part of the neck is cylindrical but close to the shaft its anterior and posterior surfaces are somewhat flattened. The upper rounded border of the neck is concave upwards. The lower border is rounded and runs obliquely downwards to join the shaft close to the lesser trochanter. The line of union of the anterior surface of the neck with the shaft is marked by a ridge, called the *intertrochanteric line*. The junction of the posterior surface of the neck with the shaft is marked by a ridge called the *intertrochanteric crest*.

The greater trochanter is a large quadrilateral projection of bone which arises from the lateral angle of the junction of the neck of the femur with the shaft. The upper border of the greater trochanter projects medially to overhang the neck, and the hollow thus produced in its medial surface is called the *trochanteric fossa*.

The lesser trochanter is a conical process which projects backwards and medially from the junction of the lower posterior part of the neck with the shaft.

The lower end of the femur

The lower end of the femur is expanded and is composed of two masses of bone called the condyles. The condyles are joined anteriorly, but are separated posteriorly by a U-shaped notch, called the *intercondylar notch*.

The posterior and inferior surfaces of each condyle bear a smooth articular surface which articulates with the tibia at the knee joint. These articular surfaces unite anteriorly, and extend onto the anterior surface of the lower end of the femur to form the articular surface for the patella.

The medial condyle projects farther downwards than the lateral condyle, but, because of the oblique position of the femur, both condyles lie in the same horizontal plane in the articulated skeleton. The tibial articular surface of the medial condyle is narrower than that of the lateral condyle.

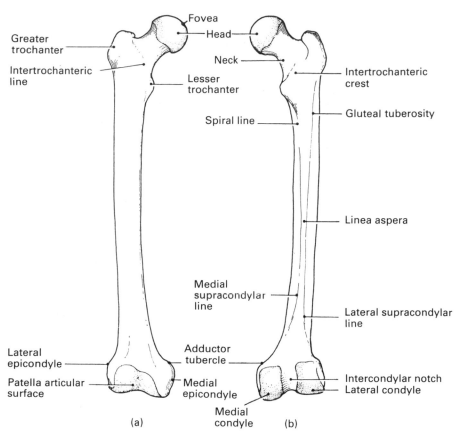

Fig. 9.6. The right femur: (a) anterior aspect, and (b) posterior aspect.

The medial surface of the medial condyle has, just behind its centre, a roughened projection called the *medial epicondyle*. Above this is a rather prominent projection, the *adductor tubercle*, which gives attachment to the adductor magnus muscle.

The lateral condyle is flattened laterally, is not so prominent as the medial condyle, and the tibial articular surface is wider and straighter than that of the medial condyle. On the lateral surface of the lateral condyle is a projection, the *lateral epicondyle*.

The patellar articular surface of the femur is situated on the anterior surface of the lower end of the bone and is concave from side to side. It extends farther upwards onto the lateral condyle than onto the medial condyle and below it is separated from the anterior part of the tibial articular surface by a faint groove.

The shaft of the femur

The shaft of the femur describes a gentle curve and is convex forwards and laterally.

The middle third of the shaft is the thinnest portion of the shaft and is roughly triangular in cross-section. The shaft widens slightly in the upper third, and expands considerably towards the lower end.

The *posterior border* of the shaft is formed in the middle third by a rough line called the *linea aspera*. The linea aspera ends above by dividing into the rough *gluteal tuberosity*, which passes upwards to the lower border of the greater trochanter, and the *spiral line*, which curves round the medial border of the shaft to join the intertrochanteric line. The linea aspera ends below by dividing into the *lateral and medial supracondylar lines* which pass to the lateral and medial condyles respectively.

The lateral and medial borders of the shaft are rounded and ill-defined.

The patella

The patella, or knee cap, is the largest sesamoid bone in the body. It is roughly triangular in shape, with the prominent rounded *apex* lying inferiorly.

The ligamentum patellae is attached to the apex of the patella.

The lateral and medial borders of the patella are rounded and well defined.

The superior border, or base, of the patella is thickened, and slopes downwards and forwards.

The anterior surface of the patella is convex and is marked by numerous longitudinal striations where fibres of the quadriceps tendon extend across it to join the ligamentum patellae.

The posterior surface of the patella is roughened, near the apex, for attachment of the ligamentum patellae. Above this is a large oval articular facet which is divided into a larger lateral and a smaller medial area for articulation with the patella articular surface of the lower end of the femur.

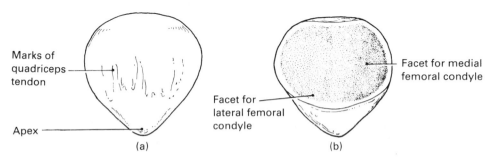

Marks of quadriceps tendon

Apex

Facet for lateral femoral condyle

Facet for medial femoral condyle

(a) (b)

Fig. 9.7. The left patella: (a) anterior aspect, and (b) posterior aspect.

The tibia

The tibia is a long bone and is the medial and larger of the two bones of the leg. It consists of an upper expanded end, a smaller lower end and a shaft.

The upper end of the tibia

The upper end of the tibia is expanded, more from side to side than from front to back, but the posterior part of the upper end curves backwards to overhang the shaft. It is composed of two masses of bone, the *lateral* and *medial condyles*, and there is a small process on the anterior surface called the tibial tuberosity. The upper surfaces of the two tibial condyles are separated in the midline by the roughened *intercondylar area.*

The intercondylar area lies between the articular surfaces of the two condyles. In the middle of the intercondylar area there is a raised portion of bone called the *intercondylar eminence,* and the lateral and medial margins of this project upwards as the small, pointed *lateral* and *medial intercondylar tubercles.*

The medial condyle is the larger of the two condyles, but it does not overhang the shaft as much as does the lateral condyle. It has a large oval articular facet on its upper surface. This facet is concave from side to side and from front to back and it extends upwards onto the medial intercondylar tubercle. The margin of this articular area is slightly elevated and the medial semilunar cartilage of the knee lies on this portion.

The lateral condyle overhangs the shaft of the tibia. The articular facet on its upper surface is roughly circular in outline and the central portion of this is concave. The lateral semilunar cartilage of the knee joint lies on the margin of the articular surface. Posterolaterally, on the under surface of the lateral condyle, is a small, circular articular facet which articulates with the head of the fibula.

The anterior surfaces of the two condyles meet at a central smooth triangular area at the apex of which lies the *tibial tuberosity.* The upper margin of the tibial tuberosity gives attachment to the ligamentum patellae.

The lower end of the tibia

The lower end of the tibia is quadrilateral in cross-section. The inferior surface of the lower end is a quadrilateral articular surface which articulates with the talus at the ankle joint. This articular surface is wider in front than behind, is concave from front to back and is slightly convex from side to side.

The medial surface of the lower end of the tibia projects downwards as a blunt process, called the *medial malleolus.* The lateral margin of the

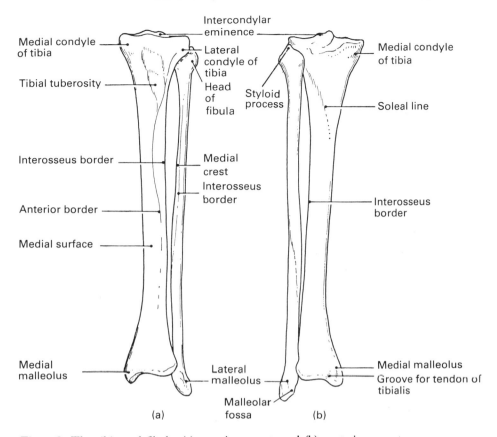

Fig. 9.8. The tibia and fibula: (a) anterior aspect, and (b) posterior aspect.

medial malleolus has a smooth crescentic articular surface, which articulates with the side of the talus and is continuous above with the articular surface on the lower end of the tibia.

The lateral surface of the lower end of the tibia has a triangular notch called the *fibular notch*. The interosseous border of the shaft of the tibia runs upwards from the apex of this notch.

The anterior surface of the lower end of the tibia is smooth. The posterior surface is grooved medially, where the tendon of the tibialis posterior muscle lies against the bone.

The shaft of the tibia

The shaft of the tibia is triangular in cross-section. It narrows slightly from the upper end to the junction of the middle and lower thirds of the bone and then expands slightly again in the lower third.

The anterior border of the shaft runs downwards from the tibial tuberosity to the medial malleolus. It is subcutaneous throughout its length but it becomes less distinct in the lower portion of the shaft. It is commonly known as the shin.

The lateral, or *interosseous border*, of the shaft begins above, just in front of the facet for the head of the fibula, and runs downwards to the apex of the fibular notch. The interosseous membrane, which connects the tibia and fibula, is attached to the interosseous border and to the anterior border of the fibular notch.

The medial border of the shaft is less well-defined and extends downwards from the medial condyle to the medial malleolus.

The lateral surface of the shaft lies between the anterior and the interosseous borders. The *medial surface* lies between the medial and the anterior borders and is smooth and subcutaneous throughout its length. The posterior *surface* lies between the medial and the anterior borders and is crossed from above downwards and medially by a rough line called the *soleal line* to which the soleus muscle is attached.

The fibula

The fibula is the lateral bone of the leg and is shorter and more slender than the tibia. It is a long bone and is composed of an upper end, or head, a lower end, or lateral malleolus, and a shaft.

The upper end of the fibula

The upper end, or **the head**, of the fibula is the expanded portion of the bone, and is roughly cylindrical in shape. The head has on its upper surface a circular articular facet which articulates with the facet on the lower surface of the lateral condyle of the tibia. Posterolaterally a blunt process, called the *styloid process*, projects upwards. The lateral ligament of the knee joint is attached to the apex of the styloid process.

The lower end of the fibula

The lower end of the fibula is the **lateral malleolus**. The posterior part of the medial surface of the lateral malleolus has a roughened depression called the *malleolar fossa*. The anterior part of the medial surface has a triangular articular facet which articulates with the talus at the ankle joint.

The lateral surface of the lateral malleolus is smooth and subcutaneous. The anterior surface merges with the lateral surface and has attached to it the anterior talofibular ligament. The posterior surface of the lateral malleolus is grooved by the tendons of the peroneal muscles.

The shaft of the fibula

The shaft of the fibula is roughly triangular in cross-section and the borders curve round the shaft in a complicated manner. The most important border is the medial or *interosseous border*. This extends upwards from a triangular area above the articular surface of the medial side of the lateral malleolus to end at the anterior aspect of the head of the fibula. Just posterior to the medial border is a line called the *medial crest* which ends at the medial side of the head.

The tarsal bones

There are seven tarsal bones all of which are of irregular shape. They correspond to the carpal bones of the hand, but they are considerably larger and are modified for bearing the weight of the body.

They are arranged in the following manner.

The *talus* lies immediately below the tibia. Its long axis runs forwards and medially so that its anterior end lies medial to the anterior end of the calcaneum.

The *calcaneum* lies below the talus and its anterior end is lateral to the anterior end of the talus.

The *cuboid* articulates with the anterior end of the calcaneum.

The *navicular* articulates with the anterior end of the talus.

The *medial, intermediate* and *lateral cuneiform bones* articulate with the anterior surface of the navicular.

The talus

The talus is the second largest of the tarsal bones. It is composed of a *body*, which lies between the lower end of the tibia and the upper surface of the calcaneum, a *head*, which is the anterior rounded portion of the bone, and a *neck*, which is the ill-defined constriction between the body and the head.

The upper surface of the body bears a large pulley-shaped articular surface, called the trochlear surface, which is concave from side to side and convex from front to back. This articulates with the lower end of the tibia. The medial surface of the body has a small comma-shaped articular surface which articulates with the medial malleolus. This articular surface is continuous above with the trochlear surface. The lateral surface of the body bears a smooth triangular articular surface which articulates with the lateral malleolus. The posterior surface of the body of the bone narrows to a pointed tubercle called the *posterior process*. Occasionally this process is represented by a separate bone which is then called the *os trigonum*.

The inferior surface of the talus bears two articular facets, separated by

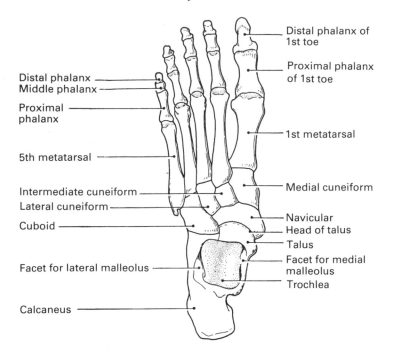

Distal phalanx of
1st toe

Proximal phalanx
of 1st toe

Distal phalanx
Middle phalanx

Proximal
phalanx

1st metatarsal

5th metatarsal

Intermediate cuneiform
Lateral cuneiform
Cuboid

Medial cuneiform

Navicular
Head of talus
Talus
Facet for medial
malleolus
Trochlea

Facet for lateral malleolus

Calcaneus

Fig. 9.9. The left foot, dorsal aspect.

a deep groove, called the *sulcus tali*, which runs forwards and laterally
across the inferior surface of the bone. The posterior facet is the larger of
the two, is oval in shape, and is concave in its long axis. It articulates with
the posterior facet on the upper surface of the calcaneum. The anterior
facet is a long concave facet which is usually divided into three by two
ridges. The most posterior area articulates with the *sustentaculum tali*,
which projects from the medial side of the calcaneum. The anterior area
articulates with the anterior end of the upper surface of the calcaneum and
is continuous with the articular surface on the head of the talus. A small
articular area on the medial side rests on the plantar calcaneonavicular
ligament (p. 257).

The head of the talus has a convex oval articular facet which faces
slightly downwards and articulates with the navicular.

The calcaneum

The calcaneum is the largest of the tarsal bones and it lies under the talus
with its long axis running forwards and laterally.

The upper surface is divided into three areas. The posterior part of the

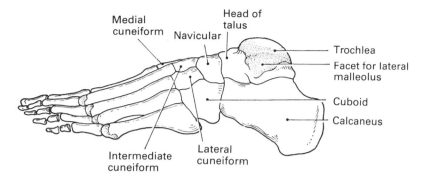

Fig. **9.10.** The left foot, lateral aspect.

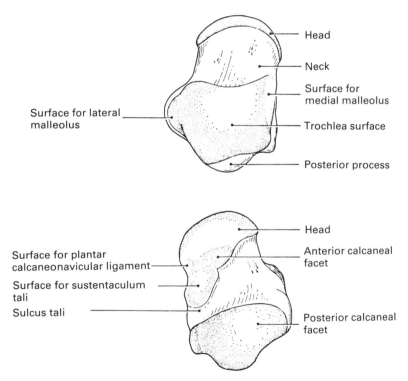

Fig. **9.11.** The left talus: (a) superior aspect, and (b) inferior aspect.

upper surface is roughened and is convex from side to side. It is covered, in the intact body, by a mass of fatty tissue which lies between the ankle joint and the tendo calcaneus (tendo achilles). The middle area is a large oval convex articular surface which articulates with the posterior facet on the inferior surface of the talus. The anterior area is grooved at its posterior end by the *sulcus calcanei*. Anteriorly and medially to this groove are two articular facets which articulate with the facets on the anterior end of the inferior surface of the talus. The most medial of these facets lies on the upper surface of the *sustentaculum tali*.

The anterior surface of the calcaneum has an oval articular facet, concave from side to side and convex from above downwards, which articulates with the cuboid.

The posterior surface of the calcaneum has a wide roughened ridge running transversely across its middle. The tendo calcaneus is attached to this ridge. The area above this ridge is smooth, and is separated from the tendo calcaneus by a bursa.

The inferior, or plantar, surface of the calcaneum is roughened and at the posterior end of this surface are two prominences called the *medial* and *lateral tubercles*.

The lateral surface of the calcaneum is rough and almost flat, but in its anterior part there is a small projection called the *peroneal tubercle*.

The medial surface of the calcaneum is concave from above downwards and from the upper border of the anterior end of this surface the shelf-like sustentaculum tali projects medially.

The navicular

The navicular lies between the talus and the cuneiform bones and is frequently described as boat-shaped.

The posterior surface of the navicular has a large oval concave articular facet which articulates with the head of the talus.

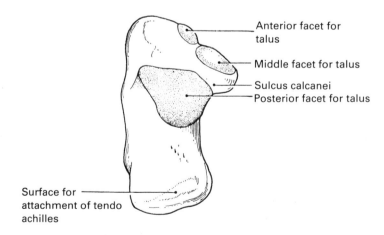

Fig. 9.12. The left calcaneus, superior aspect.

The anterior surface has a convex articular facet, which is divided by two vertical ridges into three facets for articulation with the cuneiform bones.

The medial surface of the navicular projects medially to form the tubercle of the navicular.

The lateral surface of the navicular occasionally bears a small articular facet for the cuboid.

The upper surface of the navicular is roughened and is slightly wider than the lower surface, which is also roughened and slightly concave from side to side.

The cuboid

The cuboid lies at the lateral side of the tarsus, in front of the calcaneum and behind the 4th and 5th metatarsal bones.

The posterior surface of the cuboid bears an articular facet which articulates with the anterior end of the calcaneum.

The anterior surface of the cuboid is divided by a ridge into two articular facets which articulate with the 4th and 5th metatarsal bones.

The upper surface is roughened. The lower, or plantar, surface is crossed obliquely by a groove at the lateral end of which is the tuberosity of the cuboid.

The medial surface is the largest surface of the bone and has, near its middle, an oval articular facet which articulates with the lateral cuneiform bone. Just behind this articular facet there is occasionally a small articular facet for the navicular.

The lateral surface is rough, but bears a groove which contains the tendon of the peroneus longus muscle.

The cuneiform bones

The cuneiform bones are wedge-shaped. The lateral and intermediate cuneiforms are wider above than below, but the medial cuneiform is wider below than above. The anterior ends of the lateral and medial cuneiform bones project farther forwards than the intermediate and thus form a slot into which the base of the 2nd metatarsal bone is received. The medial cuneiform is the largest of the bones and the intermediate cuneiform is the smallest.

The medial cuneiform bone has a narrow, rough, upper surface and a convex, roughened medial surface. The lateral surface is slightly concave and has an L-shaped articular facet along its posterior and upper margins. This articulates mainly with the intermediate cuneiform, but a small area anteriorly articulates with the base of the 2nd metatarsal bone. The posterior surface has a small concave articular facet which articulates with the navicular. The anterior surface has a smooth kidney-shaped convex articular surface which articulates with the base of the 1st metatarsal bone.

The intermediate cuneiform bone is wider above than below. Its upper surface is roughened. The medial surface has, along its upper and posterior borders, an L-shaped articular facet which articulates with the medial cuneiform. The lateral surface has, as its posterior end, an elongated articular surface which articulates with the lateral cuneiform. The posterior surface has a triangular articular facet which articulates with the middle facet on the anterior surface of the navicular. The anterior surface has a triangular facet for the base of the 2nd metatarsal bone.

The lateral cuneiform bone is wider above than below. The lateral surface has a large oval articular facet which articulates with the cuboid. The medial surface has two articular facets, one posteriorly which articulates with the intermediate cuneiform, and one anteriorly which articulates with the base of the 2nd metatarsal. The anterior surface has a triangular facet which articulates with the base of the 3rd metatarsal. The posterior surface has a triangular articular facet which articulates with the lateral facet on the anterior surface of the navicular.

The metatarsal bones

There are five metatarsal bones. They are long bones, having a base proximally, a head distally and a shaft. The 2nd, 3rd and 4th metatarsals are similar, but the 1st and 5th differ.

The bases of the 2nd, 3rd and 4th metatarsal bones are wedge-shaped, and are wider above than below. The bases of the 2nd and 3rd metatarsals articulate with the intermediate and lateral cuneiform bones respectively. The 2nd metatarsal bone is the longest and articulates at its base with the intermediate cuneiform bone, at the sides of its base with the lateral and medial cuneiform bones, and also, on the lateral side of its base, with the base of the 3rd metatarsal bone. The 4th metatarsal bone articulates at its base with the cuboid and, at the sides of its base, with the 3rd and 5th metatarsal bones.

The heads of the 2nd, 3rd and 4th metatarsal bones are narrower than their bases and have convex, rectangular articular facets which extend farther onto the plantar than the dorsal surface.

The shafts of the 2nd, 3rd and 4th metatarsal bones are slightly curved, being convex towards the dorsum of the foot. The shafts are somewhat compressed from side to side and have a rounded upper surface and a sharper plantar border, so that they are roughly triangular in cross-section.

The 1st metatarsal bone

The 1st metatarsal bone is the shortest but most massive metatarsal bone. The *base* has a single kidney-shaped articular surface which articulates with the anterior surface of the medial cuneiform bone. The lateral side of the base does not articulate with the 2nd metatarsal bone. The *shaft* is thicker than the shafts of the other metatarsal bones. The under surface of the *head* of the 1st metatarsal bone bears two grooves which contain sesamoid bones in the intact body.

The 5th metatarsal bone

The *base* of the 5th metatarsal bone has a triangular articular facet which articulates with the lateral facet on the anterior surface of the cuboid. The lateral side of the base of the 5th metatarsal bone bears a prominent tubercle, which projects backwards and laterally. The *shaft* and *head* of the 5th metatarsal bone are similar to those of the 2nd, 3rd and 4th metatarsal bones.

The phalanges

There are two phalanges for the 1st (big) toe, and three phalanges for each of the other toes. The middle and distal phalanges of the 5th toe are, however, frequently fused to form one bone.

The proximal phalanges

The proximal phalanx of the big toe is larger than those of the other toes. The *base* of each proximal phalanx has a concave articular facet which articulates with the head of the corresponding metatarsal bone. The *head* of each proximal phalanx has two condyles which bear a pulley-shaped articular facet which articulates with the base of the middle phalanx. The *shaft* is slightly curved, being convex downwards.

The middle phalanges

The middle phalanges are shorter and thicker in comparison to their length than are the proximal phalanges. Their *bases* bear a concave articular facet for the head of the corresponding proximal phalanx, and their *heads* have a convex articular facet which articulates with the corresponding distal phalanx.

The distal phalanges

The distal phalanges have concave facets on their proximal ends which articulate with the corresponding phalanx. Their distal ends are expanded to support the nail and the soft tissue of the toes.

The arches of the foot

The foot is built in a similar manner to the hand, but it is modified to support the weight of the body. When the bones are articulated the foot is seen to have longitudinal and transverse arches.

The **longitudinal arch** is considerably more marked on the medial side of the foot, so that the transverse arch is, in reality, half an arch.

The medial part of the longitudinal arch is formed posteriorly by the calcaneum, the apex of the arch is the head of the talus and the anterior part of the arch is formed by the navicular, the cuneiform bones and the 1st, 2nd and 3rd metatarsal bones.

The lateral part of the longitudinal arch is formed posteriorly by the calcaneum and anteriorly by the cuboid and the 4th and 5th metatarsal bones.

The **transverse arch** of the foot is most marked in the region of the bases of the metatarsal bones.

Arteries and nerves related to the bones of the lower limb

Certain arteries and nerves are related to the bones of the lower limb and are liable to damage if the bone is fractured (see Fig. 11.22, p. 299; Fig. 11.23, p. 300; and Fig. 16.21, p. 443). The most important are:

1   The femoral artery which passes through a fibrous arch in the adductor magnus muscle, at the junction of the middle and lower thirds of the shaft of the femur. At this point the artery lies close to the medial side of the femur and may be damaged by fractures in this region.

2   The popliteal artery, which lies close to the posterior surfaces of the lower end of the femur and the upper end of the tibia. It may be damaged by fractures of the lower end of the femur or of the upper end of the tibia.

3   The anterior tibial artery which grooves the medial side of the neck of the fibula as it passes forwards through the interosseous membrane which joins the tibia and fibula. It may be damaged by fractures of the upper end of the fibula.

4   The common peroneal nerve which lies on the posterior aspect of the head of the fibula, and then winds round the lateral side of the neck of the fibula. It is liable to damage in injuries involving the lateral ligament of the knee joint and the upper end of the fibula.

Ossification of the bones of the lower limb

The femur

The femur is ossified from five centres of ossification.

The shaft of the femur is ossified from one centre which appears in the 7th week of intrauterine life.

The upper end of the femur ossifies from three secondary centres:

1 a centre for the *head*, which appears during the first six months of life;

2 a centre for the *greater trochanter*, which appears during the 4th year of life; and

3 a centre for the *lesser trochanter* which appears during the 13th or 14th year of life.

The neck of the femur is ossified as an extension of the shaft.

The lower end of the femur is ossified from one secondary centre which appears during the 9th month of intrauterine life.

The centres for the greater and lesser trochanters fuse with the shaft soon after puberty and the centres for the head and the lower end fuse with the shaft between the 17th and the 20th year.

The patella

The patella is ossified from a single centre of ossification which appears between the 3rd and the 6th years of life.

The tibia

The tibia is ossified from three centres of ossification.

The shaft of the tibia is ossified from one centre which appears during the 7th week of intrauterine life.

The upper end of the tibia is ossified from one secondary centre which appears just before, or just after, birth.

The lower end of the tibia is ossified from one secondary centre which appears during the 1st year of life.

These secondary centres fuse with the shaft during the 17th or 18th years of life.

The fibula

The fibula is ossified from three centres of ossification.

The shaft of the fibula is ossified from one primary centre of ossification which appears during the 8th week of intrauterine life.

The upper end of the fibula is ossified from one secondary centre which appears during the 4th or 5th years of life.

The lower end of the fibula is ossified from one secondary centre of ossification which appears during the 1st year of life.

The two secondary centres fuse with the shaft between the 15th and 17th years of life.

The bones of the foot

The tarsal bones all ossify from a single centre of ossification with the exception of the calcaneum, which has a secondary centre of ossification for the posterior part of the bone.

The dates of appearance of the centres are:

The calcaneum	6th month of intrauterine life
The talus	7th month of intrauterine life
The cuboid	9th month of intrauterine life
The lateral cuneiform	1st year of life
The medial cuneiform	3rd year of life
The intermediate cuneiform	4th year of life
The navicular	4th year of life

The secondary centre for the calcaneum appears during the 6th to 8th year of life and unites with the rest of the bone during the 14th to 16th years of life.

The metatarsal bones each ossify from two centres. The 2nd to 5th metatarsal bones have a primary centre for the shaft and a secondary centre for the head of the bone. The first metatarsal bone has a primary centre for the shaft and a secondary centre for the base of the bone.

The primary centres for the *shafts* appear during the 9th to 10th weeks of intrauterine life.

The secondary centres appear during the 3rd and 4th years of life and they all unite with the shafts during the 17th to 20th years of life.

The phalanges all ossify from two centres.

The primary centres for the *shafts* appear during the 9th to 16th weeks of intrauterine life.

The secondary centres for the *bases* appear between the 2nd and 8th years of life and fuse with the shaft by the 18th year of life.

CHAPTER 10

The joints and the muscular system

A joint is formed where two or more bones of the skeleton meet. Joints are classified according to their structure and this is largely dependent on the function of the joint. Thus at joints where there is little movement, or no movement at all, the bones forming the joint are united by fibrous tissue or by cartilage. Where more movement is required the ends of the bones forming the joint are united by a fibrous capsule which contains fluid. This type of joint is called a synovial joint.

Movement of a joint is produced by the contraction of muscles. There are three types of muscle found in the human body (p. 20), but in this chapter we are concerned with the voluntary or skeletal muscles which are under the control of conscious will.

Voluntary muscles are composed of a mass of striped muscle fibres which are attached at their ends to other structures such as bones, cartilage or skin. The muscle fibres may be attached directly to these structures or they may be attached through the medium of fibrous bands which are called *tendons*. Some muscles have wide attachments to fibrous bands which are called *aponeuroses* (sing. aponeurosis). The proximal attachment of the muscle is usually called the *origin* of the muscle and the distal attachment is usually called the *insertion* of the muscle.

Muscular contraction is initiated by an impulse which passes out from the central nervous system in a nerve fibre. When a nerve fibre reaches a muscle it divides into a variable number of branches which pass to individual muscle cells. Thus each individual nerve fibre usually supplies several muscle cells. When the stimulus initiating contraction reaches a muscle cell that cell contracts fully and the strength of contraction of a muscle as a whole depends on the number of its component muscle cells which are contracting at that time.

When the cells of a muscle contract, the two attachments of the muscle are pulled towards each other and, if these two attachments lie on opposite sides of a joint, movement of that joint occurs.

Types of joints

Fibrous joints

In fibrous joints there is virtually no movement and the bones are joined by fibrous tissue. There are three types of fibrous joints:

1   Sutures, which are found only between the bones of the skull. The bones are joined by a thin layer of fibrous tissue and no movement occurs.
2   Gomphoses, which are peg and socket joints. The articulations between the teeth and the jaws are of this type.
3   Syndesmoses, in which the bones are joined by an interosseous membrane of fibrous tissue, as in the tibiofibular syndesmoses.

Cartilaginous joints

In cartilaginous joints the bones are joined by a layer of cartilage and slight movement can occur. There are two types of cartilaginous joints:
1   Synchondroses, or primary cartilaginous joints, in which the cartilage joining the two bones is of the hyaline type and is later replaced by bone. The joint between the diaphysis and epiphysis of a growing long bone is of this type.
2   Symphyses, or secondary cartilaginous joints, in which the ends of the articulating bones are covered by hyaline cartilage and are joined by a disc of fibrocartilage. The joints between the bodies of the vertebrae are of this type.

Synovial joints

The majority of the joints of the body and all the joints of the limbs, apart from the tibiofibular syndesmoses, are synovial joints. Synovial joints have the following characteristics:
1   The articular surfaces of the bones involved in the formation of the joints are covered with *hyaline cartilage.*
2   There is a *joint cavity* which contains synovial fluid.
3   The joint is surrounded by a fibrous capsule, called the *articular capsule.*
4   The inner surface of the joint, with the exception of the articular cartilage, is covered by a *synovial membrane* which secretes the synovial fluid. Pouches of synovial membrane may protrude from this joint to form sacs called *bursae.* These bursae lie between tendons, or muscles, and bones where there is friction. Some bursae which lie in the region of synovial joints do not communicate with the joint however.
5   The joint capsule is strengthened by fibrous bands which are called *ligaments.*
6   *Movement* occurs at synovial joints, but the extent of this movement varies considerably.
7   The joint receives a *nerve supply* and a *blood supply.*

Types of synovial joints

1   Hinge joints. In this type of joint, movement is allowed on one plane

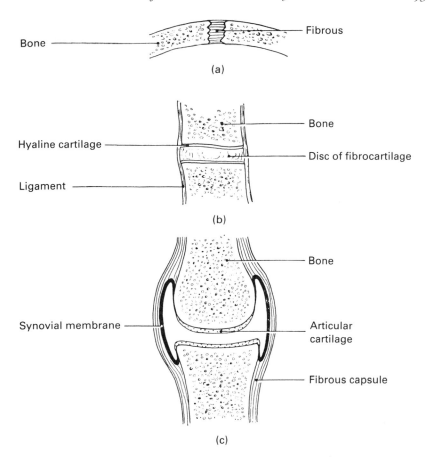

Bone — Fibrous

(a)

Hyaline cartilage —

Ligament —

Bone

Disc of fibrocartilage

(b)

Synovial membrane —

Bone

Articular cartilage

Fibrous capsule

(c)

Fig. 10.1. Types of joints: (a) fibrous joint (suture); (b) cartilaginous joint (symphysis); and (c) synovial joint.

only, and the bones are joined by strong ligaments on each side of the joint. The elbow joint and the interphalangeal joints are of this type.

2 Condylar joints. In this type of joint there are two pairs of articular surfaces with their long axes in line. This restricts the movements mainly to one plane, but a small amount of movement can occur in a second plane. The knee joint is of this type.

3 Ellipsoid joints. In this type of joint one articular surface is elliptical and convex and the other is elliptical and concave. In this type of joint, movement can occur in two planes. The wrist joint and the metacarpophalangeal joints are of this type.

4 Saddle joints. In this type of joint the articular surfaces are saddle-shaped and the movements occur in two planes, but some rotation of one

bone on the other can also occur. The carpometacarpal joint of the thumb
is of this type.

5 Pivot joint. In this type of joint the only movement which can occur is
rotation. The superior and inferior radioulnar joints are of this type.

6 Ball-and-socket joints. In this type of joint the end of one bone is
hemispherical and articulates with a cup-shaped depression in the other
bone. The range of movement in this type of joint is very free. The hip
joint and the shoulder joint are of this type.

7 Plane joints. In this type of joint the articular surfaces are flat and the
only movement that is allowed is slight gliding movement. The joints
between the articular processes of the vertebrae are of this type.

The movements of joints

Certain terms are used to describe the movements of joints and these terms
need some explanation.

Flexion is bending of the joint towards the front of the body. In the
knee joint, however, flexion is backward bending.

Extension is the opposite movement to flexion.

Abduction is movement away from the midline of the body.

Adduction is movement towards the midline of the body.

Circumduction is a combination of all these movements producing a
circular movement.

Rotation is movement around the long axis of the limb. An example of
this is the movement of the forearm when a screwdriver is used.

Inversion occurs at the foot when the sole of the foot is turned inwards.

Eversion is the opposite movement to inversion.

The muscles and joints of the head

The muscles of facial expression

There are numerous muscles which are responsible for producing the full
range of facial movement of which we are capable, but only the main
muscles will be described (see Fig. 10.3, p. 198). All the muscles of facial
expression are supplied by the *facial nerve* (p. 437).

The occipitofrontalis muscle is composed of an anterior muscular
part and a posterior muscular part which are separated by a large aponeurosis which extends over the top of the skull. The occipital part of the
muscle arises from the superior nuchal line of the occipital bone and is
inserted into the posterior edge of the aponeurosis. The frontal part arises
from the anterior edge of the aponeurosis and passes downward to blend

with the muscles around the orbits. The aponeurosis can slide freely over the surface of the skull but the skin of the scalp is firmly attached to it.

The two parts of the muscle can thus produce forward or backward movements of the scalp and the skin of the forehead. They can thus raise the eyebrows, as occurs in surprise, or wrinkle the skin of the forehead, as occurs in frowning.

The orbicularis oculi surrounds the eye in a circular manner. Gentle contraction of this muscle closes the eyes and stronger contraction 'screws up' the eyes.

The orbicularis oris is a circular muscle which surrounds the mouth. Gentle contraction of this muscle closes the lips and stronger contraction brings the lips closely together and causes them to protrude.

The buccinator is a flat quadrilateral muscle of the cheek and it occupies the interval between the maxilla and the mandible. Its action is to compress the cheek against the teeth during chewing.

The temporomandibular joint

The temporomandibular joint is a **synovial condylar joint**.

The articular surfaces involved in the formation of the joint are the anterior part of the mandibular fossa of the temporal bone and the head of the mandible.

The joint is divided into two halves by an *articular disc* and each half of the joint has its own synovial membrane.

The ligaments of the joint are: the *fibrous capsule*, the *temporomandibular ligament*, the *sphenomandibular ligament*, and the *stylomandibular ligament*.

The fibrous capsule of the joint is attached above to the margins of mandibular fossa and below to the neck of the mandible. The circumference of the articular disc is attached to the capsule. Above the disc the capsule is loose, but below it is tight, and holds the disc firmly to the head of the mandible.

The temporomandibular ligament is a thickened band which lies on the lateral side of the joint. It is attached above to the articular tubercle of the temporal bone and below to the posterior border of the neck of the mandible.

The sphenomandibular ligament lies on the medial side of the joint, and is attached above to the spine of the sphenoid bone and below to the lingula of the mandibular foramen.

The stylomandibular ligament is an accessory ligament of the joint. It is attached above to the tip of the styloid process of the temporal bone and below to the posterior border of the ramus of the mandible.

The articular disc is an oval fibrous disc. Its upper surface is convex

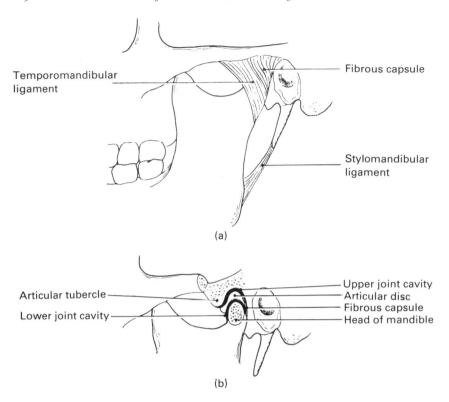

Temporomandibular ligament

Fibrous capsule

Stylomandibular ligament

(a)

Articular tubercle

Lower joint cavity

Upper joint cavity
Articular disc
Fibrous capsule
Head of mandible

(b)

Fig. 10.2. The left temporomandibular joint: (a) lateral aspect, and (b) section though the joint.

from front to back and its lower surface is concave. It is attached by its circumference to the fibrous capsule. The disc moves with the head of the mandible, and it moves forwards onto the articular tubercle when the mandible is protruded.

Blood supply

The temporomandibular joint is supplied with blood by branches of the superficial temporal and the maxillary arteries.

Nerve supply

The nerves of the joint are branches of the mandibular division of the trigeminal nerve.

Movements

The movements of the temporomandibular joints are produced mainly by a group of muscles called *the muscles of mastication*, but some of the superficial muscles of the neck are also involved. The muscles of mastication are the temporalis, the masseter, and the lateral and medial pterygoid muscles. The muscles of the neck which are involved are the suprahyoid muscles and the platysma. The movements which occur at the joint are:

1 *depression*, or opening the mouth, which is produced by the suprahyoid muscles and the platysma;

2 *elevation*, or closing the mouth, which is produced by the temporalis and the masseter;

3 *protrusion*, or forward movement, which is produced by the medial and lateral pterygoid muscles of both sides;

4 *retraction*, or backward movement, which is produced by the temporalis muscle, and

5 *side to side movement*, as occurs for example in chewing, which is produced by the medial and lateral pterygoid muscles of each side acting alternately.

The muscles which move the temporomandibular joint

The temporalis is a fan-shaped muscle which arises from the temporal fossa and passes deep to the zygomatic arch to be inserted into the coronoid process of the mandible. It elevates and retracts the lower jaw.

The masseter arises from the lower surface of the zygomatic arch and is inserted into the body of the mandible. It assists the temporalis to elevate the lower jaw.

The lateral and medial pterygoid muscles arise from the lateral and medial pterygoid plates of the sphenoid bone and are inserted into the ramus of the mandible. The two muscles, acting with the corresponding muscles of the opposite side, protrude the lower jaw. The muscles of each side, acting alternately, produce the side to side movements of the lower jaw which occur during chewing.

The platysma is a thin sheet of muscle which extends upwards from the fibrous tissue covering the front of the chest to be inserted into the body of the mandible. Its fibres also blend with the fibres of the facial muscles. It helps to depress the lower jaw.

The suprahyoid and infrahyoid muscles are small muscles which lie at the front of the neck. They extend from the mandible and the styloid process of the temporal bone to the hyoid bone and from the hyoid bone to the thyroid cartilage of the larynx and the sternum. Their action is to

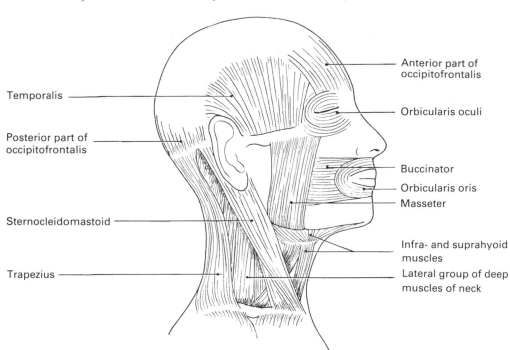

Temporalis

Posterior part of occipitofrontalis

Sternocleidomastoid

Trapezius

Anterior part of occipitofrontalis

Orbicularis oculi

Buccinator

Orbicularis oris

Masseter

Infra- and suprahyoid muscles

Lateral group of deep muscles of neck

Fig. 10.3. The muscles of the head and neck.

depress the lower jaw, and, when the mandible is held in an elevated position, these muscles elevate the larynx and the hyoid bone during the act of swallowing.

The joints and muscles of the neck and trunk

The joints of the vertebral column

Adjacent vertebrae articulate with each other by an anterior cartilaginous joint between the vertebral bodies, and four posterior synovial joints between the articular processes of the vertebral arches.

The joints of the vertebral bodies

The joints between the bodies of the vertebrae are **symphyses**.

The upper and lower surfaces of the bodies of the vertebrae are covered by a thin layer of hyaline cartilage. The bodies of the vertebrae are then united by *intervertebral discs* of fibrocartilage, which are closely adherent to the hyaline cartilage, and by ligaments. The discs, which increase in thickness from the upper thoracic region downwards, are composed of an outer fibrous layer, called the *annulus fibrosus*, and an inner layer of gelatinous material, called the *nucleus pulposus*.

The ligaments uniting the vertebral bodies are called the *anterior* and *posterior longitudinal ligaments.*

The anterior longitudinal ligament extends along the anterior surface of the vertebral bodies from the base of the occipital bone to the first part of the sacrum. It is closely adherent to the anterior surfaces of the intervertebral discs and to the upper and lower margins of the anterior surfaces of the vertebral bodies.

The posterior longitudinal ligament lies within the vertebral canal on the posterior surfaces of the bodies of the vertebrae. It is attached above to the posterior surface of the body of the axis, and continues downwards to the sacrum. Like the anterior longitudinal ligament, it is closely adherent to the posterior surfaces of the intervertebral discs and to the adjacent margins of the bodies of the vertebrae.

The joints of the vertebral arches

The joints between the articular processes of the vertebral arches of adjacent vertebrae are **synovial plane joints**. Each is surrounded by a loose *fibrous capsule*.

The spines, the laminae and the transverse processes of the vertebral arches are joined by ligaments.

The ligaments are: the *ligamenta flava*, the *supraspinous ligament*, the *ligamentum nuchae*, the *interspinous ligaments*, and the *intertransverse ligaments*.

The ligamenta flava are short ligaments of yellow elastic tissue which are attached above to the lower part of the anterior surface of each lamina and below to the upper margin and the posterior surface of the lamina of the vertebra below.

The supraspinous ligament is a strong fibrous band which connects the tips of the spinous processes of all the vertebrae from the 7th cervical vertebra to the sacrum.

The ligamentum nuchae is a fibrous sheet, the outer margin of which extends from the external occipital protuberance to the spine of the 7th cervical vertebrae. The inner margin of the ligament is attached to the occipital bone, below the external occipital protuberance, and to the spines of the cervical vertebrae.

The interspinous ligaments connect adjoining spinous processes and extend from the lower border of one spinous process to the upper border of the process below.

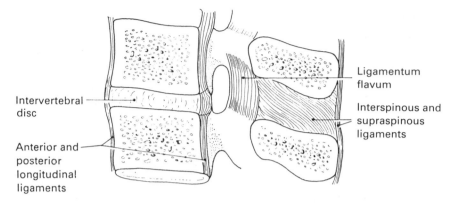

Intervertebral disc

Anterior and posterior longitudinal ligaments

Ligamentum flavum

Interspinous and supraspinous ligaments

Fig. 10.4. Vertical section of the vertebral column to show the principal ligaments.

The intertransverse ligaments extend between the transverse processes of adjacent vertebrae.

The sacrococcygeal joint

The sacrococcygeal joint is a *symphysis*. The apex of the sacrum and the upper surface of the coccyx are united by a disc of fibrocartilage.

The atlantoaxial joints

There are three **synovial joints** between the atlas and the axis.

There are two **lateral atlantoaxial joints** which are formed between the superior articular facets of the axis and the inferior articular facets of the atlas, and a third joint is formed between the **dens** of the axis and the facet on the posterior surface of the **anterior arch of the atlas**.

The lateral atlantoaxial joints are **plane joints**. The joints are each surrounded by a *fibrous capsule* which is strengthened posteriorly by an *accessory ligament*. The atlas and the axis are joined anteriorly by the anterior longitudinal ligament and posteriorly by the ligamentum flavum.

The joint between the dens and the anterior arch of the atlas is a **pivot joint**. The joint is surrounded by a thin *fibrous capsule*. The *transverse ligament* of the atlas, which is attached to the medial surface of each lateral mass of the atlas, keeps the dens in position against the anterior arch of the atlas.

The axis is joined to the occipital bone by three ligaments:

The alar ligaments, which are attached to each side of the dens and run upwards and laterally to the medial surface of each occipital condyle.

The apical ligament, which is attached to the apex of the dens and runs upwards to the anterior margin of the foramen magnum.

The atlanto-occipital joints

There are two joints between the atlas and the occiput. Each is a **synovial ellipsoid joint** between the occipital condyle and the superior articular facet of the atlas.

The occipital condyles have oval convex articular facets the long axes of which pass forwards and medially. The superior articular facets of the atlas are oval and concave and almost exactly correspond to the facets on the occipital condyles.

A *fibrous capsule* surrounds each joint.

Two ligaments unite the atlas with the occipital bone:

The anterior atlanto-occipital membrane, which is a broad fibrous sheet that extends between the anterior margin of the foramen magnum and the upper border of the anterior arch of the atlas.

The posterior atlanto-occipital membrane, which is a broad thin fibrous sheet that extends between the posterior margin of the foramen magnum and the upper border of the posterior arch of the atlas.

Movements of the vertebral column and the atlanto-occipital joints

The range of movement between individual vertebrae is limited and this is mainly due to the presence of the intervertebral discs. The sum of the movements between individual vertebrae, however, amounts to a considerable range of movement for the vertebral column as a whole. The movements which occur in the vertebral column are flexion, extension, lateral flexion, circumduction and rotation.

Flexion is most extensive in the cervical region, occurs to a lesser extent in the lumbar region and is almost absent in the thoracic region.

In the cervical region flexion is produced by the action of the anterior group of deep muscles of the neck and by the sternocleidomastoid muscles of each side acting together. In the lumbar region flexion is mainly produced by the action of the muscles of the anterior abdominal wall.

Extension is most extensive in the cervical and lumbar regions where it is mainly produced by the action of the sacrospinalis muscle.

Lateral flexion is most extensive in the cervical and lumbar regions.

In the cervical region it is produced by the action of the lateral group of deep muscles of the neck assisted by the sternocleidomastoid muscle of that side acting alone. In the lumbar region lateral flexion is produced by the action of the psoas and quadratus lumborum muscles of the side to which the lateral flexion is occurring.

Circumduction is a combination of the three previous movements.

Rotation mainly occurs in the thoracic region and is produced by the action of the multifidus and rotatores.

At the atlantoaxial joints the only movement which can occur is rotation, and at the atlanto-occipital joints the only movements which can occur are flexion and extension with a minor degree of lateral movement. This means that **rotation** of the head occurs at the atlantoaxial joint and is produced by the action of the sternocleidomastoid muscle of the opposite side to which the face is turning. **Flexion** of the head is produced by the action of the deep anterior muscles of the neck and **extension** is produced by the action of the upper fibres of the trapezius assisted by a group of deep muscles which lie on the posterior aspect of the upper cervical spine.

The muscles which move the vertebral column and the atlanto-occipital joints

The sternocleidomastoid arises from the manubrium sterni and the inner end of the clavicle and passes obliquely upwards and laterally to be inserted into the mastoid process of the temporal bone. When acting with its fellow of the opposite side the sternocleidomastoid flexes the cervical spine. When acting alone the sternocleidomastoid rotates the head towards the opposite side and it also assists the lateral muscles of the neck to laterally flex the cervical spine.

The **trapezius** is also concerned with movements of the shoulder girdle and is described on p. 216.

The **deep muscles of the neck** are divided into an anterior group which lie in front of the cervical spine and a lateral group which lie at the side of the cervical spine. The *anterior group of muscles* flex the cervical spine and the *lateral group of muscles*, which are called the scalene muscles, laterally flex the cervical spine.

The **muscles of the abdomen**, besides providing a firm but flexible support for the abdominal viscera, also play quite an important part in producing movements of the lumbar vertebral column. The muscles of the abdomen can be divided into two groups, the anterolateral muscles of the abdomen and the posterior muscles of the abdomen.

The anterolateral muscles of the abdomen

The rectus abdominis muscle is a long, flat muscle which lies close to the midline of the anterior abdominal wall. It is separated from the corresponding muscle of the opposite side by a fibrous cord called the linea alba. The rectus abdominis arises from the front of the pubic symphysis and the pubic crest and is inserted into the cartilages of the 5th, 6th and 7th ribs. It is enclosed in a fibrous sheath which is formed from the aponeuroses of the other muscles of the anterior abdominal wall. Medially the sheath is attached to the linea alba.

The external oblique muscle of the abdomen lies on the anterior

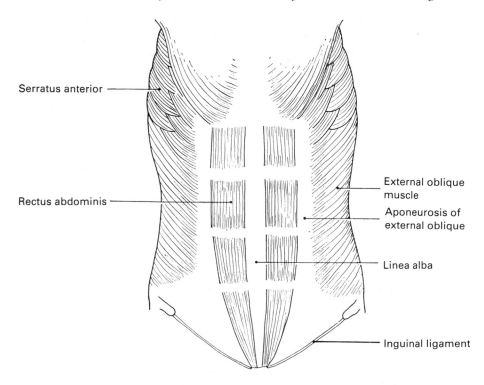

Serratus anterior

Rectus abdominis

External oblique muscle

Aponeurosis of external oblique

Linea alba

Inguinal ligament

Fig. 10.5. The muscles of the anterior abdominal wall.

and lateral aspects of the abdomen. It arises from the external surfaces of the lower eight ribs and its fibres run downwards and forwards to be inserted into the anterior half of the iliac crest, and into a wide aponeurosis which passes in front of the rectus abdominis to be attached to the linea alba. The lower margin of the aponeurosis forms the inguinal ligament which extends from the anterior superior iliac spine to the pubic crest.

The internal oblique muscle of the abdomen lies deep to the external oblique muscle. It arises from the lateral two-thirds of the inguinal ligament, from the anterior two-thirds of the iliac crest and from a sheet of fibrous tissue, which connects it with the lumbar vertebrae. The fibres of the muscle pass upwards and forwards to be inserted into the lower four ribs and into a wide aponeurosis. This aponeurosis divides into two layers at the lateral border of the rectus abdominis. These two layers then enclose the rectus abdominis like a sheath and reunite at its medial border, where they are attached to the linea alba.

The transversus abdominis is the innermost muscle of the anterior and lateral walls of the abdomen. It arises from the lateral third of the

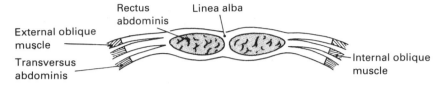

Fig. 10.6. The arrangement of the aponeuroses of the muscles of the anterior abdominal wall.

inguinal ligament, from the lateral two-thirds of the iliac crest, from a sheet of fibrous tissue, which connects it with the lumbar vertebrae, and from the inner surface of the lower six costal cartilages. Its fibres pass horizontally to be inserted into an aponeurosis which passes behind the rectus muscle to be attached to the linea alba.

Actions of the anterolateral muscles of the abdomen

The anterolateral muscles of the abdomen provide a firm but flexible support for the abdominal viscera. Acting together, when the spine remains fixed, they increase the intra-abdominal pressure in forced expiration, in defaecation and in micturition.

The rectus abdominis muscles, aided by the oblique muscles, act to flex the lumbar part of the vertebral column. The muscles of one side of the abdomen, contracting alone, laterally flex the lumbar vertebral column.

The posterior muscles of the abdomen

There are three main muscles which form the posterior abdominal wall. Two of these, the psoas major and the iliacus muscle, are flexors of the hip joint, but they lie mainly in the posterior part of the abdomen.

The psoas major arises from the transverse processes and the sides of the bodies of the lumbar vertebrae. The muscle runs downwards, along the brim of the true pelvis, passes beneath the inguinal ligament and is inserted into the lesser trochanter of the femur.

The iliacus is a triangular-shaped muscle which arises from the upper two-thirds of the iliac fossa of the hip bone. The fibres of the muscle converge to be inserted, with the psoas major muscle, into the lesser trochanter of the femur.

The psoas major and iliacus muscles flex the thigh on the trunk. The psoas muscle of one side, acting alone, laterally flexes the lumbar vertebral column.

The quadratus lumborum is a quadrilateral-shaped muscle which arises from the posterior part of the iliac crest and from the ilio-lumbar

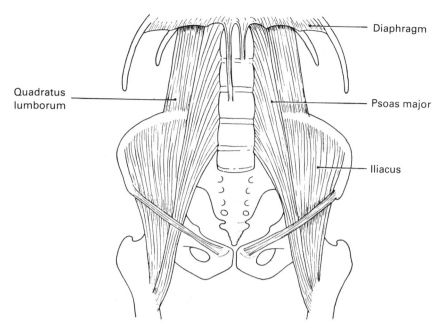

Quadratus lumborum

Diaphragm

Psoas major

Iliacus

Fig. 10.7. The posterior muscles of the abdomen.

ligament. Its fibres pass upwards to be inserted into the transverse processes of the upper four lumbar vertebrae and into the medial half of the lower border of the 12th rib.

The quadratus lumborum muscles of both sides, acting together, extend the lumbar part of the vertebral column. The muscle of one side, acting alone, laterally flexes the lumbar vertebral column.

The costovertebral joints

Each rib articulates with the vertebral column at two joints. One joint is formed between the head of the rib and the bodies of the vertebrae and the other is formed between the tubercle of the rib and the transverse process of the appropriate vertebrae.

The joints of the heads of the ribs

The head of each typical rib articulates with the facet on the upper part of the lateral surface of the thoracic vertebra from which it takes its number, with the facet on the lower part of the lateral surface of the vertebra above and with the intervening intervertebral disc. The 1st, 10th, 11th and 12th ribs articulate with one vertebral body only, that is the body of the vertebra from which they take their number.

The joints between the heads of the ribs and the vertebral bodies are **synovial plane joints**.

The crest of the head of each typical rib is united to the intervertebral disc by a ligament which divides each of the joints into two halves. Each half has its own synovial membrane. Each joint is surrounded by a *fibrous capsule* which is attached laterally to the head of the rib and medially is attached to the circumference of the articular area on the vertebral column. The capsule is thickened anteriorly to form **the radiate ligament**.

The joints between the tubercle of the ribs and the transverse processes of the vertebrae

The tubercles of the upper ten ribs articulate at **synovial plane joints** with the transverse process of the vertebra from which they take their number.

Each joint has a *fibrous capsule* which is lined by synovial membrane.

The ribs are joined to the transverse processes by three ligaments:

The superior costotransverse ligament joints the neck of each rib to the transverse process of the vertebra above.

The costotransverse ligament runs between the back of the neck of each rib and the anterior surface of the transverse process of the vertebra of the same level.

The lateral costotransverse ligament runs between the tubercle of each rib and the apex of the transverse process of the vertebra of the same level.

The sternocostal joints

The 1st costal cartilage articulates with the sternum at a **synchondrosis**.

The costal cartilages of the remaining true ribs usually articulate with the sternum at small **synovial joints**. Thin *fibrous capsules* surround these joints and the capsules are thickened anteriorly and posteriorly to form ligaments.

The interchondral joints

The adjacent borders of the 6th and 7th, the 7th and 8th, and the 8th and 9th costal cartilages articulate with each other at small **synovial joints**.

The movements of the costovertebral and sternocostal joints

Limited gliding movement is the only movement which can occur at the costovertebral and the sternocostal joints. Movements of these joints are

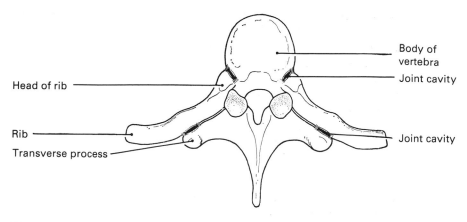

Fig. 10.8. The costovertebral joints.

produced by the action of the muscles of respiration and the joints are so arranged that the gliding movements which do occur result in an increase in the volume of the thoracic cavity during inspiration and a decrease during expiration. The movements of respiration are discussed in more detail on p. 355.

The muscles of respiration

The muscles of respiration are the external and internal intercostal muscles and the diaphragm.

The external intercostal muscles. There are eleven external intercostal muscles on each side of the thorax, and they lie in the spaces between the ribs. Each arises from the lower border of one rib, from the tubercle to the anterior end of that rib, and its fibres pass obliquely downwards and forwards to be inserted into the upper border of the rib below. The action of the external intercostal muscles is to elevate the ribs to produce inspiration.

The internal intercostal muscles. There are eleven intercostal muscles on each side of the thorax and they lie between the ribs, deep to the external intercostal muscles. They arise from the lower border of the costal cartilage of each rib and from the lower border of the rib, from the anterior end to the angle. Their fibres pass obliquely downwards and backwards to be inserted into the upper border of the rib below. their fibres thus lie at right angles to the fibres of the external intercostal muscles. The action of the internal intercostal muscles is to depress the ribs in forced expiration.

The diaphragm is a fibromuscular dome, which is convex upwards, and which separates the thoracic and abdominal cavities. It is composed of a peripheral muscular portion and a central tendinous portion.

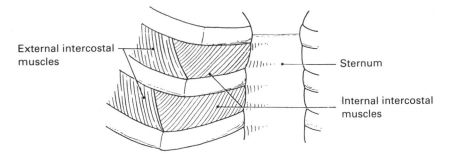

Fig. 10.9. The intercostal muscles.

The muscular fibres are divided into three groups:

1   *Sternal fibres*, which arise from the back of the xiphisternum.

2   *Costal fibres*, which arise from the costal cartilages, and the adjoining shafts of the lower six ribs.

3   *Lumbar fibres*, which arise from two tendinous arches, called the arcuate ligaments, and by two bundles, called the crura, from the upper lumbar vertebrae. The arcuate ligaments extend from the transverse process of the 1st lumbar vertebra medially to the tip of the 12th rib laterally. The right crus arises from the anterior surface of the upper three lumbar vertebrae and the left crus from the anterior surface of the upper two lumbar vertebrae.

These muscular fibres pass inwards, from their origins, to be inserted into the central tendon of the diaphragm.

The diaphragmatic openings. There are three openings in the diaphragm which transmit structures passing between the abdominal and thoracic cavities.

1   *The aortic opening* lies posteriorly at the level of the lower border of the 12th thoracic vertebra. It lies between the vertebral column and the diaphragm, slightly to the left of the midline, and is thus, strictly, behind the diaphragm. It transmits the aorta and the thoracic duct.

2   *The oesophageal opening* lies above and to the left of the aortic opening, at the level of the 10th thoracic vertebra. The opening is surrounded by the muscular fibres of the right crus. It transmits the oesophagus and the vagus nerve. It is often called the oesophageal hiatus.

3   *The vena caval opening* lies to the right of the oesophageal opening at the level of the 9th thoracic vertebra. It lies in the central tendon of the diaphragm and is roughly quadrilateral in shape. It transmits the inferior vena cava.

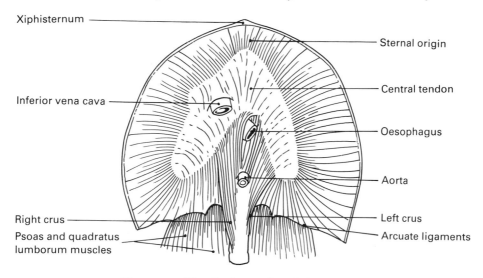

Xiphisternum

Sternal origin

Central tendon

Inferior vena cava

Oesophagus

Aorta

Right crus

Left crus

Psoas and quadratus lumborum muscles

Arcuate ligaments

Fig. 10.10. The diaphragm from below.

The muscular fibres of the diaphragm contract during inspiration and pull the central tendon downwards. This downward movement increases the volume of the thorax. The diaphragm is supplied by the right and left phrenic nerves, which are branches of the cervical plexus (p. 439).

The joints of the pelvis

The sacroiliac joint

The sacroiliac joint is a **synovial plane joint**.

The articular surfaces involved in the formation of the joint are the auricular surface of the iliac bone and the auricular surface of the sacrum. Unlike most synovial joints the articular surfaces are irregular, but the irregularities in one bone closely approximate to those in the other and thus the bones interlock.

The ligaments of the joint are: the *fibrous capsule*, the *interosseous sacroiliac ligament*, the *posterior sacroiliac ligament*, and the *anterior sacroiliac ligament*.

The fibrous capsule surrounds the joint and is attached to the margins of the auricular surfaces of the sacrum and the iliac bone.

The interosseous sacroiliac ligament is a strong ligament which binds the two bones together. It occupies the interval above and behind the joint where the two bones are in close proximity.

The **posterior sacroiliac ligament** lies over the posterior surface of the interosseous ligament. It extends from the transverse tubercles of the sacrum to the posterior superior iliac spine.

The **anterior sacroiliac ligament** is a thickening in the anterior part of the fibrous capsule.

Movements of the sacroiliac joint

Movement at the sacroiliac joint is restricted to slight **anterior and posterior rotation** which occurs during flexion and extension of the trunk.

The vertebropelvic ligaments

Three ligaments unite the pelvis and the vertebral column and are, therefore, accessory ligaments to the sacroiliac joints.

The **iliolumbar ligament** is attached medially to the tip of the transverse process of the 5th lumbar vertebra and laterally to the iliac crest in front of the sacroiliac joint.

The **sacrotuberous ligament** is attached medially to the posterior iliac spines, the transverse tubercles of the sacrum and the lateral margins of the sacrum and coccyx. Its fibres run obliquely downwards and laterally to be attached to the medial margin of the ischial tuberosity.

The **sacrospinous ligament** is a triangular ligament which is attached, by its base, to the lateral margins of the sacrum and coccyx in front of the sacrotuberous ligament, and by its apex to the spine of the ischium.

The sacrotuberous and sacrospinous ligaments convert the greater and lesser sciatic notches into foramina, through which vessels and nerves leave the pelvis to pass into the gluteal region.

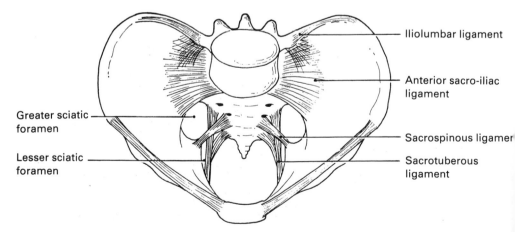

Fig. 10.11. The ligaments of the pelvis.

The pubic symphysis

The pubic bones articulate in the midline at the pubic symphysis.

The articular surfaces of the two pubic bones are covered by a thin layer of hyaline cartilage and are united by a **disc of fibrocartilage**.

Two ligaments strengthen the joint.

The superior pubic ligament unites the upper surfaces of the pubic bones.

The arcuate pubic ligament unites the lower surfaces of the pubic bones.

The muscles of the pelvis

The floor of the true pelvis is a muscular diaphragm, which is formed, on each side, by two main muscles, the levator ani and the coccygeus.

The levator ani is a wide muscle which arises from the anterior and lateral walls of the true pelvis. The fibres of the muscle pass towards the midline and unite with the muscle of the opposite side, enclosing the anus.

The coccygeus is a small triangular muscle which lies behind the levator ani. It arises from the spine of the ischium and is inserted into the base of the coccyx.

These two muscles are covered by a sheet of fibrous tissue, called the pelvic fascia and are perforated, in the female, by the urethra, the vagina and the anus. In the male they are perforated by the urethra and the anus.

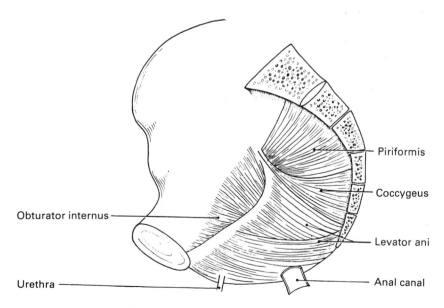

Fig. **10.12.** The muscles of the pelvis.

Arising from the lateral and posterior walls of the true pelvis are two muscles, the **obturator internus** and the **piriformis**, which leave the pelvis through the lesser and greater sciatic notches to be attached to the femur.

The joints and muscles of the upper limb

The sternoclavicular joint

The sternoclavicular joint is a **synovial saddle joint**.

The articular surfaces forming the joint are the sternal end of the clavicle, the clavicular notch of the manubrium sterni and the upper surface of the cartilage of the 1st rib. The joint is divided into two parts by an articular disc.

The ligaments of the joint are:

1　**The fibrous capsule** which surrounds the joint.
2　**The anterior** and **posterior sternoclavicular ligaments** which lie at the front and back of the joint respectively.
3　**The interclavicular ligament** which extends between the sternal ends of the two clavicles.
4　**The costoclavicular ligament** which extends between the upper surface of the 1st rib and the under-surface of the medial end of each clavicle.

The articular disc is a flat, circular disc of fibrous tissue which is attached, by its circumference, to the fibrous capsule of the joint.

The acromioclavicular joint

The acromioclavicular joint is a **synovial plane joint**.

The articular surfaces involved in the formation of the joint are the acromial end of the clavicle and the medial border of the acromion process of the scapula.

The ligaments of the joint are:

1　**The fibrous capsule**, which surrounds the joint and is attached to the articular margins.
2　**The acromioclavicular ligament**, which extends across the superior surface of the joint. It is attached to the upper surface of the acromial end of the clavicle and the upper surface of the acromion process.
3　**The coracoclavicular ligament**, which extends between the coracoid process of the scapula and the clavicle. It is divided into a conoid and a trapezoid part. The *conoid part* is attached above to the conoid tubercle of

the clavicle and below to the medial side of the root of the coracoid process. The *trapezoid part* is attached above to the trapezoid line of the clavicle and below to the upper surface of the coracoid process.

The shoulder joint

The shoulder joint is a **synovial ball-and-socket joint**.

The articular surfaces involved in the formation of the joint are the head of the humerus and the shallow glenoid cavity of the scapula. The glenoid cavity is deepened by a fibrous rim, called the labrum glenoidale, which is attached to its margin.

The articular surfaces are covered by *hyaline cartilage*.

The ligaments of the joint are: the *fibrous capsule*, the *glenohumeral ligaments*, the *coracohumeral ligament*, the *transverse humeral ligament* and *the labrum glenoidale*.

The fibrous capsule encloses the joint. It is attached medially to the margins of the glenoid and laterally to the anatomical neck of the humerus, except on the medial side of the humerus where it extends for a short distance onto the shaft of the bone. The capsule is lax but it is supported by the muscles which surround the joint. There are usually two openings in the capsule. One communicates anteriorly with the *subscapularis bursa*, which lies between the joint and the tendon of the *subscapularis* muscle, and the second allows the tendon of the long head of the biceps to leave the joint. The *tendon of the long head of the biceps* arises from the upper surface of the glenoid, passes over the head of the humerus, within the shoulder joint, and finally turns downwards to lie in the intertubercular sulcus of the humerus.

The glenohumeral ligaments are three bands which strengthen the capsule. They are attached medially to the upper part of the glenoid, and they spread out as they pass laterally to be attached to the lesser tubercle and anatomical neck of the humerus.

The coracohumeral ligament strengthens the upper part of the capsule. It arises from the root of the coracoid process and passes downwards and laterally to the front of the greater tubercle of the humerus.

The transverse humeral ligament extends from the lesser to the greater tubercles of the humerus and retains the tendon of the long head of the biceps in the intertubercular sulcus.

The labrum glenoidale is a fibrocartilaginous rim which is attached to the margins of the glenoid, and serves to deepen the socket for articulation with the head of the humerus.

The synovial membrane lines the fibrous capsule and extends over the anatomical neck of the humerus to the margins of the articular cartilage.

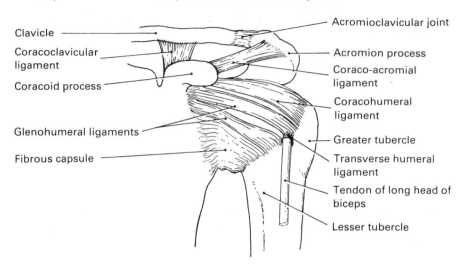

Fig. 10.13. The left shoulder joint, anterior aspect.

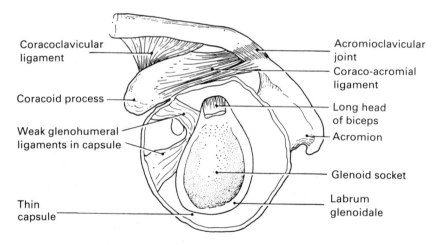

Fig. 10.14. The left shoulder joint after removal of the humerus.

The tendon of the long head of the biceps is enclosed in a sheath of synovial membrane.

Bursae

There are several bursae around the shoulder joint, but apart from the subscapularis and the subacromial bursae, they are of limited importance.

The *subscapularis bursa* has already been mentioned (see above). The *subacromial bursa* lies between the shoulder joint and the acromion process beneath the deltoid muscle and the tendon of the supraspinatus. It does not communicate with the joint.

The blood supply to the shoulder joint

The shoulder joint receives its blood supply from branches of the axillary artery and the subclavian artery.

The nerve supply to the shoulder joint

The shoulder joint receives its nerve supply from the circumflex humeral nerve and from other branches of the posterior cord of the brachial plexus (p. 000).

Muscles related to the shoulder joint

Above: the supraspinatus.
Below: the long head of the triceps.
Anteriorly: the subscapularis.
Posteriorly: the infraspinatus and the teres minor.

The deltoid muscle covers the shoulder joint laterally, in front and behind.

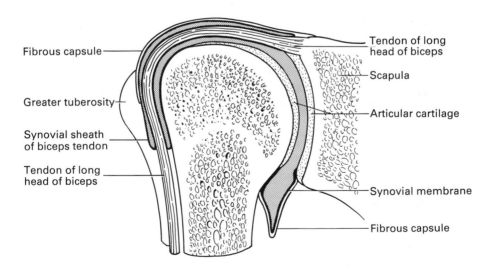

Fig. 10.15. Cross-section of the shoulder joint.

The muscles which move the shoulder girdle and the shoulder joint

The trapezius is a large muscle which arises from the external occipital protuberance, the superior nuchal line of the occipital bone, the ligamentum nuchae, which attaches it to all the spines of the cervical vertebrae, and from the spines of all the thoracic vertebrae. It is inserted into the spine and the acromion process of the scapula and into the posterior border of the lateral one-third of the clavicle. The upper fibres of the muscle help to pull the head backwards, or, if the head is fixed, to elevate the shoulder girdle. They also help to rotate the scapula when the arm is elevated above the horizontal (p. 221). The whole muscle helps to pull the scapula backwards, as in bracing the shoulders.

The latissimus dorsi is a large triangular muscle which arises from the spines of the lower six thoracic vertebrae, from the spines of the lumbar and sacral vertebrae, to which it is attached by a sheet of fibrous tissue, and from the posterior part of the iliac crest. The muscular fibres converge from this wide origin to join a narrow tendon which is inserted into the floor of the intertubercular sulcus of the humerus. The latissimus dorsi

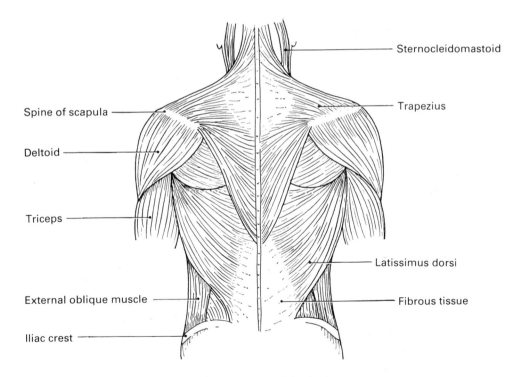

Fig. 10.16. The superficial muscles of the back.

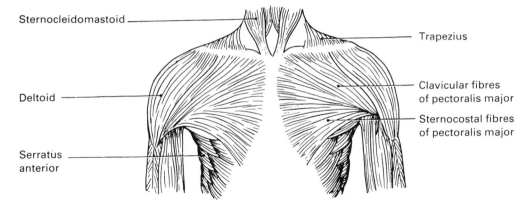

Sternocleidomastoid

Trapezius

Deltoid

Clavicular fibres
of pectoralis major

Sternocostal fibres
of pectoralis major

Serratus
anterior

Fig. 10.17. The superficial muscles of the anterior chest and shoulder region.

adducts and medially rotates the humerus, and also helps to depress the raised arm against resistance.

The pectoralis major is a broad triangular muscle which covers the front of the upper part of the chest. It arises from the front of the sternal half of the clavicle, from the anterior surface of the sternum and from the costal cartilages of the 2nd to 6th ribs. Its fibres converge to join a flat tendon by which it is inserted into the lateral lip of the intertubercular sulcus of the humerus.

The pectoralis major produces adduction and medial rotation of the humerus. In addition, the fibres which arise from the clavicle, acting independently of the rest of the muscle, help to flex the arm, and the fibres which arise from the sternum and the costal cartilages help to depress the arm against resistance.

The pectoralis minor is a small muscle which lies deep to the pectoralis major. It arises from the 2nd, 3rd and 4th ribs and is inserted into the coracoid process of the scapula. It assists in pulling the scapula forwards round the chest wall, as, for example, during the action of pushing with the arm forward.

The serratus anterior is a flat muscular sheet which arises from the outer surfaces of the upper eight ribs on the lateral side of the chest. The fibres of the muscle then run backwards round the chest, and pass behind the scapula, to be inserted into the costal surface of the medial border of that bone. The serratus anterior assists the pectoralis minor muscle in pulling the scapula forwards round the chest wall, and it also helps the trapezius to rotate the scapula upwards when the arm is raised above the horizontal position.

The deltoid is a thick triangular muscle which overlies the shoulder joint. It arises from the lateral third of the clavicle and from the acromion

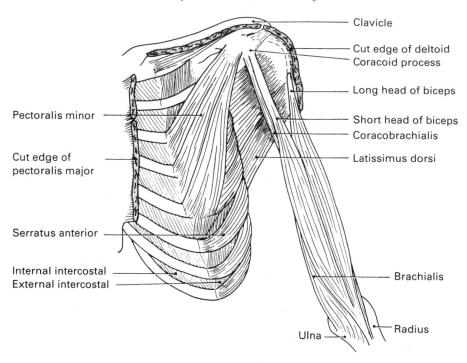

Clavicle
Cut edge of deltoid
Coracoid process
Long head of biceps
Short head of biceps
Coracobrachialis
Latissimus dorsi
Brachialis
Radius
Ulna

Pectoralis minor
Cut edge of
pectoralis major
Serratus anterior
Internal intercostal
External intercostal

Fig. 10.18. The anterior muscles of the upper arm, and the deep muscles of the anterior chest wall. The pectoralis major and the deltoid have been removed.

process and the lower part of the spine of the scapula. The fibres of the muscle converge to be inserted into the deltoid tuberosity of the humerus.

The fibres of the deltoid muscle which arise from the acromion process abduct the arm, while the anterior and posterior fibres steady the arm during abduction. The fibres which arise from the clavicle assist the clavicular fibres of the pectoralis major in flexing the arm, and the fibres which arise from the spine of the scapula assist the latissimus dorsi and teres major in pulling the arm backwards.

The subscapularis arises from the medial two-thirds of the subscapular fossa and is inserted into the lesser tubercle of the humerus.

The supraspinatus arises from the supraspinous fossa of the scapula and its fibres are attached to a tendon which passes over the upper surfaces of the shoulder joint to be inserted into the greater tubercle of the humerus.

The infraspinatus arises from the infraspinous fossa of the scapula and its fibres are attached to a tendon which passes behind the shoulder joint to be inserted into the greater tubercle of the humerus.

The teres minor arises from the upper two-thirds of the lateral border

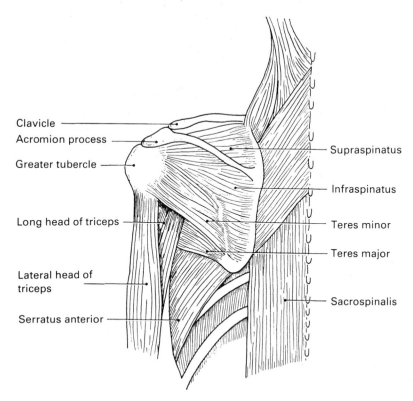

Clavicle

Acromion process

Greater tubercle

Supraspinatus

Infraspinatus

Long head of triceps

Teres minor

Teres major

Lateral head of triceps

Sacrospinalis

Serratus anterior

Fig. 10.19. The posterior muscles of the upper arm and the deep muscles of the posterior aspect of the chest. The deltoid and the trapezius have been removed.

of the scapula. Its fibres pass obliquely upwards and laterally to be inserted by a tendon into the greater tubercle of the humerus.

The subscapularis, the supraspinatus, the infraspinatus and the teres minor all help to keep the head of the humerus against the glenoid during movements of the shoulder joint. The supraspinatus helps to initiate abduction of the arm, the infraspinatus and teres minor laterally rotate the humerus and the subscapularis medially rotates the humerus.

The teres major is a thick muscle which arises from the inferior angle of the scapula. The fibres of the muscle pass upwards and laterally to be attached, by a tendon, into the medial lip of the intertubercular sulcus of the humerus. The teres major pulls the humerus backwards and medially and also rotates it medially.

The coracobrachialis arises from the coracoid process of the scapula and is inserted into the middle of the medial border of the humerus. The coracobrachialis flexes the arm and also draws it medially.

The movements of the shoulder girdle and the shoulder joint

The shoulder girdle is attached to the trunk by one joint, the sternoclavicular joint, and by several muscles. Movements of the shoulder girdle, however, involve both the sternoclavicular joint and the acromioclavicular joint. Movements of the shoulder girdle and movements of the shoulder joint usually occur together and the movements of the shoulder girdle contribute considerably to the range of movement which is possible with the upper limb.

A wide range of movement is possible at the shoulder joint but because of this wide range of movement the shoulder joint itself is rather weak. The shoulder joint is strengthened, however, by:

1   The strong muscles which surround the joint.
2   The tendon of the long head of the biceps muscle which passes through the joint and helps to keep the head of the humerus against the glenoid cavity.
3   The coracoacromial ligament which, together with the coracoid and acromion processes, forms an arch over the shoulder joint.

The movements which occur at **the shoulder girdle** are:

1   *Elevation*, as in shrugging the shoulders. This movement is produced by the upper fibres of the trapezius.
2   *Depression*, which is the reverse of elevation. This movement is produced by the action of the serratus anterior and pectoralis minor.
3   *Forward movement of the scapula* round the chest wall, as in the action of pushing. This forward movement is produced by the serratus anterior and the pectoralis minor.
4   *Backward movement of the scapula* round the chest wall, as in the action of bracing the shoulders. This backward movement is mainly produced by the whole of the trapezius muscle.
5   *Upward rotation of the scapula*, so that the glenoid faces upwards. This is produced by the action of the trapezius and serratus anterior and is a necessary part of the movement by which the arm is abducted above the horizontal (see below).
6   *Downward rotation of the scapula*, which is the reverse of the above movement is produced by the action of the pectoralis minor.

A wide range of movement is possible at **the shoulder joint** and the movements which occur are flexion, extension, abduction, adduction, circumduction, medial rotation and lateral rotation.

Flexion of the shoulder joint is produced by the action of the anterior fibres of the deltoid muscle, the coracobrachialis and the fibres of the pectoralis major muscle which arise from the clavicle.

When the arm is flexed from the extended position the sternocostal

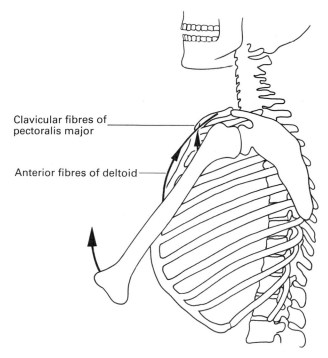

Clavicular fibres of
pectoralis major

Anterior fibres of deltoid

Fig. 10.20. Flexion of the shoulder joint is mainly produced by the clavicular fibres of the pectoralis major, and the anterior fibres of the deltoid.

fibres of the pectoralis major muscle play a large part in bringing the arm back to the side of the trunk.

Extension of the shoulder joint is produced by the action of the posterior fibres of the deltoid.

When the arm is extended from the flexed position against resistance, the sternocostal fibres of the pectoralis major and the latissimus dorsi draw the arm back to the plane of the trunk.

Abduction of the shoulder joint is produced by the action of the deltoid muscle assisted by the supraspinatus. The middle fibres of the deltoid, which arise from the acromion process, actually produce the abduction but the anterior and posterior fibres, together with the subscapularis, the infraspinatus and the teres minor, help to steady the head of the humerus against the glenoid.

Abduction at the shoulder joint can only occur to a point slightly above the horizontal plane. Elevation of the arm beyond this level can only occur if there is upward rotation of the scapula so that the glenoid faces upwards.

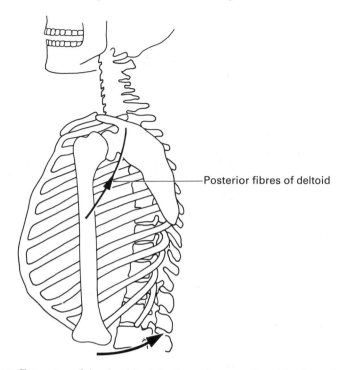

Posterior fibres of deltoid

Fig. 10.21. Extension of the shoulder joint from the neutral position is produced by the posterior fibres of the deltoid.

This upward rotation is produced by the action of the trapezius and the serratus anterior.

Adduction of the shoulder joint is produced by the action of the pectoralis major and the latissimus dorsi.

Circumduction of the shoulder joint is a combination of the last four movements.

Medial rotation of the shoulder joint is produced by the pectoralis major, the latissimus dorsi, the teres major, and the anterior fibres of the deltoid.

Lateral rotation is produced by the action of the infraspinatus, the teres minor and the posterior fibres of the deltoid.

The elbow joint

The elbow joint is a **synovial hinge joint**.

The articular surfaces involved in the formation of the joint are the trochlea of the humerus, which articulates with the trochlear notch of the ulna, and the capitulum of the humerus, which articulates with the facet on

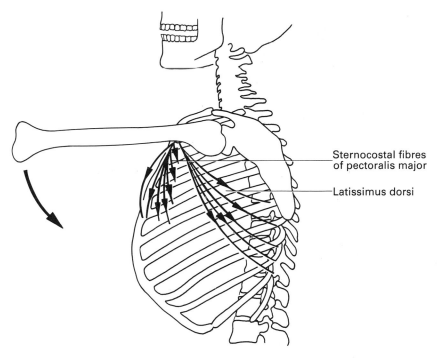

Fig. 10.22. Extension of the shoulder joint from the flexed position is produced by the latissimus dorsi and the sternocostal fibres of the pectoralis major.

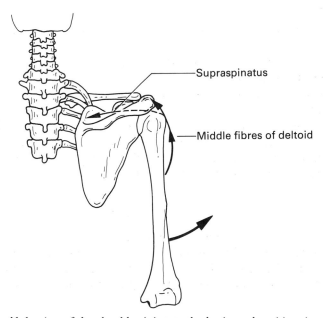

Fig. 10.23. Abduction of the shoulder joint, to the horizontal position, is produced by the supraspinatus and the middle fibres of the deltoid.

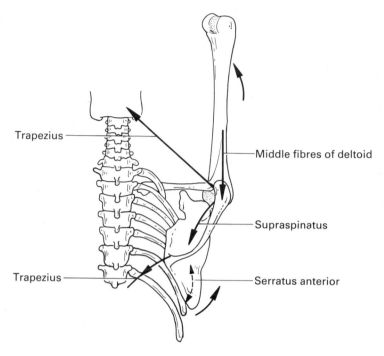

Trapezius

Middle fibres of deltoid

Supraspinatus

Trapezius

Serratus anterior

Fig. 10.24. Abduction of the shoulder joint above the horizontal can only occur if the scapula is rotated upwards by the trapezius and the serratus anterior.

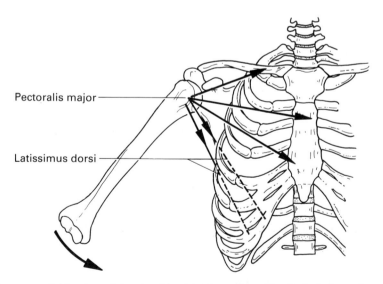

Pectoralis major

Latissimus dorsi

Fig. 10.25. Adduction of the shoulder joint is produced by the latissimus dorsi and the pectoralis major.

the upper surface of the head of the radius. The synovial membrane and the fibrous capsule of the elbow joint are continuous with the superior radioulnar joint, but this joint is not strictly part of the elbow joint.

The articular surfaces are covered with *hyaline cartilage*.

The ligaments of the joint are: the *fibrous capsule*, the *ulnar collateral ligament*, and the *radial collateral ligament*.

The fibrous capsule is attached anteriorly, by its upper margin, to the medial epicondyle of the humerus and to the lower end of the humerus just above the radial and coronoid fossae and, by its lower margin, to the anterior surface of the coronoid process of the ulna and the annular ligament of the superior radioulnar joint. Posteriorly the capsule is attached, by its upper margin, to the medial epicondyle of the humerus, to the margins of the olecranon fossa and to the lower end of the humerus above the capitulum; and, by its lower margin, to the lateral and upper borders of the olecranon process and to the capsule of the superior radioulnar joint. At the sides of the elbow joint the capsule is continuous with the ulnar and radial collateral ligaments.

The ulnar collateral ligament is a triangular ligament which is attached above to the medial epicondyle of the humerus and below to the medial side of the olecranon and coronoid processes of the ulna.

The radial collateral ligament is attached above to the lateral epicondyle of the humerus and below to the annular ligament of the superior radioulnar joint.

The synovial membrane of the elbow joint is quite extensive. It commences at the articular surfaces of the humerus, passes upwards to line the coronoid, radial and olecranon fossae and then passes onto the deep surface of the capsule. It extends into the superior radioulnar joint and lines the lower part of the annular ligament.

The blood supply to the elbow joint

The elbow joint receives its blood supply from branches of the brachial, ulnar and radial arteries which form a network of vessels in the elbow region.

The nerve supply to the elbow joint

The elbow joint receives its nerve supply from the musculocutaneous and radial nerves, and it also receives small branches from the ulnar and median nerves.

Muscles related to the elbow joint

Anteriorly: the brachialis.

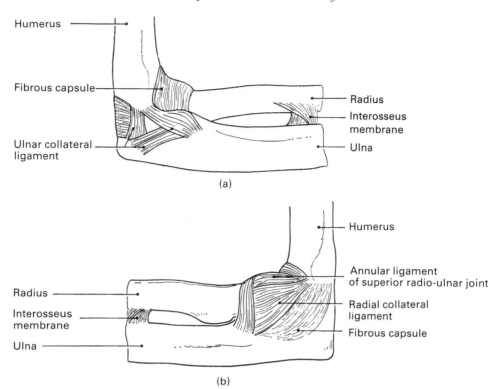

Fig. 10.26. The left elbow joint: (a) medial aspect, and (b) lateral aspect.

Posteriorly: the triceps.
Laterally: the supinator and the common tendon of the extensor muscles of the forearm.
Medially: the common tendon of the flexor muscles of the forearm.

The muscles which move the elbow joint

The biceps brachii arises from two different sites and is thus said to have two heads of origin, a short head and a long head. *The short head* arises from the coracoid process of the scapula and the long head arises by a tendon from the upper surface of the glenoid of the scapula. The tendon of *the long head* of the biceps passes through the shoulder joint, over the head of the humerus, and then runs downwards in the intertubercular sulcus. The fibres from the two heads unite to form a fleshy mass of muscle which lies on the anterior aspect of the forearm. The muscle is inserted, by a long tendon, into the radial tuberosity. The biceps brachii flexes the elbow joint and supinates the forearm. It also helps to flex the shoulder joint.

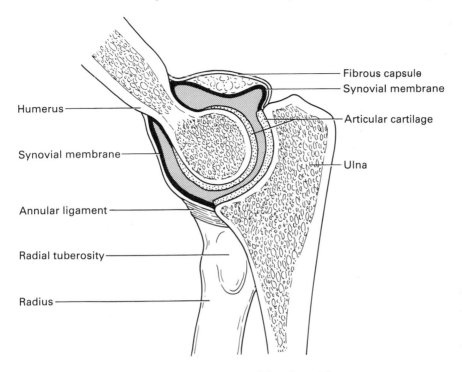

Fig. **10.27.** Cross-section of the elbow joint.

The brachialis lies deep to the biceps brachii on the anterior aspect of the humerus. It arises from the lower half of the anterior aspect of the humerus and is inserted, by a short tendon, into the coronoid process of the ulna. The brachialis flexes the elbow joint.

The triceps is a large muscle which lies on the posterior aspect of the upper arm. It has three heads of origin, a long head, a lateral head and a medial head. The *long head* arises by a tendon from the lower surface of the glenoid of the scapula and the *lateral and medial heads* arise from the posterior surface of the humerus. The fibres from the three heads unite to be inserted, by a tendon, into the upper surface of the olecranon process of the ulna. The triceps extends the elbow joint.

The brachioradialis arises from the lateral epicondyle of the humerus and is inserted by a tendon into the lateral side of the radius just above the styloid process. It flexes the elbow joint.

The movements of the elbow joint

The elbow joint is a hinge joint and the only movements that can occur are flexion and extension.

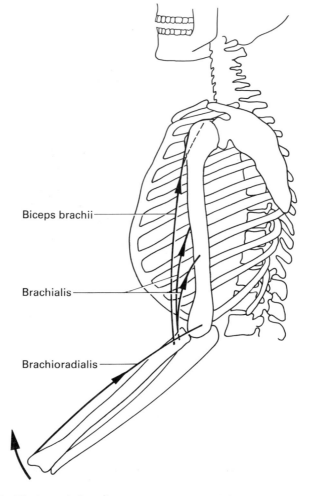

Biceps brachii

Brachialis

Brachioradialis

Fig. 10.28. Flexion of the elbow joint is produced by the biceps brachii, the brachialis and the brachioradialis.

Flexion is produced by the action of the brachialis, the biceps and the brachioradialis.

Extension is produced by the action of the triceps.

The radioulnar joints

There are two joints between the radius and the ulna, the *superior radioulnar joint* and the *inferior radioulnar joint*, and the shafts of the two bones are, in addition, united by an *interosseous membrane*.

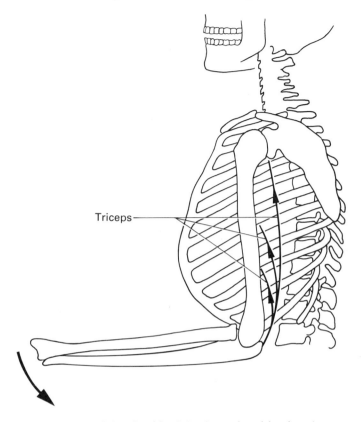

Triceps

Fig. 10.29. Extension of the shoulder joint is produced by the triceps.

The interosseous membrane is a fibrous sheet, the fibres of which run obliquely downwards and medially from the interosseous border of the radius to the interosseous border of the ulna. At the wrist the force exerted on the hand is transmitted to the radius. The interosseus membrane transfers the force from the radius to the ulna and thus to the humerus.

The superior radioulnar joint is a *synovial pivot joint*.

The articular surfaces involved in the formation of the joint are the circumference of the head of the radius and the radial notch of the ulna.

The capsule and synovial membrane of the joint are continuous with those of the elbow joint.

The annular ligament is a strong band which runs round the head of the radius and retains it against the radial notch of the ulna. It is attached to the anterior and posterior margins of the radial notch and its inner surface has a covering of cartilage where it is in contact with the head of the radius.

The inferior radioulnar joint is a *synovial joint*.

The articular surfaces involved in the formation of the joint are the head of the ulna and the ulnar notch of the radius.

The joint is enclosed in a *fibrous capsule* which is thickened anteriorly and posteriorly.

An articular disc helps to bind the two bones together. It is triangular in shape, and is attached by its base to the medial border of the inferior surface of the radius and, by its apex, to the inferior surface of the lower end of the ulna close to the styloid process. The edges of the disc are attached to the fibrous capsule of the wrist joint.

The muscles which move the radioulnar joints

Three muscles of the forearm and one of the upper arm are involved in movements of the radioulnar joints. The muscle of the upper arm, which has already been described, is the biceps brachii. The muscles of the forearm involved are:

The pronator teres, which passes obliquely across the upper third of the forearm from the medial epicondyle of the humerus to the lateral surface of the shaft of the radius.

The pronator quadratus, which is a quadrilateral muscle and which extends across the lower anterior surfaces of the shafts of the radius and ulna.

The supinator surrounds the upper third of the radius. It arises from the lateral epicondyle of the humerus and the upper end of the ulna and it is inserted into the lateral surface of the radius. It supinates the forearm.

Movements of the radioulnar joints

The movements which occur at the radioulnar joints are pronation and supination and these movements involve both joints.

In the anatomical position, with the arm by the side, the palm of the hand faces forwards and the shafts of the radius and ulna are parallel. This is the supinated position. When the forearm is pronated the hand turns through an angle of 150°, and the lower end of the radius rotates round the lower end of the ulna to lie first in front of and then medial to the lower end of the ulna.

Pronation is produced by the pronator quadratus and the pronator teres. Quick pronation is mainly produced by the pronator quadratus and pronation against resistance is mainly produced by the pronator teres.

Supination is produced by the action of the supinator but when the elbow is flexed the biceps brachii plays an important part.

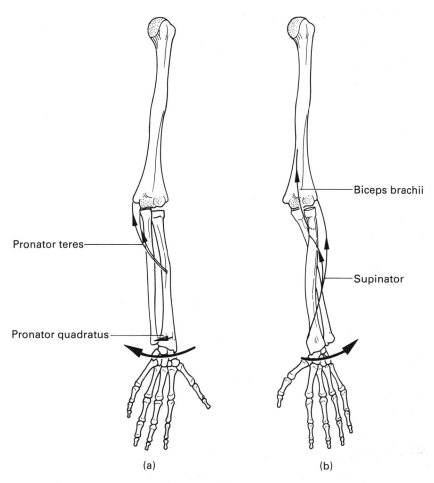

Biceps brachii

Pronator teres

Supinator

Pronator quadratus

(a)

(b)

Fig. 10.30. Pronation and supination. (a) The forearm and hand are in the supine position. The pronator quadratus and pronator teres contract to rotate the distal end of the radius and the hand round the head of the ulna to reach the pronated position. (b) The reverse movement, supination, is produced by the supinator and when the elbow is flexed the supinator is assisted by the biceps.

The wrist joint (radiocarpal joint)

The wrist is a **synovial ellipsoid joint**.

The articular surfaces involved in the formation of the joint are the distal end of the radius and the distal surface of the articular disc of the inferior radioulnar joint, which articulate with the proximal surfaces of the scaphoid, lunate and triquetral bones. The distal end of the radius and the distal surface of the articular disc form a concave elliptical surface into which the proximal surfaces of the scaphoid, lunate and triquetral fit.

The ligaments of the joint are: the *fibrous capsule*, the *palmar radiocarpal ligament*, the *palmar ulnocarpal ligament*, the *ulnar collateral ligament*, the *radial collateral ligament* and the *dorsal radiocarpal ligament*.

The fibrous capsule surrounds the joint and is strengthened by the ligaments.

The palmar radiocarpal ligament is attached proximally to the anterior surface of the distal end of the radius and distally to the palmar surfaces of the scaphoid, lunate and triquetral bones.

The palmar ulnocarpal ligament is a rounded band which extends from the styloid process of the ulna to the lunate and triquetral bones.

The ulnar collateral ligament extends from the styloid process of the ulna to the medial side of the triquetral and pisiform bones.

The radial collateral ligament runs from the styloid process of the radius to the lateral side of the scaphoid bone.

The dorsal radiocarpal ligament runs from the posterior border of the distal end of the radius to the dorsal surfaces of the scaphoid, lunate and triquetral bones.

The synovial membrane of the wrist joint lines the articular capsule and is usually separate from that of the inferior radioulnar joint.

The blood supply to the wrist joint

The wrist joint receives its blood supply from branches of the radial and ulnar arteries.

The nerve supply to the wrist joint

The wrist joint receives its nerve supply from branches of the radial and median nerves.

The intercarpal joints

Each carpal bone articulates with its neighbours at synovial joints. These joints can be divided into three groups; the *joints of the proximal* and *distal rows of the carpus,* and the *midcarpal joint*.

The joints of the proximal row of the carpus

The scaphoid and triquetral bones articulate with the lateral and medial sides of the lunate respectively and the three bones are united by interosseous ligaments and palmar and dorsal ligaments. The pisiform bone articulates with the palmar surface of the triquetral bone at a separate synovial joint.

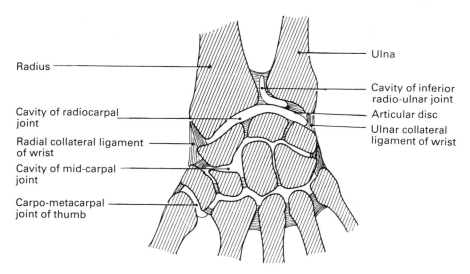

Radius

Cavity of radiocarpal joint

Radial collateral ligament of wrist

Cavity of mid-carpal joint

Carpo-metacarpal joint of thumb

Ulna

Cavity of inferior radio-ulnar joint

Articular disc

Ulnar collateral ligament of wrist

Fig. 10.31. Section through the wrist, carpal and carpometacarpal joints.

The joints of the distal row of the carpus

The bones of the distal row of the carpus articulate with each other at synovial joints and are united by interosseous ligaments and by palmar and dorsal ligaments.

The midcarpal joint

The midcarpal joint is the joint between the proximal and distal rows of the carpus. It is divided into an ellipsoid joint between the hamate and capitate distally and the scaphoid, lunate and triquetral bone proximally, and a **plane joint** between the trapezium and trapezoid distally and the scaphoid proximally. The midcarpal joint is strengthened by *dorsal* and *palmar ligaments*, and by *collateral ligaments*, which lie on each side of the carpus.

The **synovial membrane** of the carpus extends between the bones of the carpus and frequently is continuous with the synovial membrane of the carpometacarpal joints. The joint between the pisiform bone and the triquetral bone, however, has a synovial membrane which is separate from that of the rest of the carpus.

The carpometacarpal and intermetacarpal joints

The 2nd to 5th metacarpal bones articulate with the bones of the distal row of the carpus at **synovial plane joints**. The 2nd metacarpal articulates with the trapezium, the trapezoid and the capitate, the 3rd metacarpal with the capitate, the 4th metacarpal with the capitate and hamate, and the 5th metacarpal with the hamate. The **synovial membrane** of these joints is frequently continuous with that of the carpal joints and of the intercarpal joints.

The 2nd to 5th metacarpal bones articulate with each at their bases.

The 1st metacarpal bone articulates with the trapezium at a **synovial saddle joint**. The **synovial membrane** of this joint is separate from that of the other carpometacarpal joints.

The metacarpophalangeal joints

The metacarpophalangeal joints are **ellipsoid synovial joints**.

The articular surfaces involved in the formation of these joints are the convex heads of the metacarpal bones and the concave bases of the proximal phalanges.

The joints are surrounded by **fibrous capsules** and are strengthened by **palmar ligaments** and **collateral ligaments**. In addition the heads of the 2nd to 5th metacarpal bones are united by ligaments called the *deep transverse metacarpal ligaments*.

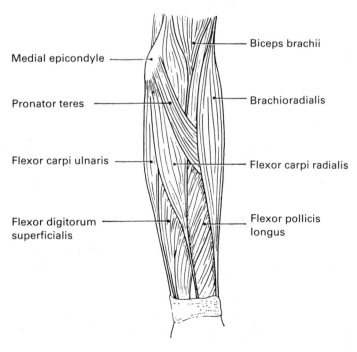

Medial epicondyle

Biceps brachii

Pronator teres

Brachioradialis

Flexor carpi ulnaris

Flexor carpi radialis

Flexor digitorum superficialis

Flexor pollicis longus

Fig. 10.32. The muscles of the left forearm, anterior aspect.

The interphalangeal joints

The interphalangeal joints are **synovial hinge joints**.

The joints are surrounded by **fibrous capsules** and have **palmar and collateral ligaments**.

The muscles which move the wrist joint and the joints of the hand

The muscles which move the wrist joint and the joints of the hand are the muscles of the forearm and the small muscles of the hand. The muscles of the forearm are divided into an **anterior group**, which flex the wrist joint and the fingers and pronate the hand, and a **posterior group**, which extend the wrist joint and the fingers and supinate the hand. The two pronators and supinator have already been described in the section on the radioulnar joints.

The flexors, or **the anterior muscles of the forearm**, are divided into a superficial group, which arise by a common tendon from the medial epicondyle of the humerus, and a deep group which arise from the anterior surfaces of the shafts of the radius and ulna and from the interosseous membrane.

The superficial flexors of the forearm are:

The flexor carpi radialis which flexes the wrist joint and adducts the hand.

The flexor carpi ulnaris which flexes the wrist joint and abducts the hand.

The flexor digitorum superficialis which has four tendons. These tendons are inserted into each of the fingers and the muscle flexes the fingers and the wrist joint.

The deep flexors of the forearm are:

The flexor digitorum profundus which also has four tendons, which are inserted into each of the fingers. The muscle flexes the fingers and the wrist joint.

The flexor pollicis longus which is inserted, by a tendon, into the thumb. The muscle flexes the thumb.

The extensors, or the **posterior muscles of the forearm**, are divided into a superficial group which arise from the lateral side of the lower end of the humerus and by a common tendon from the lateral epicondyle of the humerus, and a deep group which arise from the posterior surfaces of the shafts of the radius and ulna.

The superficial extensors are:

The extensors carpi radialis longus and *brevis* which are inserted by tendons into the bases of the 2nd and 3rd metacarpal bones respectively. They extend the wrist joint and abduct the hand.

The extensor carpi ulnaris which is inserted by a long tendon into the base of the 5th metacarpal bone. It extends the wrist joint and adducts the hand.

The extensor digitorum which has four tendons, which are inserted into each of the fingers. It extends the fingers and wrist joint.

The extensor digiti minimi which is inserted into the little finger. It extends the little finger.

The deep extensors are:

The extensors pollicis longus and *brevis* which are inserted into the thumb. They extend the thumb.

The extensor indicis which is inserted by a tendon into the index finger. It extends the index finger.

The abductors pollicis longus and *brevis* which are inserted by tendons into the thumb. They abduct the thumb.

The muscles of the hand are divided into three groups:

1   *The thenar muscles* which form the prominence at the base of the thumb on the palmar surface of the hand. They produce movements of the thumb.

2   *The hypothenar muscles* which form the prominence at the base of the little finger on the palmar surface of the hand. They produce movements of the little finger.

3   *The interosseous muscles* which lie between the metacarpal bones. They are divided into a dorsal group, which abduct the fingers from the midline, and a palmar group which adduct the fingers towards the midline.

The movements of the wrist joint and the joints of the hand

The range of movement of the **wrist joint** is increased by the movements which occur at the **midcarpal joint** and the movements of the two joints are considered together. The movements which occur are:

Flexion of the wrist joint is produced by the flexors carpi ulnaris and radialis, assisted by the long flexors of the fingers. The range of flexion at the wrist joint is not as great as it might appear, as a large amount of the apparent flexion in fact occurs at the midcarpal joint.

Extension is produced by the extensors carpi radialis longus and brevis and the extensor carpi ulnaris, assisted by the extensors of the fingers.

Abduction is produced by the flexor carpi radialis and the extensors carpi radialis longus and brevis. Abduction at the wrist joint is limited and most of the apparent movement occurs at the midcarpal joint.

Adduction is produced by the flexor carpi ulnaris and the extensor carpi ulnaris.

Circumduction is a combination of these four movements.

At the 2nd to the 5th **carpometacarpal joints** and the **intermetacarpal joints** only slight *gliding movements* can occur.

At **the 1st carpometacarpal joint**, however, there is a considerable

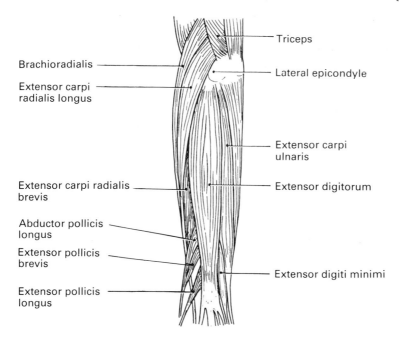

Triceps

Brachioradialis

Lateral epicondyle

Extensor carpi
radialis longus

Extensor carpi
ulnaris

Extensor carpi radialis
brevis

Extensor digitorum

Abductor pollicis
longus

Extensor pollicis
brevis

Extensor digiti minimi

Extensor pollicis
longus

Fig. 10.33. The muscles of the left forearm, posterior aspect.

range of movement and this accounts for the mobility of the thumb. The movements which occur are *flexion, extension, abduction, adduction, circumduction* and *opposition*. The positions of the articular surfaces of the joint are such that flexion and extension take place in a plane which is parallel to the palm of the hand. Opposition is a composite movement by which the tip of the thumb can be brought into contact with the tip of each of the fingers and it is this movement which accounts for the manipulative ability of the human hand.

The movements which occur at the **metacarpophalangeal joints** are *flexion* and *extension* and a limited range of *abduction* and *adduction*. The only movements which can occur at the **interphalangeal joints** are *flexion* and *extension*.

The joints and muscles of the lower limb

The hip joint

The hip joint is a **synovial ball-and-socket joint**.

The **articular surfaces** involved in the formation of the joint are the head of the femur and the acetabulum of the hip bone.

The head of the femur is completely covered by *hyaline cartilage* apart from a small roughened pit, called the *fovea*, to which the ligament of the head of the femur is attached. The articular surface of the acetabulum is a horeshoe-shaped area which is covered with hyaline cartilage. The central area of the acetabulum is depressed and non-articular and this area is occupied by a pad of fat. The articular area of the acetabulum is also deficient below at the *acetabular notch*. The depth of the acetabulum is increased by a fibrocartilaginous rim, called the **acetabular labrum**.

The **ligaments** of the joint are: the *fibrous capsule*, the *iliofemoral ligament*, the *ischiofemoral ligament*, the *pubofemoral ligament*, the *ligament of the head of the femur*, the *acetabular labrum*, and the *transverse acetabular ligament*.

The **fibrous capsule** of the joint is thick and strong. It is attached medially to the acetabulum, the acetabular labrum and the transverse acetabular ligament. It surrounds the neck of the femur and is attached, laterally, to the base of the neck.

The **iliofemoral ligament** is a strong triangular ligament which is attached by its apex to the lower part of the anterior inferior iliac spine. From the apex the fibres of the ligament run downwards and laterally, across the front of the capsule of the joint, to be attached to the trochanteric line of the femur.

The **ischiofemoral ligament** is attached medially to the body of the ischium, below and behind the acetabulum. The fibres of the ligament then run, in a spiral fashion, upwards and laterally, round the back of the neck of the femur, to be attached to the anterior part of the greater trochanter deep to the iliofemoral ligament.

The **pubofemoral ligament** is a triangular ligament which is attached by its base to the iliopectineal eminence and the superior pubic ramus. The fibres of the ligament converge to join the fibres of the joint capsule and the deep surface of the iliofemoral ligament.

The **ligament of the head of the femur** is a triangular flattened ligament which is attached, by its apex, to the fovea of the head of the femur. The base of the ligament is attached to the margins of the acetabular notch and to the transverse acetabular ligament. It is enclosed in a sheath of synovial membrane.

The **acetabular labrum** is a fibrocartilaginous rim which is attached to the margins of the acetabulum. It is triangular in cross-section, being thick at its attachment and thin at its free edge.

The **transverse acetabular ligament** completes the acetabular labrum over the acetabular notch.

The **synovial membrane** of the hip joint is extensive. It commences at the margin of the articular cartilage, which covers the head of the femur, and ensheaths the portion of the neck of the femur which lies within the hip joint. It is then reflected onto the deep surface of the capsule and

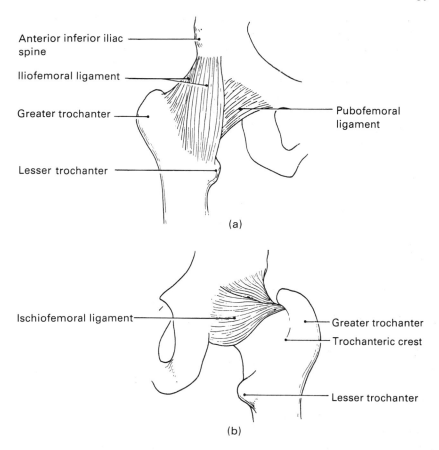

Anterior inferior iliac spine

Iliofemoral ligament

Greater trochanter

Lesser trochanter

Pubofemoral ligament

(a)

Ischiofemoral ligament

Greater trochanter

Trochanteric crest

Lesser trochanter

(b)

Fig. 10.34. The right hip joint: (a) anterior aspect, and (b) posterior aspect.

extends medially to cover both sides of the acetabular labrum. It ensheaths the ligament of the head of the femur and covers the pad of fat in the acetabular fossa.

The blood supply to the hip joint

The hip joint receives its blood supply from braches of the internal iliac and femoral arteries.

The nerve supply to the hip joint

The hip joint receives its nerve supply from branches of the femoral, obturator and sciatic nerves.

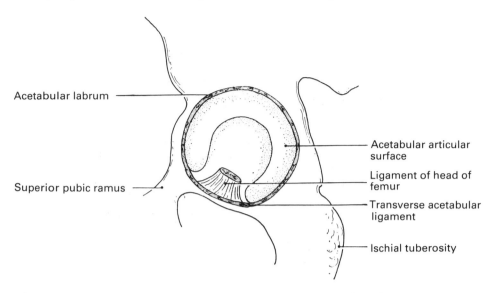

Fig. **10.35.** The interior of the hip joint after removal of the femur.

The relations of the hip joint

Anteriorly: the pectineus and ilio-psoas, which separate the femoral vessels and the femoral nerve from the joint.

Posteriorly: the obturator internus and externus, which separate the sciatic nerve from the joint.

Laterally: the gluteus medius and minimus.

Superiorly: the rectus femoris and gluteus minimus cover the joint.

Inferiorly: the lateral part of the pectineus and the obturator externus are in contact with the joint.

The muscles which move the hip joint

The muscles which move the hip joint and the muscles which move the knee joint are mainly the muscles of the thigh and some of the muscles have actions which involve both joints. The muscles involved in the movements of the two joints will therefore be described together (p. 248).

The knee joint

The knee joint is a **synovial condylar joint**.

The articular surfaces involved in the formation of the joint are the lateral and medial condyles of the femur, which articulate with the lateral

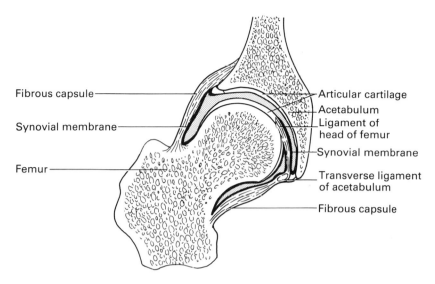

Fibrous capsule

Synovial membrane

Femur

Articular cartilage

Acetabulum

Ligament of head of femur

Synovial membrane

Transverse ligament of acetabulum

Fibrous capsule

Fig. 10.36. Cross-section of the hip joint.

and medial condyles of the tibia and with the semilunar cartilages. In addition, the patella articulates with the patellar articular area of the lower end of the femur.

The ligaments of the joint are: the *fibrous capsule*, the *ligamentum patellae*, the *oblique and arcuate popliteal ligaments*, the *medial ligament* (tibial collateral), the *lateral ligament* (fibular collateral), the *cruciate ligaments*, and the *transverse ligament*.

The fibrous capsule of the knee joint is somewhat complicated as it is absent in some parts and in others it is replaced by fibrous expansions from the tendons which surround the joint.

Posteriorly it is attached above to the femoral condyles and below to the tibial condyles. The posterior part of the capsule blends on either side with the origins of the lateral and medial heads of the gastrocnemius.

Medially the capsule blends with the medial ligament of the knee and is attached above to the medial femoral condyle and below to the medial tibial condyle.

Laterally the capsule is attached above to the lateral femoral condyle and below to the lateral tibial condyle and the head of the fibula. The lateral ligament of the knee joint, is, however, separate from the capsule.

Anteriorly the capsule is absent above and over the patellar area and elsewhere it blends with expansions from the tendons of the vastus medialis and lateralis.

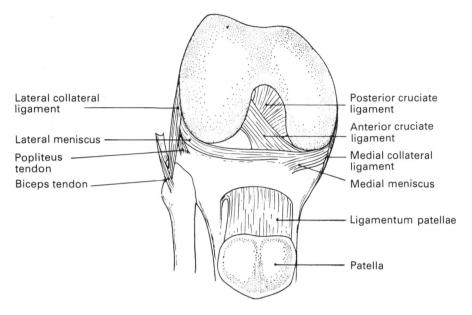

Lateral collateral
ligament

Lateral meniscus

Popliteus
tendon

Biceps tendon

Posterior cruciate
ligament

Anterior cruciate
ligament

Medial collateral
ligament

Medial meniscus

Ligamentum patellae

Patella

Fig. 10.37. The knee joint, anterior view. The patella has been turned downwards.

The ligamentum patellae is attached above to the lower part of the posterior surface of the patella and below to the tibial tuberosity.

The oblique popliteal ligament is a flat band of fibrous tissue which passes upwards and laterally from the insertion of the semimembranosus, on the medial side of the tibia, to be attached to the lateral femoral condyle. It blends with the posterior aspect of the capsule.

The arcuate popliteal ligament is a Y-shaped ligament which is attached below to the head of the fibula and above to the lateral femoral condyle and the posterior part of the intercondylar area of the tibia.

The medial ligament lies nearer to the back of the joint than the front. It is a wide flat band which is attached above to the medial epicondyle of the femur and below to the medial condyle and medial surface of the shaft of the tibia.

The lateral ligament is a strong cord which is attached above to the lateral epicondyle of the femur and below to the head of the fibula.

The cruciate ligaments are two strong ligaments which lie within the knee joint. *The anterior cruciate ligament* arises from the anterior part of the intercondylar area of the tibia and passes upwards, backwards and laterally to be attached to the medial surface of the lateral femoral condyle. *The posterior cruciate ligament* arises from the posterior part of the intercondylar area of the tibia and passes upwards, forwards and medially to be attached to the lateral surface of the medial femoral condyle. The cruciate ligaments cross each other like the limbs of the letter X and it is from this that they receive their names.

The synovial membrane of the knee joint is extensive. It extends upwards from the upper border of the patella to form a large pouch, called the suprapatellar bursa, which lies between the lower end of the femur and the quadriceps femoris. At the sides of the joint it passes downwards, on the inside of the capsule as far as the menisci. Posteriorly the synovial membrane is reflected forwards to cover the anterior and lateral surfaces of the cruciate ligaments and thus the central posterior portion of the fibrous capsule has no synovial covering.

The semilunar cartilages (menisci)

The semilunar cartilages are two crescentic pieces of cartilage, which lie between the femur and the tibia, and serve to deepen the articular surfaces of the tibial condyles. They are roughly triangular in cross-section, the peripheral margin of each being thick, and the inner border thin. The upper surface of each meniscus is smooth and concave and articulates with the femoral condyle. The lower surface of each meniscus is smooth and flat and lies on the upper surface of the tibia.

The medial meniscus is roughly semicircular, whereas the lateral is more nearly circular. Each is attached, by its periphery, to the fibrous capsule and, by its anterior and posterior ends, to the anterior and posterior parts of the intercondylar area of the tibia. The **transverse ligament** runs between the anterior margins of the two menisci.

Bursae around the knee joint

There are a number of bursae in the region of the knee joint. The three important bursae, however, lie in front of the joint. They are:

1   **The suprapatellar bursa**, which lies between the lower end of the femur and the quadriceps femoris.

2   **The prepatellar bursa**, which lies between the anterior surface of the patella and the skin.

3   **The infrapatellar bursa**, which lies between the lower part of the tibial tuberosity and the skin.

The relations of the knee joint

Anteriorly: the quadriceps femoris.
Posterolaterally: the tendon of the biceps femoris.
Posteromedially: the sartorius and the tendon of the gracilis.

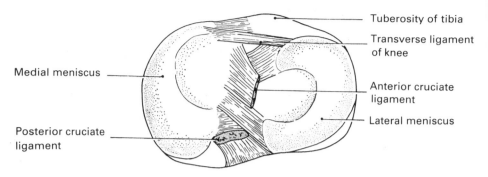

Medial meniscus

Posterior cruciate ligament

Tuberosity of tibia

Transverse ligament of knee

Anterior cruciate ligament

Lateral meniscus

Fig. 10.38. The knee joint in transverse section.

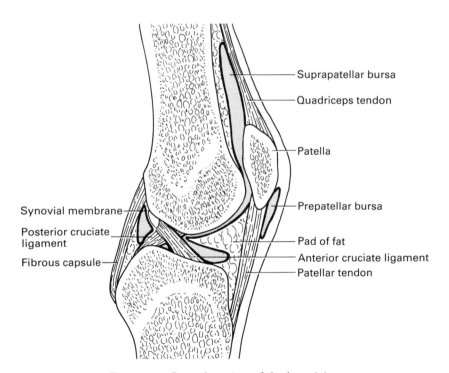

Suprapatellar bursa

Quadriceps tendon

Patella

Prepatellar bursa

Synovial membrane

Posterior cruciate ligament

Pad of fat

Fibrous capsule

Anterior cruciate ligament

Patellar tendon

Fig. 10.39. Sagittal section of the knee joint.

Posteriorly: the popliteal vessels and the medial popliteal nerve. They are overlapped below by the gastrocnemius.

The blood supply to the knee joint

The knee joint receives its blood supply from branches of the femoral artery, the popliteal artery and the anterior tibial artery.

The nerve supply to the knee joint

The knee joint receives its nerve supply from branches of the obturator nerve, the femoral nerve and the medial and lateral popliteal nerves.

The muscles which move the hip joint and the knee joint

The muscles involved in movements of the hip joint and the knee joint are the muscles of the thigh and the iliacus and psoas major muscles which have already been described (p. 204).

The muscles of the thigh

The muscles of the thigh are divided into four groups, the anterior muscles, the medial muscles, the gluteal muscles and the posterior muscles.

The anterior muscles of the thigh

The tensor fasciae latae arises from the anterior part of the iliac crest and is inserted into a sheet of fibrous tissue which is attached below to the lateral condyle of the tibia. It helps to extend the knee and abduct the thigh.

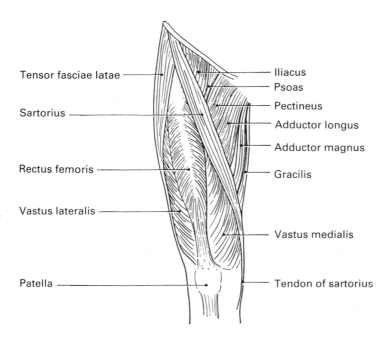

Fig. 10.40. The anterior and medial muscles of the right thigh.

The sartorius is a long, strap-like muscle and is the longest muscle in the body. It arises from the anterior superior iliac spine and curves obliquely downwards and medially, across the front of the thigh, to be inserted into the upper part of the medial surface of the shaft of the tibia. It flexes the hip joint and the knee joint, and it helps to abduct the thigh.

The quadriceps femoris makes up the mass of muscle at the front and sides of the thigh. It is composed of four parts:

The rectus femoris which arises from the anterior superior iliac spine.

The vastus lateralis, the vastus medialis and *the vastus intermedius* which arise from the anterior and lateral surfaces of the shaft of the femur and the linea aspera of the femur.

The muscle fibres of these four parts unite to join a strong tendon which is inserted into the base of the patella, and through the patellar tendon, into the tibial tuberosity.

The quadriceps femoris extends the knee joint, and the rectus femoris, in addition, plays a small part in flexing the hip joint.

The medial muscles of the thigh

The gracilis is a thin muscle which arises from the body of the pubis, the inferior pubic ramus and the ischial ramus. It is inserted into the shaft of the tibia just below the medial condyle. The gracilis flexes the knee joint and also acts as an adductor of the thigh.

The pectineus arises from the body of the pubis and is inserted into the linea aspera of the femur. It adducts and flexes the thigh.

The adductor muscles. There are three adductor muscles, the *adductor longus*, the *adductor brevis* and the *adductor magnus*. They arise from the body of the pubis, the inferior pubic ramus and the ischial tuberosity, and they are inserted into the linea aspera of the femur. The adductor muscles adduct and laterally rotate the femur.

The gluteal muscles

The gluteus maximus is the largest muscle of the gluteal region and it forms the prominence of the buttock. It arises from the posterior part of the outer surface of the iliac bone and from the sacrum, and it is inserted into the gluteal tuberosity of the femur. The gluteus maximus extends the hip joint and it is important both for maintaining the extension of the hip joint in the erect position and for the returning the trunk to the erect position after stooping.

The gluteus medius and minimus. The gluteus medius arises from the outer surface of the iliac bone and lies partially underneath the gluteus maximus. The gluteus minimus also arises from the outer surface of the iliac bone but it lies beneath the gluteus medius. Both muscles are inserted into the greater trochanter of the femur. Both abduct and medially rotate the thigh. They are important during the act of walking when they tilt the pelvis slightly towards their own side when the opposite leg is removed from the ground.

The obturator muscles. The obturator internus and externus arise from the inner and outer walls of the true pelvis and are inserted into the greater trochanter and the trochanteric fossa of the femur. They laterally rotate the femur.

The posterior muscles of the thigh

The posterior femoral muscles are three in number and are often referred to as the hamstring muscles.

The biceps femoris has two heads of origin, a long head and a short head. The long head arises from the ischial tuberosity and the short head from the linea aspera of the femur. The two heads unite and are inserted by a tendon into the lateral condyle of the tibia and the head of the fibula.

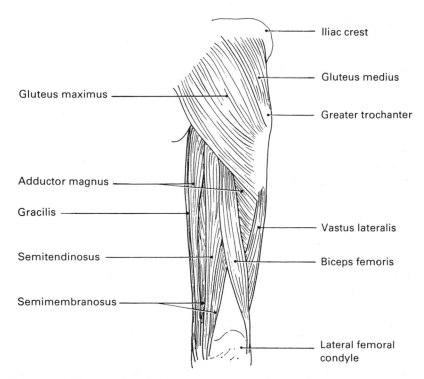

Fig. 10.41. The muscles of the posterior aspect of the right thigh.

The semitendinosus arises from the ischial tuberosity and is inserted, by a long tendon, into the upper end of the medial surface of the shaft of the tibia.

The semimembranosus arises from the ischial tuberosity and its fibres converge to a tendon which is inserted into the posterior aspect of the medial condyle of the tibia.

The hamstring muscles flex the knee joint and also help to draw the trunk upright from the stooping position. The biceps femoris is, in addition, a lateral rotator of the leg and the semitendinosus and semimembranosus are medial rotators of the leg.

The movements of the hip joint

The movements which occur at the hip joint are flexion, extension, abduction, adduction, circumduction, medial rotation and lateral rotation.

Flexion of the hip joint is mainly produced by the psoas and iliacus but the rectus femoris muscle plays a small part. In the erect position the weight of the trunk is transferred to the pelvis at the sacroiliac joints and to the lower limbs at the hip joint. Since the sacroiliac joint lies behind the hip joints there is a tendency for the pelvis to rotate backwards on the hip joints producing hyperextension and this tendency is resisted by reflex contraction of the psoas and iliacus.

Extension of the hip joint is produced by the action of the gluteus maximus assisted by the hamstring muscles. The hamstrings are particularly important in producing extension of the hip when the body is returning to the erect position from the stooping position.

Abduction of the hip is mainly produced by the action of the gluteus medius and minimus, assisted by the tensor fascia lata. During walking, the weight of the body is alternately transferred from one leg to the other and the gluteus medius and minimus produce slight abduction at the hip joint on the side to which the weight is being transferred in order to allow the opposite leg to move forwards clear of the ground.

Adduction of the hip joint is produced by the action of the adductor muscles, the pectineus and the gracilis.

Circumduction is a combination of the previous four movements.

Medial rotation of the hip joint is produced by the action of the tensor fascia lata and the anterior fibres of the gluteus medius and minimus.

Lateral rotation of the hip joint is produced by the action of the obturatores internus and externus and the gluteus maximus.

The movements of the knee joint

The movements which occur at the knee joint are flexion and extension and, to a very limited extent, medial and lateral rotation. Extension of the

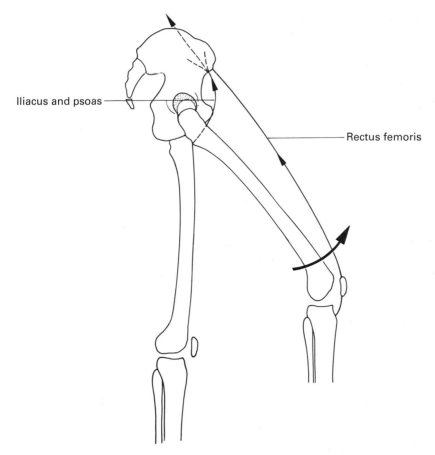

Iliacus and psoas

Rectus femoris

Fig. 10.42. Flexion of the hip joint is mainly produced by the psoas and the iliacus but the rectus femoris plays a small part.

knee joint involves a small degree of medial rotation and flexion involves an equal degree of lateral rotation. Medial and lateral rotation, independent of flexion or extension, can only occur when the knee joint is not fully extended.

Flexion of the knee joint is mainly produced by the action of the hamstring muscles but they are assisted by the gracilis, the sartorius and the gastrocnemius (p. 258).

Extension of the knee joint is produced by the action of the quadriceps femoris.

Medial rotation is produced by the action of the popliteus, the semimembranosus and the semitendinosus.

Lateral rotation is produced by the action of the biceps femoris.

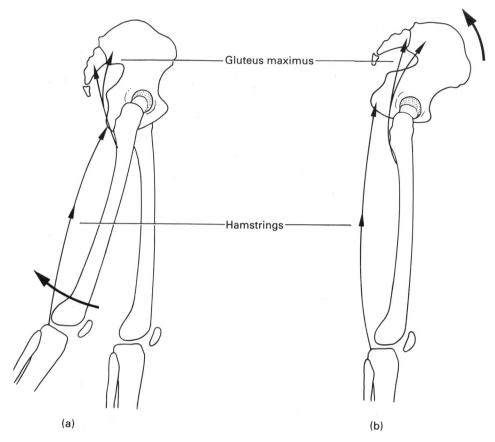

(a)   (b)

Fig. 10.43. Extension of the hip joint (a) is produced by the gluteus maximus and the hamstrings. The hamstrings are particularly important in producing extension of the hip joint when the body is returning to the erect position from the stooping position (b).

The tibiofibular joints

There are two joints between the tibia and fibula, and the shafts of the two bones are, in addition, joined by an *interosseous membrane*, the fibres of which run downwards and laterally from the interosseous border of the tibia to the interosseous border of the fibula.

The superior tibiofibular joint

The superior tibiofibular joint is a **synovial plane joint**. The **articular surfaces** involved in the formation of the joint are the articular facet on the head of the fibula and the articular facet on the under-surface of the lateral

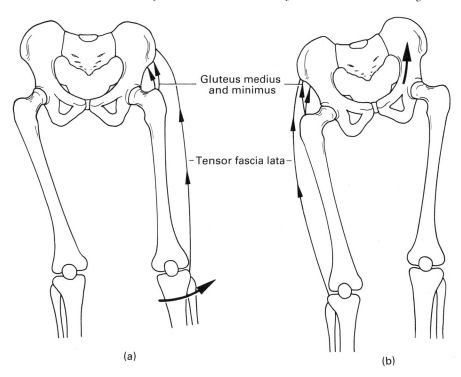

Fig. **10.44.** (a) Abduction of the hip joint is produced by the action of the gluteus medius and minimus assisted by the tensor fascia lata. (b) During walking, abduction at the hip joint occurs on the side of the leg which remains on the ground in order to tilt the pelvis to allow the opposite leg to move forwards clear of the ground.

tibial condyle. The joint is surrounded by a **fibrous capsule** which is strengthened by **anterior** and **posterior ligaments**.

The inferior tibiofibular joint

The inferior tibiofibular joint is a **syndesmosis**. The **articular surfaces** involved in the formation of the joint are the rough convex area on the medial surface of the lower end of the fibula and the rough concave fibular notch on the lateral surface of the lower end of the tibia. The two surfaces are united by an **interosseous ligament** which is continuous, above, with the interosseous membrane.

The ankle (talocrural) joint

The ankle joint is a **synovial hinge joint**.
    The articular surfaces involved in the formation of the joint are the

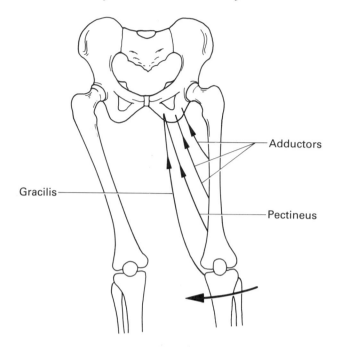

Fig. 10.45. Adduction of the hip joint is produced by the adductor muscles and the pectineus and gracilis.

articular surface of the medial malleolus, the inferior surface of the tibia and the articular surface of the lateral malleolus of the fibula, which together form a mortice for articulation with the articular area of the body of the talus.

The articular surfaces are covered with *hyaline cartilage.*

The ligaments of the joints are: the *fibrous capsule,* the *deltoid ligament,* the *anterior* and *posterior talofibular ligaments,* and the *calcaneofibular ligament.*

The fibrous capsule surrounds the joint. It is attached above to the margins of the articular surfaces of the tibia and malleoli, and below to the posterior and lateral margins of the articular surface of the talus and to the neck of the talus anteriorly.

The deltoid ligament lies on the medial side of the joint. It is a strong triangular ligament which is attached above to the medial malleolus, and below to the tuberosity of the navicular, the sustentaculum tali and the medial surface of the talus.

The anterior and **posterior talofibular ligaments** lie on the lateral side of the joint. The anterior talofibular ligament is attached above to the anterior margin of the lateal malleolus and passes forwards and medially to

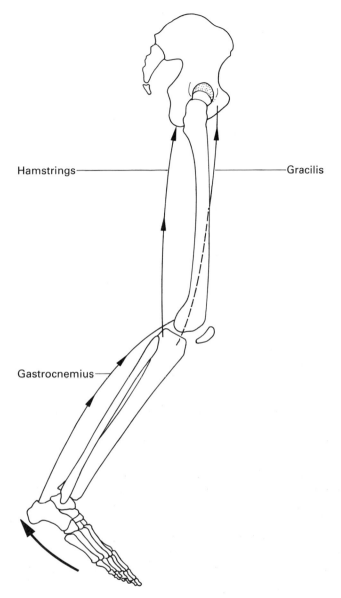

Hamstrings

Gracilis

Gastrocnemius

Fig. 10.46. Flexion of the knee joint is mainly produced by the hamstrings but they are assisted by the gracilis and the gastrocnemius.

be attached to the lateral side of the neck of the talus. The posterior talofibular ligament passes laterally from the lower part of the malleolar fossa of the fibula to be attached to the posterior process of the talus.

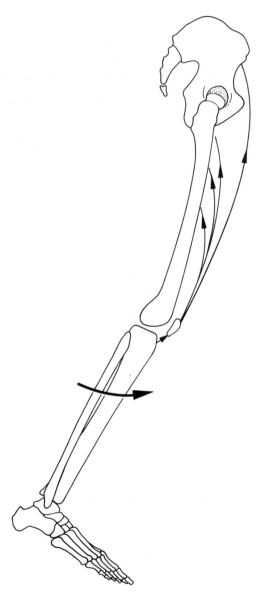

Fig. 10.47. Extension of the knee joint is produced by the quadriceps femoris.

The calcaneofibular ligament is a rounded cord which runs downwards and forwards from the apex of the lateral malleolus to be attached to the lateral side of the calcaneus.

The synovial membrane lines the fibrous capsule and frequently

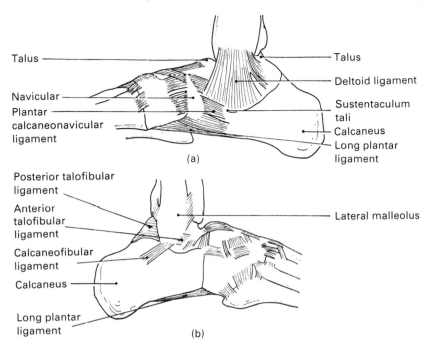

Talus

Navicular

Plantar
calcaneonavicular
ligament

Talus

Deltoid ligament

Sustentaculum
tali

Calcaneus

Long plantar
ligament

(a)

Posterior talofibular
ligament

Anterior
talofibular
ligament

Calcaneofibular
ligament

Calcaneus

Long plantar
ligament

Lateral malleolus

(b)

Fig. 10.48. The right ankle joint: (a) medial aspect, and (b) lateral aspect.

passes upwards for a short distance between the lower ends of the tibia and fibula.

The blood supply to the ankle joint

The ankle joint receives its blood supply from branches of the anterior tibial and peroneal arteries.

The nerve supply to the ankle joint

The ankle joint receives its nerve supply from the lateral popliteal and posterior tibial nerves.

The intertarsal joints

1 The subtalar joint is composed of two articulations. One, called the **talocalcanean joint,** is formed between the posterior facet on the lower surface of the talus and the posterior facet on the upper surface of the calcaneus. At the second joint, which is called the **talocalcaneonavicular joint**, the head of the talus articulates with the proximal surface of the navicular, the middle and anterior facets for the talus on the upper surface of the calcaneus and the upper surface of the plantar calcaneonavicular ligament. Each joint is surrounded by a fibrous capsule which is lined with synovial membrane.

2 The calcaneocuboid joint is a **synovial saddle joint**. The **articular surfaces** involved in the formation of the joint are the distal surface of the calcaneus and the proximal surface of the cuboid.

3 The cuboidonavicular joint. The joint between the cuboid and the navicular is either a **synovial plane joint**, or a **syndesmosis** in which case the two bones are united by an interosseous ligament.

4 The cuneonavicular joint. The joint between the three cuneiform bones and the navicular is a **synovial joint** which is continuous with the intercuneiform and the cuneocuboid joints.

5 The intercuneiform and the cuneocuboid joints. The joints between the cuneiform bones and the lateral cuneiform bone and the cuboid are all **synovial plane joints**.

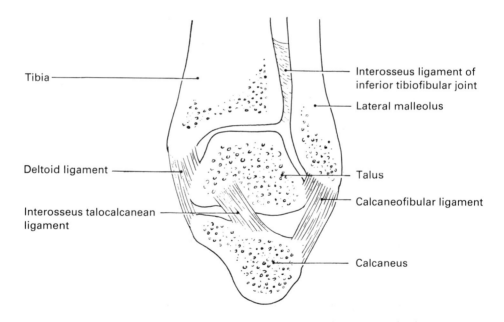

Fig. 10.49. Coronal section through the ankle and subtalar joints.

The ligaments of the tarsus

There are numerous short ligaments which connect the bones of the tarsus and most of these are of relatively little importance. There are several ligaments, however, which are of greater importance.

1 **The interosseous talocalcanean ligament** lies in the sinus tarsi and connects the talus and the calcaneus.

2 **The long plantar ligament** is the longest ligament of the tarsus. It is attached, posteriorly, to the plantar surface of the calcaneus close to the posterior border of the bone. It extends forwards to be attached to the plantar surface of the cuboid and to the bases of the 2nd, 3rd and 4th metatarsal bones.

3 **The short plantar ligament** lies deep to the long plantar ligament and extends from the plantar surfaces of the calcaneus to the plantar surface of the cuboid.

The long and short plantar ligaments are important for maintaining the longitudinal arch of the foot.

4 **The plantar calcaneonavicular ligament**, or the spring ligament, connects the sustentaculum tali to the plantar surface of the navicular. The upper surface of this ligament forms part of the articular surface for the head of the talus at the anterior part of the subtalar joint. It helps to maintain the medial side of the longitudinal arch of the foot.

The tarsometatarsal joints

The tarsometatarsal joints are **plane synovial joints**. The 1st tarsometatarsal joint is a separate joint which has its own synovial membrane and joint capsule. The other tarsometatarsal joints are continuous with the intercuneiform and cuneocuboid joints.

The metatarsophalangeal joints

The metatarsophalangeal joints are **ellipsoid synovial joints**. Each joint is surrounded by a fibrous capsule which is strengthened by **plantar and collateral ligaments**. The heads of the metatarsal bones are united by ligaments called the **deep transverse metatarsal ligaments**.

The interphalangeal joints

The interphalangeal joints are **synovial hinge joints** and each has an **articular capsule** and two **collateral ligaments**.

The muscles which move the ankle joint and the joints of the foot

The muscles which are involved in movements of the ankle joint and the foot are the muscles of the leg and the small muscles of the foot.

The muscles of the leg are divided into anterior, lateral and posterior groups.

The anterior muscles of the leg

The tibialis anterior arises from the lateral condyle and the upper half of the lateral surface of the shaft of the tibia. It is inserted by a long tendon into the medial cuneiform bone and the base of the 1st metatarsal bone. The tibialis anterior dorsiflexes the foot.

The extensor hallucis longus arises from the medial surface of the shaft of the fibula and from the interosseous membrane, which lies between the tibia and fibula. It is inserted, by a long tendon, into the big toe. It extends the big toe and assists in dorsiflexion of the foot.

The extensor digitorum longus arises from the lateral condyle of the tibia, the medial surface of the shaft of the fibula and the interosseous membrane. It is inserted by four tendons into the 2nd to 5th toes. It extends the toes and assists in dorsiflexion of the foot.

The lateral muscles of the leg

The peroneus longus and brevis both arise from the lateral surface of the shaft of the fibula. The tendons of bone muscles pass behind the lateral malleolus at the ankle joint. The peroneus longus tendon is inserted into the lateral side of the medial cuneiform bone and the base of the 1st metatarsal bone and the peroneus brevis tendon is inserted into the base of the 5th metatarsal bone. Both muscles evert the foot.

The posterior muscles of the leg

The posterior muscles of the leg are divided into a superficial group and a deep group.

The superficial muscles are the gastrocnemius and the soleus, which are plantarflexors of the foot.

The deep muscles are the flexor hallucis longus, the flexor digitorum longus, the tibialis posterior and the popliteus.

The gastrocnemius is the most superficial muscle of this group, and it forms the main mass of the calf. It arises by lateral and medial heads from the lateral and medial condyles of the femur. The two heads are attached to an aponeurosis, which narrows in the lower part of the leg, and unites with the tendon of the soleus, to form the tendo calcaneus (Achilles tendon) which is attached to the posterior aspect of the calcaneum.

The *soleus* lies beneath the gastrocnemius and arises from the upper ends of the tibia and fibula. The tendon of the soleus unites with the tendon of the gastrocnemius to form the tendo calcaneus.

The gastrocnemius and the soleus are the main plantarflexors of the foot and they are important for providing the plantarflexion necessary for running and walking.

The popliteus is a triangular muscle which lies at the upper part of the posterior aspect of the leg. It arises from the lateral condyle of the femur and is inserted into the posterior surface of the upper part of the shaft of the tibia. It is a medial rotator of the knee.

The flexor digitorum longus arises from the posterior surface of the shaft of the tibia. The fibres of the muscle are attached to a tendon which passes behind the medial malleolus and, on the sole of the foot, it divides into four tendons which are inserted into the 2nd to 5th toes. It flexes the toes and helps to plantarflex the foot.

The flexor hallucis longus arises from the posterior surface of the shaft of the fibula and from the interosseus membrane which lies between the tibia and fibula. The tendon of the muscle passes behind the medial malleolus and is inserted into the 1st toe. It flexes the 1st toe and helps to plantarflex the foot.

The tibialis posterior lies between the preceding two muscles and arises

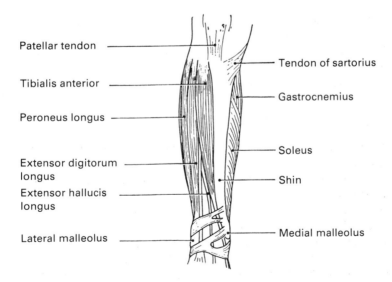

Fig. 10.50. The muscles of the right leg, anterior aspect.

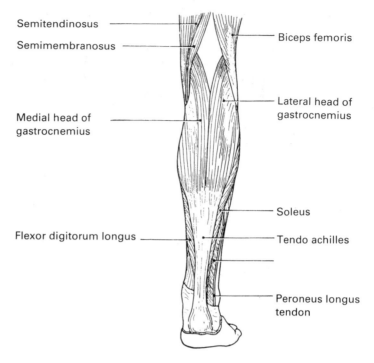

Fig. 10.51. The muscles of the right leg, posterior aspect.

from the interosseous membrane and the adjacent portions of the shafts of the tibia and fibula. The fibres join a tendon which passes round the medial side of the ankle joint and is attached to the navicular. The tibialis posterior inverts the foot.

The muscles of the foot

There are numerous small muscles in the foot which assist in producing movements of the toes and which are important for maintaining the arches of the foot.

The movements of the ankle joint and the joints of the foot

The ankle joint is a synovial hinge joint so the only movements which can occur are dorsiflexion and plantarflexion. Plantarflexion increases the angle between the dorsum of the foot and the anterior aspect of the leg and dorsiflexion decreases the angle.

Plantarflexion is produced by the action of the gastrocnemius and the soleus.

Dorsiflexion is produced by the tibialis anterior assisted by the extensors of the toes.

The subtalar joint is formed by the talocalcanean joint and the talocalcaneonavicular joint and these two form a functional unit at which a considerable range of movement can occur. The calcaneus and the navicular, together with the rest of the foot, can move medially on the talus to produce the movement of inversion of the foot. The reserve movement, called eversion, is more limited.

Inversion of the foot is produced by the tibialis anterior and the tibialis posterior.

Eversion of the foot is produced by the peroneus longus and the peroneus brevis.

During inversion and eversion of the foot slight gliding movements occur at the calcaneocuboid joint.

At the **metatarsophalangeal joints** there are limited movements and these are *flexion, extension, abduction* and *adduction.*

At the **interphalangeal joints** the only movements which occur are *flexion* and *extension.*

CHAPTER 11

The circulatory system

The blood

Blood is a fluid which circulates in the closed system of vessels that, with the heart, make up the circulatory system. It is responsible for supplying the requirements of the body cells for oxygen and nutritive materials, and it is also important for the removal of the products of metabolism. This transport function of the blood is very important.

The physical and chemical properties of haemoglobin allow the blood to transport large amounts of oxygen. In addition, the haemoglobin acts as a buffer helping to maintain a constant blood pH, and, in conjuction with the lungs and the kidneys, a constant body fluid pH. The colour of the blood varies with its oxygen contents; thus in the arteries it is bright scarlet and in veins it is dark red or purple.

Blood consists of a fluid called *plasma*, in which are suspended a large number of cells: red cells (erythrocytes or red corpuscles), white cells (leucocytes or white corpuscles), and platelets (thrombocytes). Plasma accounts for 55% of the blood volume and the suspended cells for 45% (packed cell volume). The total blood volume of an average adult is about 5 l.

The distribution of water in the body

Water constitutes about 60% of total body weight in men and 50% in women. The difference between the proportions of water is due to the difference in lean body mass, which is the total weight of fat-free tissue, and this is greater in men than in women. Fat is stored free of water and more obese individuals, therefore, have a lower percentage of total body weight as water.

The proportion of water in different tissues varies; bone has only 20% while skin has 70% and brain, lung and heart have nearly 80%.

Water is necessary as a solvent for most of the metabolic activities of cells and it is also important for the passage of foodstuffs, gases and waste products between cells and their surroundings. Every cell is surrounded by fluid which allows these exchanges to take place.

The total body water is divided between two main compartments:

1 **Intracellular fluid** which is the fluid within the cells: this constitutes 70% of the total body water.

2 **Extracellular fluid** which constitutes about 30% of the total body water and is contained within:

a) *the interstitial fluid* which is the fluid surrounding the cells (about 20% of total body water);

b) *the plasma* which constitutes about 10% of the total body water; and

c) *transcellular fluid* which includes synovial fluid, urine, the digestive juices in the alimentary canal and the cerebro-spinal fluid (only about 150 ml).

Water is taken in through the mouth and passes into the blood stream from the gastrointestinal tract. Water is lost from the body via the kidneys as urine, via the lungs as water vapour and via the skin as sweat. There are also small losses in the tears.

The water balance of the body is maintained by the rate of water excretion by the kidneys. This rate of excretion is controlled by antidiuretic hormone secreted from the posterior pituitary gland (see Chapter 17).

The plasma

Plasma is a slightly alkaline yellowish fluid which consists of the following:

Water	90%
Proteins	Albumin and globulin
Inorganic salts	Sodium chloride, sodium bicarbonate and smaller amounts of potassium, calcium, magnesium, phosphorous (as inorganic phosphate), sulphate and iron. These are sometimes referred to as plasma electrolytes
Foodstuffs (nutrients)	Glucose, amino acids, fats and vitamins
Waste products	Urea, uric acid and creatinine
Fibrinogen *Prothrombin*	Necessary for the clotting of blood
Hormones	
Enzymes	

The plasma proteins are generally of too large a molecular size to leave the blood vessels. They are thus important for maintaining the osmotic pressure of blood and preventing loss of fluid from the circulation. At the arterial end of the capillary the hydrostatic force within the vascular compartment is sufficient to cause fluid to pass into the interstitial space. However, at the venous end of the capillary the osmotic pressure of the plasma proteins exceeds the hydrostatic pressure and most of the fluid is returned to the vascular compartment. The residual fluid is usually drained away by the lymphatic system. This cycle of fluid exchange at the capillaries between the vascular and interstitial spaces is important, and is the main function of the capillaries, which have a vast capacity. In some kidney disorders protein is lost in the urine in large amounts, the osmotic pressure of the blood falls and fluid gathers in the intercellular tissue spaces causing swelling, known as *oedema*. Plasma proteins have some buffering capacity, that is they help to maintain a constant blood pH.

The globulin fraction of the plasma proteins contains the antibodies which are important in the body's defence against infection (see Chapter 3).

The inorganic salts are important for the normal functioning of cells and for maintenance of the normal electrical potential difference between the inside and outside of all cells. This is particularly important for the normal functioning of nerve and muscle (both skeletal and cardiac) cells. In order to maintain the gradient of sodium, which is at a greater concentration in the extracellular than in the intracellular fluid, energy expenditure must take place. The mechanism by which sodium is transported out of the cells is known as the '*sodium pump*'.

Fibrinogen and prothrombin are necessary for the normal clotting of blood.

The blood cells

There are three main types of blood cells: red cells (*erythrocytes*), white cells (*leucocytes*) and platelets (*thrombocytes*).

The erythrocytes, or red cells, are by far the most numerous of the blood cells. They are biconcave, non-nucleated discs about 7.2 μm in diameter. The biconcave shape gives a larger surface area for oxygen and carbon dioxide exchange than a spherical shape. The cell membrane is elastic and deformable allowing the red cell to be distorted to pass through narrow capillaries. There are approximately $5 \times 10^{12}/l$ of blood (5 000 000 ml).

The eythrocytes contain a special protein called *haemoglobin*. Haemoglobin consists of an iron containing portion, called *haem*, and the protein *globin*. There are about 14.8 g of haemoglobin per dl (100 ml) of blood. Each red cell contains about 30 pg of haemoglobin and this index is known as the *mean corpuscular haemoglobin* or *MCH*. Another index of the amount of haemoglobin in each red cell is the *mean corpuscular haemoglobin concentration*, or *MCHC*. This is normally about 33 g/dl of blood. Red blood counts and indices of the red cell haemoglobin are now measured in automated machines which also count the numbers of white cells and platelets. Haemoglobin has a marked affinity for oxygen with which it combines reversibly in the lungs where the oxygen tension is high. In the tissues, where cell metabolism produces carbon dioxide, the haemoglobin delivers up its attached oxygen. The containment of haemoglobin within the red cell prevents loss of the pigment in the urine and destruction by the macrophage (reticulo-endothelial) system as well as preventing the pigment from causing undesirable effects on the physical characteristics of the blood.

The formation of erythrocytes. Red blood cells are formed in the red bone marrow from large nucleated cells, called pronormoblasts (proeryth-roblasts). These cells divide to form progressively smaller cells and as the erythrocytes mature they lose their nucleus. Normal red cell formation (erythropoiesis) is called *normoblastic*. In the absence of sufficient Vitamin B12 or folic acid erythropoiesis is *megaloblastic*, the red cells produced being abnormally large.

A normal red cell usually has a life span of about 120 days in the circulation. After this time the red cell is destroyed in the macrophage system (see Chapter 12). The haemoglobin is released, and iron and globin are separated from the haem portion, which is converted to form the bile pigment bilirubin. This is excreted by the liver in the bile. The iron is stored as ferritin until needed for new haemoglobin formation.

Erythrocyte sedimentation rate, or *ESR*, is a measure of the rate at which erythrocytes will settle if an anticoagulated sample of blood is allowed to stand in a narrow vertical tube. The rate is increased in various diseases and though this not a diagnostic test it does allow disease activity to be measured during treatment.

Anaemia is the condition in which the haemoglobin concentration is below normal. It can occur as a result of:

1   *loss of blood*, either acute or chronic.

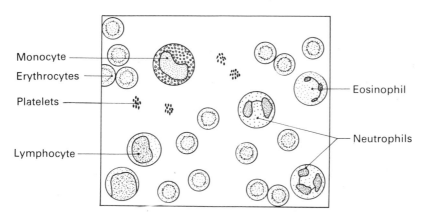

Monocyte

Erythrocytes

Platelets

Eosinophil

Neutrophils

Lymphocyte

Fig. 11.1. The various types of blood corpuscles.

2 *Inadequate red cell production.* This may be due to lack of essential factors such as iron, Vitamin B12 or folate. It may also be caused by destruction of the bone marrow by certain chemicals, X-rays, drugs, secondary carcinoma invasion or leukaemia.

3 *Increased red cell destruction,* known as *haemolysis.* Haemolytic anaemias may be due to abnormalities of the red cells, which are therefore destroyed more quickly than normal red cells, or to factors (such as malaria, drugs or antibodies) which lead to premature destruction of normal red cells.

The leucocytes, or white blood cells, are nucleated cells which are present in far smaller numbers than the red cells. The number ranges from 4 to 11 × 10^9/l and even in health varies widely from hour to hour. There are three main types of white cell: *granulocytes, lymphocytes* and *monocytes.*

Granulocytes (polymorphonuclear leucocytes) are cells with prominent granules in their cytoplasm. The nuclei have several lobes and stain variably with dyes according to the type of granulocyte. Three types are recognized:

1 **neutrophils** which are 10−15 μm in diameter and are important in the body's defence against infection. At sites of injury or infection neutrophils are present in large numbers. These cells have the property of devouring bacteria by phagocytosis. They are motile and can pass out of the capillaries into the tissues;

2 **eosinophils,** which are phagocytic but less mobile than neutrophils and are found at sites of allergic reactions; and

3 **basophils** are the least numerous of the granulocyte series. They are similar to tissue mast cells and are important in certain hypersensitivity reactions.

Lymphocytes are formed in the bone marrow. They have a single large nucleus. Some migrate to the thymus where they are converted to T-lymphocytes while others become B-lymphocytes or K- ('killer') lymphocytes (see Chapter 12).

Monocytes are larger than granulocytes and are about 10−18 μm in diameter. They are phagocytic. Some circulate in the blood stream while others leave to join the macrophage system.

The proportions of the various white cell types in the blood are as follows: granulocytes 70% (neutrophils 40−75%, eosinophils 1−6%, basophils 0−1%), lymphocytes 20−45% and monocytes 2−10%. These are the ranges in health. The numbers and proportions of white cell type are altered by illness, especially infective and allergic conditions.

Platelets, or thrombocytes, are colourless cells which vary in shape and are small (only 2−4 μm in diameter). They have no nucleus and they number 150−400 × 10^9/l. Platelets are formed in the bone marrow from much larger cells called megakaryocytes. The main function of platelets is the clotting of blood.

The bone marrow

Bone marrow is found in the centre of all bones and is of two types: *red marrow* and *yellow marrow*.

Red marrow is usually present throughout the skeleton at birth, but after the age of five years it is gradually replaced in the long bones by yellow marrow. By the age of 20–25 years the red marrow is present only in the ribs, the sternum, the vertebrae, the skull, the pelvis and the upper ends of the femora and humeri.

Red marrow consists of a framework of connective tissue within which are large numbers of blood sinusoids lined with endothelial tissue. These are supplied with arterial blood from the nutrient artery to the bone, and blood from the sinusoids drains into adjacent veins. Within the sinusoids are large numbers of mature and developing red cells, white cells and platelets as well as their precursor cells.

Yellow marrow consists of connective tissue containing a large number of fat cells. At times when increased blood formation is required, for example following severe haemorrhage or haemolysis, yellow marrow can revert to red marrow and become active in the production of blood cells. Because of the widespread red marrow, children have little reserve under such circumstances. The liver and spleen can also be the sites of blood cell formation in certain diseases.

Blood groups

Blood loss, if significant, is often treated by *transfusion*. Blood is removed from the vein of one individual, called the *donor*, and is given by intravenous infusion to the *recipient*. The early use of blood transfusion showed that there were four main blood groups, and that the blood of one group could not be given with safety to people with all of the other blood groups. These blood groups are called **A**, **B**, **AB** and **O**. When blood of one group is mixed with the blood of another group with which it is incompatible, the red cells become stuck together in clumps, or *agglutinated*. If this happens during a blood transfusion serious symptoms or even death can result. The agglutinated red cells are haemolysed. Fever and jaundice result and severe reactions may cause renal failure.

The process of agglutination is caused by a substance, called an *agglutinogen*, which is present in the red cells of one group reacting with a substance, called an *agglutinin*, which is present in the serum of another group. The distribution of agglutinogens and agglutinins in people with different ABO blood groups is shown in table 11.1.

Since blood group AB has no agglutinins in the serum, persons of this blood group can often be transfused with blood of the other groups without

Table 11.1. The distribution of agglutinins in ABO blood groups

Blood Group	Agglutinogen	Agglutinin
A	A	β (anti-B)
B	B	α (anti-A)
AB	A & B	none
O	none	α & β

agglutination resulting. Persons of blood group AB are referred to as *universal recipients*.

Blood group O has no agglutinogens present in its red cells and blood of this group can often be transfused to people of other blood groups without agglutination or transfusion reaction. People with blood group O have been described as *universal donors*.

The terms universal recipient and universal donor are not truly valid since they ignore complications due to the presence of rhesus factors. The only test of compatibility is to test the donor's red cells against the recipient's serum. Figure 11.2 illustrates the classical interactions of the various ABO blood groups.

The distribution of blood groups varies from one part of the world to another. In Western Europe the distribution is:

Group A	42%
Group B	9%
Group AB	3%
Group O	46%

The rhesus factor. A factor called the rhesus factor has been found in 85% of Europeans irrespective of their ABO blood group. The rhesus factor is so-called because the factor present in the blood of rhesus positive individuals resembles closely a factor present in the red cells of the rhesus monkey. If blood containing the rhesus factor (rhesus positive blood) is transfused to a patient who does not possess the rhesus factor (a rhesus negative individual) no immediate reaction occurs. Over a period of about ten days, however, the rhesus negative recipient develops antibodies against the rhesus factor. A second transfusion of rhesus positive blood will precipitate a reaction which may be severe or even fatal (see above) due to destruction of the rhesus positive red cells by the antibodies.

In addition to its significance in blood transfusion, the rhesus factor is also important in the conception of rhesus positive children by a rhesus negative mother. A rhesus negative mother may conceive a rhesus positive child if the father is rhesus positive. Assuming that the mother has at no

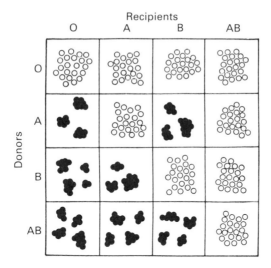

Fig. 11.2. The interactions of the ABO blood groups. Clumping of the red cells indicates an incompatible transfusion.

stage been given a rhesus positive blood transfusion, the birth of the first rhesus positive child does not carry any problem from the blood group point of view. At some stage in the pregnancy rhesus positive fetal blood cells may escape into the mother's circulation causing the mother's immune system to form antibodies against rhesus positive cells. As little as 0.5 ml of blood may be enough to cause antibody formation. If a further rhesus positive fetus is conceived these maternal antibodies can cross the placenta and destroy the fetal red cells. The effects of this *rhesus incompatibility* may be slight or may be severe enough to cause the death of the fetus *in utero*, or of the neonate soon after birth.

The ABO and rhesus blood group systems are the main groups but there are others such as the M, N, Kell, Luther and Duffy, which may be occasional causes of blood transfusion reactions.

The clotting of blood

When blood escapes from a blood vessel there is a natural tendency for the escaped blood to coagulate, or clot, in an attempt to prevent further blood loss. Whether or not this is effective depends upon a number of factors, including the blood pressure within the vessel. If a large artery is cut, for example in a stab wound, blood loss cannot be prevented by the actual clotting mechanism. Blood loss from arterioles can be stopped by reactive vasoconstriction.

The mechanisms involved in blood clotting are complex and only a brief account will be given here. The process involves a number of steps in a so-called cascade system (Fig. 11.3). The clotting cascade is a chain reaction initiated either by contact of platelets with a damaged vascular endothelium or by contact with tissue fluid when the vessel wall is breached. Initially a platelet plug is formed as a result of platelet aggregation (clumping). Subsequent release of substances from disintegration of aggregated platelets initiates the clotting cascade.

The initial steps involve the production of *thrombin* from its inactive precursor, *prothrombin*, which is formed in the liver. Thrombin, an enzyme, then reacts with *fibrinogen*, a normal constituent of plasma, to form the insoluble *fibrin*. The fine fibrin threads, by adhering to each other, form a network entangling blood cells. After a while the fresh clot will retract and become firmer as further chemical changes of the fibrin render it stable.

There are a total of twelve known factors which must be present in blood for a normal clot to be formed and stabilized. A full description is not possible here but two factors which are of particular importance and interest will be mentioned.

1 *Anti-haemophilic globulin* (AHG or factor VIII). This is a protein which is normally present in blood but is absent from the blood of patients suffering from the disease *haemophilia*. This is an inherited disorder in

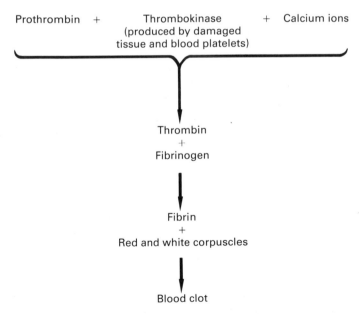

Fig. **11.3.** Scheme of the cascade system of blood coagulation.

...ssive bleeding due to impairment of the ... is an X-linked recessive disorder, women ... from haemophilia, whilst males carrying

... nother protein normally present in the ... blood of patients suffering from Christmas ..., a disorder very similar to haemophilia and which has the same mode of inheritance. (The disease is so-called because the first patient in whom the disease was described was called Mr Christmas.)

Thrombosis and fibrinolysis

Normally blood does not clot within blood vessels but, occasionally, this does happen and is called *thrombosis*. A clot formed within one vessel sometimes becomes dislodged and moves around the circulation until it lodges within another vessel. It is then called an *embolus*. Drugs called anticoagulants delay the process of clotting and are sometimes used in the treatment or prevention of thrombosis.

There is a natural system which opposes the clotting of blood. Plasminogen is converted to the active plasmin, which digests the insoluble fibrin. In healthy individuals the activities of coagulation and fibrinolytic systems are balanced.

Bone marrow transplantation

Bone marrow transplantation is now being used to replace diseased marrow with healthy bone marrow cells obtained from a suitably matched donor. The donor may or may not be related to the recipient, but the ideal donor is clearly an identical twin. Occasionally marrow tissue is taken from a patient with a healthy marrow prior to treatment which may damage the blood cell precursors. After treatment has been given the stored marrow is given back to the same individual. This is known as autografting.

The blood vessels

The blood is pumped by the action of the heart through a closed system of vessels. These vessels are of three types: *arteries*, *capillaries*, and *veins*.

The structure of the arteries

The arteries carry the blood away from the heart to the tissues. The exact structure of the arteries varies with the size of the lumen but they all consist of three coats.

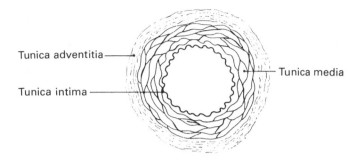

Tunica adventitia

Tunica media

Tunica intima

Fig. 11.4. Cross-section of an artery.

1 The outer coat is called the *tunica adventitia*. It is composed of fibrous tissue containing a few elastic fibres.

2 The middle coat is called the *tunica media*. It is composed of smooth muscle fibres which run in a circular direction. Interspersed with the smooth muscle fibres are yellow elastic fibres.

3 The inner coat is called the *tunica intima*. It is composed of a single layer of endothelial cells with underlying elastic tissue, which produces a series of small folds.

The main difference between large and small arteries is in the tunica media. In large arteries there is more elastic tissue in the tunica media than in small arteries. The very small arteies are called *arterioles*.

The structure of the capillaries

The arterioles divide into very small vessels called capillaries. The wall of the capillaries is composed of a single layer of endothelial cells which is continuous with the endothelial lining of the arterioles. It is through the wall of the capillaries that the exchanges between plasma and interstitial fluid take place. Water, and substances of small molecular size, can pass through the capillary wall, but the plasma proteins and red blood cells are unable to pass through. The diameter of a capillary is about the same as that of a red blood cell.

The structure of the veins

Veins, like arteries, have a wall composed of three coats. The middle coat, however, is much thinner than in arteries and the lumen is larger as compared to the thickness of the wall.

1 *The tunica adventitia* is composed of connective tissue and smooth muscle fibres.

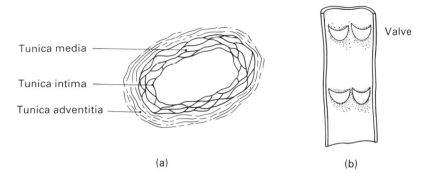

Tunica media

Tunica intima

Tunica adventitia

Valve

(a)

(b)

Fig. **11.5.** The structure of a vein: (a) transverse section, and (b) longitudinal section.

2 *The tunica media* is composed of connective tissue and elastic fibres and in some veins there are some circular smooth muscle fibres.
3 *The tunica intima* is composed of a single layer of endothelial cells, which are shorter and broader than those in arteries, and there is an underlying layer of elastic fibres.

Valves

Most veins have, on their inner surface, valves which serve to prevent backflow of blood. The valves consist of a fold of the tunica intima strengthened by connective tissue and elastic fibres. They are semilunar in shape and allow blood to flow in one direction only, that is towards the heart. They are of particular importance in the lower part of the body when the blood is returning towards the heart against gravity.

The heart

The heart is a muscular pump, consisting of right and left sides with no direct communication between the two sides. Blood is returned to the heart from the body through the *superior* and *inferior venae cavae* which open into a thin-walled chamber called the *right atrium*. The right atrium pumps the blood through a valve, called the *tricuspid valve*, into a thicker-walled chamber called the *right ventricle*. The right ventricle in its turn pumps the blood through the *pulmonary artery* into the lungs. In the capillaries of the lungs the blood takes up oxygen and releases carbon dioxide. The freshly oxygenated blood then passes through the *pulmonary veins* into a thin-walled chamber called the *left atrium*. The left atrium pumps the blood through a valve, called the mitral valve, into the thickest-walled chamber of the heart, the *left ventricle*. The left ventricle in turn pumps the blood into the *aorta* from whence it is distributed to the body. These series of events occur in parallel so that the left and right atria contract together and the left and right ventricles contract together.

The general circulation of the body is called the *systemic circulation* and the circulation of the lungs is called the *pulmonary circulation*. Greater pressure is required to pump the blood through the systemic circulation

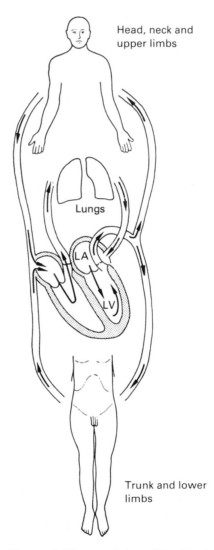

Head, neck and upper limbs

Lungs

LA

LV

Trunk and lower limbs

Fig. 11.6. The circulation of the blood.

than is required to pump the blood through the pulmonary circulation, thus the left atrium is slightly thicker-walled than the right atrium and the left ventricle is considerably thicker-walled than the right ventricle.

The heart lies obliquely in the middle mediastinum, (p. 70) more to the left than to the right, and it is roughly cone-shaped, having an apex, a base and four surfaces.

The *apex of the heart* is formed by the left ventricle and points forwards, downwards and to the left, being overlapped by the left lung and pleura.

The *base of the heart* faces upwards, backwards and to the right and is mainly formed by the left atrium. It is separated from the thoracic vertebrae by the oesophagus and the descending thoracic aorta.

The *anterior*, or *sternocostal*, surface of the heart is formed by the anterior part of the right atrium, the right ventricle and part of the left ventricle and it lies behind the sternum and the costal cartilages of the 3rd to 6th ribs. The line of division between the right and left ventricles is marked by the *interventricular groove* and running in this groove is a branch of the left coronary artery.

The *diaphragmatic surface of the heart* is formed by the left ventricle and a small part of the right ventricle and it lies mainly on the central tendon of the diaphragm.

The *left surface of the heart* is formed by the left ventricle and a small part of the left atrium and it faces upwards, backwards and to the left.

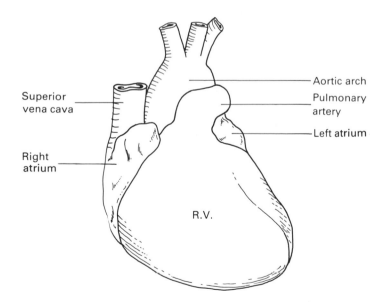

Fig. 11.7. The heart, anterior view.

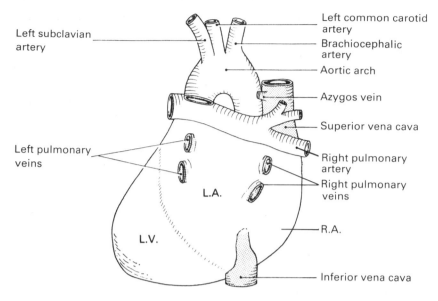

Left subclavian artery

Left common carotid artery

Brachiocephalic artery

Aortic arch

Azygos vein

Superior vena cava

Left pulmonary veins

Right pulmonary artery

Right pulmonary veins

L.A.

R.A.

L.V.

Inferior vena cava

Fig. 11.8. The heart, posterior view.

The *right surface of the heart* is formed by the right atrium and it is almost vertical.

The chambers of the heart

The right atrium forms the right surface of the heart, lying in front of, and to the right of, the left atrium. The cavities of the right and left atria are separated by the *interatrial septum*, a thin muscular partition on the lower part of which there is a shallow depression called the *fossa ovalis*. This marks the site of a communication between the two atria which is patent during fetal life. Projecting from the upper anterior surface of the right atrium is a small conical pouch called the *right auricle*. The *superior vena cava* opens into the upper posterior part of the right atrium and the *inferior vena cava* opens into the lower posterior part of the right atrium. Also opening into the right atrium close to the site of entry of the inferior vena cava is the *coronary sinus*, through which the venous blood from the heart muscle is returned to the heart. The orifice of the *triscuspid valve* lies in the lower anterior part of the right atrium.

   The right ventricle forms most of the sternocostal surface and part of the diaphragmatic surface of the heart, and extends from the anterior border of the right atrium almost to the apex of the heart. The cavities of the right and left ventricles are separated by the *interventricular septum* which forms the posterior wall of the right ventricle. The cavity of the right ventricle is almost crescentic in cross-section because the interventricular septum bulges into it due to the greater thickness of the walls of the left ventricle. The *tricuspid valve* opens into the right side of the base of the right ventricle. The valve consists of three triangular cusps, or flaps, which close under pressure when the right ventricle contracts and they thus prevent the backflow of blood into the right atrium. Protruding from the walls of the right ventricle are small muscular bundles, called the *papillary muscles* and these are attached to the margins of the cusps of the tricuspid valve by tendinous threads called the *chordae tendinae*. During right ventricular contraction the papillary muscles also contract and thus prevent the cusps everting under pressure. The *pulmonary artery* arises from the upper anterior part of the right ventricle and the smooth-walled area just below its orifice is called the *infundibulum*. Backflow of blood from the pulmonary artery into the right ventricle is prevented by the *pulmonary valve* which is located in the base of the pulmonary artery and is composed of three semilunar cusps.

The left atrium forms most of the base of the heart and it lies behind and to the left of the right atrium. The walls of the left atrium are rather

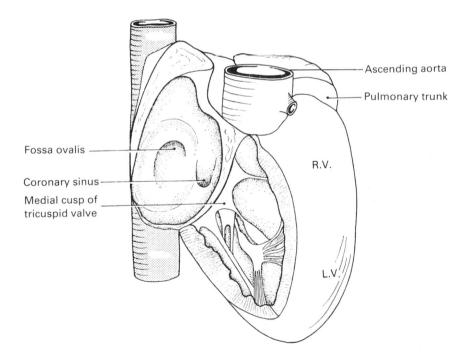

Fossa ovalis

Coronary sinus

Medial cusp of
tricuspid valve

Ascending aorta

Pulmonary trunk

R.V.

L.V.

Fig. 11.9. The interior of the right atrium and right ventricle.

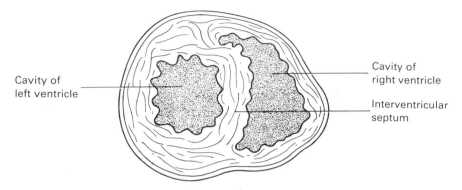

Cavity of
left ventricle

Cavity of
right ventricle

Interventricular
septum

Fig. 11.10. Transverse section through the ventricles as seen from above.

thicker than those of the right atrium. Projecting upwards from the upper left corner of the left atrium and curving forwards to lie on the left side of the base of the pulmonary artery is a small conical pouch called the *left auricle*. Four *pulmonary veins*, two on each side, open into the upper posterior part of the left atrium. In the lower anterior part of the left atrium is the orifice of the *mitral valve*.

The left ventricle forms the left surface of the heart, most of the diaphragmatic surface of the heart and part of the sternocostal surface of the heart. It is considerably thicker-walled than the right ventricle from which it is separated by the *interventricular septum*. The *mitral valve* lies in the upper posterior part of the left ventricle and is composed of two triangular cusps. As with the right ventricle, *papillary muscles* protrude from the walls of the left ventricle and are attached to the margins of the cusps by *chordae tendinae* which are thicker and stronger than those of the tricuspid valve. The *ascending aorta* arises from the left ventricle in front of, and to the right of, the *mitral valve*. Backflow of blood into the left ventricle is prevented by the *aortic valve* which is composed of three semilunar cusps.

Surface anatomy of the heart

The apex of the heart lies in the 5th intercostal space about 9 cm to the left of the midline. A line running vertically downwards from the middle of the clavicle crosses the apex of the heart.

The right border of the heart describes a gentle curve about 2.5 cm to the right of the right border of the sternum from the 3rd to the 6th costal cartilages.

The lower border of the heart extends from the lower border of the right atrium, at the level of the 6th right costal cartilage, to the apex.

The left border of the heart extends from the apex upwards and to

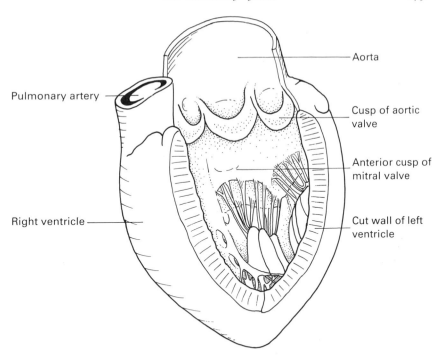

Aorta

Pulmonary artery

Cusp of aortic valve

Anterior cusp of mitral valve

Right ventricle

Cut wall of left ventricle

Fig. 11.11. The interior of the left ventricle.

the right to end at the lower border of the 2nd left costal cartilage about 1 cm from the left sternal edge.

The tricuspid valve lies sloping downwards and to the right at about the midline of the sternum opposite the 4th intercostal space.

The mitral valve slopes slightly more obliquely downwards and to the right than the tricuspid valve, and lies behind the left half of the sternum opposite the 4th costal cartilage.

The pulmonary valve lies horizontally behind the upper border of the junction of the 3rd left costal cartilage and the sternum.

The aortic valve lies below and to the right of the pulmonary valve at the level of the 3rd intercostal space.

The structure of the heart

The outer surface of the heart is enclosed in a sac called the *pericardium*. The pericardium is actually composed of two sacs, an outer sac called the *fibrous pericardium* which consists of fibrous tissue, and an inner sac called the *serous pericardium* which is composed of a serous membrane.

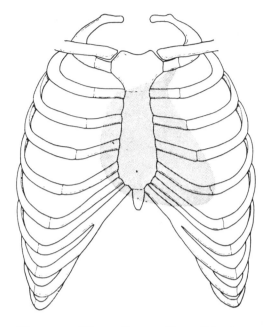

Fig. 11.12. The surface markings of the heart.

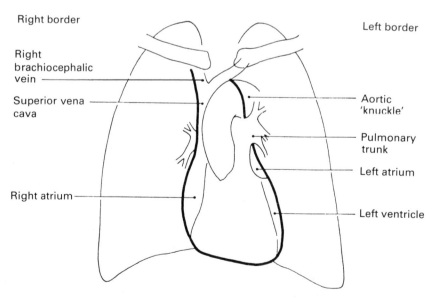

Right border

Left border

Right
brachiocephalic
vein

Superior vena
cava

Aortic
'knuckle'

Pulmonary
trunk

Left atrium

Right atrium

Left ventricle

Fig. 11.13. A tracing of a PA radiograph of the chest.

The fibrous pericardium is roughly cone-shaped. Its apex is attached to the outer coats of the great vessels and its base is attached to the central tendon of the diaphragm.

The serous pericardium is a closed sac which lines the fibrous pericardium and covers the surface of the heart. The layer lining the fibrous pericardium is called the parietal pericardium and the layer lining the surface of the heart is called the visceral pericardium. Between the two layers is a very thin film of fluid, which allows easy movement between the heart and the fibrous sac during contraction of the heart muscle.

The muscle of the heart is called the *myocardium*. It is composed of cardiac muscle fibres, the structure of which has already been described (p. 21).

The inner layer of the heart is called the *endocardium*. The endocardium is a layer of flat endothelial cells with an underlying layer of connective tissue and yellow elastic fibres. The endocardium is continuous with the tunica intima of the great vessels which enter and leave the heart.

The blood supply of the heart

The heart muscle is supplied with arterial blood by the right and left coronary arteries, which arise from the aorta just above the aortic valve.

The coronary veins drain the blood from the heart muscle and they mostly drain into the coronary sinus, which opens into the right atrium.

The nerve supply to the heart

The heart is supplied by branches of the vagus nerve and by branches from the sympathetic nervous system.

The functions of the heart and circulation

The function of the cardiovascular system is to maintain an adequate supply of blood to all the tissues appropriate to their metabolic needs. Arterial blood carries oxygen, glucose and other nutrients to organs and tissues. Venous blood transports the 'waste' products of tissue metabolism such as carbon dioxide and lactic acid away from these tissues for eventual excretion or metabolism.

The blood supply to vital organs, such as the brain and kidneys, is particularly favoured, even in the face of a diminished cardiac output such as may occur after blood loss. The blood pressure in the arteries is in excess of atmospheric pressure which ensures that organs are not deprived of blood because of gravity. This is particularly important in the case of the brain.

The flow of blood to exercising muscle is increased. This is important in ensuring an adequate supply of nutrients and also adequate removal of the products of metabolism. The circulation of blood through the skin is important in the regulation of temperature. Constriction of skin blood vessels decreases heat loss via this route, whereas an increase in skin blood flow enhances loss of blood heat via the skin.

The supply of blood to the arterial tree is maintained by rhythmic contractions of the heart muscle, which occur at a rate of about 72/min in the average healthy adult. In the resting state, the output of the right and left ventricles over a 70 year life span exceeds 400 million litres.

Atrial contraction occurs first, followed by ventricular contraction and there is then a short period of time when both chambers relax and fill with blood. This series of events is known as the *cardiac cycle*.

The following events occur in the cardiac cycle:
1 At first both the atria and ventricles are in a relaxed state and are filled by the venous return. The atrio-ventricular valves are open, and the atria and ventricles are in direct communication with each other.
2 The atria contract and empty blood into the ventricles through the atrio-ventricular (mitral and tricuspid) valves. This phase is known as *atrial systole* and lasts about 0.1 s.
3 As atrial contraction ceases the ventricles begin to contract. When the pressure in the ventricles rises above that in the atria, the mitral and tricuspid valves close and pressure rises further within the ventricles. When the pressure in the ventricles exceeds that in the aorta and pulmonary artery the aortic and pulmonary valves open, and blood is pumped into these two great vessels. This phase is known as *ventricular systole* and lasts about 0.3 s.
4 The ventricular muscle then relaxes and the aortic and pulmonary valves close as pressure in the ventricles falls below that in the great vessels. This prevents backflow of blood from the aorta and pulmonary artery into the ventricles. The atrio-ventricular valves open and the cycle is repeated, as blood from the superior and inferior venae cavae flows into the right atrium and blood from the pulmonary veins flows into the left atrium. *Ventricular diastole* lasts about 0.5 s in a total cardiac cycle of 0.8 s at a heart rate of 72/min.

Cardiac output

Each ventricular systole ejects about 70 ml of blood into the aorta. This is known as the stroke volume. At a normal resting heart rate of around 70/min the cardiac output is approximately 5 l/min. The ventricles do not empty completely of blood, about 50 ml remaining within the ventricular

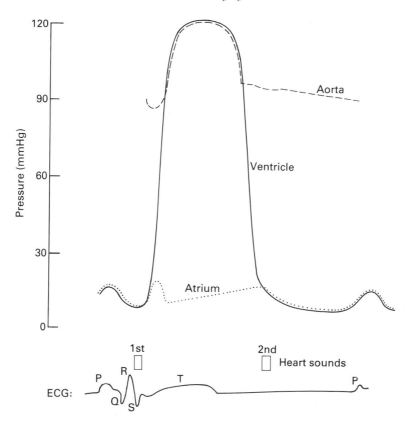

Fig. 11.14. The cardiac cycle: the relationships between the pressures in the left atrium, the left ventricle and the aorta as well as the heart sounds and the ECG.

cavity at the end of ventricular systole. Thus the total ventricular volume during diastole is about 120 ml.

During exercise, both the stroke volume and the heart rate rise to provide an increased cardiac output of blood to meet the increased metabolic demand of contracting muscle.

Heart sounds and murmurs

If a stethoscope is placed on the chest wall, two sounds, called the heart sounds, can be heard.

The first sound is due to closure of the mitral and tricuspid valves and the second sound is due to closure of the pulmonary and aortic valves. The closure of the two valves, which cause each of the heart sounds, is usually synchronous. A delay in closure of one of the valves causes the sound to be split—that is each component of the heart sound is heard separately. This usually indicates disease of the heart, as do the presence of the third and fourth heart sounds. The third heart sound is caused by vibrations in the cardiac muscle during diastole and the fourth heart sound is caused by abnormally forceful atrial systole.

Cardiac murmurs are noises which are heard when abnormal turbulence occurs within the heart. This may be due to increased blood flow through the heart and may not necessarily indicate heart disease. However, valve narrowing, or failure of closure of a valve, due to disease, causes murmurs at various time points during the cardiac cycle. The timing and the character of the murmurs varies according to the valve affected and the type of disturbance of valve function.

The conducting system of the heart

The conducting system of the heart consists of a network of specialized conducting fibres which coordinate atrial and ventricular contraction.

The impulse for contraction originates in the *sinu-atrial node*, which is a group of specialized cells situated close to the junction of the right atrium and the superior vena cava. It has a rich blood supply and receives fibres from the sympathetic and vagal parasympathetic nervous systems.

A wave of contraction spreads from the sinu-atrial node through the walls of the atria to stimulate the *atrio-ventricular node*. This node is a group of cells in the interatrial septum near the mouth of the coronary sinus. The atrio-ventricular node is also supplied by a branch of the vagus nerve and has a structure identical to that of the sinu-atrial node. When stimulated the atrio-ventricular node initiates an impulse in the *atrio-ventricular bundle* (also known as the bundle of His).

The atrio-ventricular bundle is a bundle of specialized fibres which passes down the interventricular septum, dividing into left and right branches, which conduct the impulses to the respective ventricles. The left and right main branches break up into smaller branches to the ventricular muscle.

The control of heart rate

The sinu-atrial node initiates the impulse which causes first the atria and then the ventricles to contract. The heart has an inherent rhythm of contraction, the rate varying from one part of the heart to another. Since the sinu-atrial node normally has the fastest intrinsic rate it is the so-called *pacemaker*.

The rate of impulses generated by the sinu-atrial node can be altered

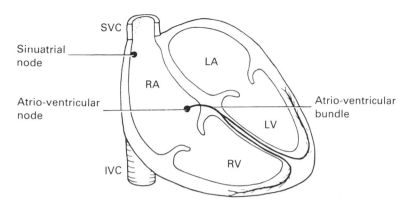

Fig. 11.15. Diagrammatic representation of the conducting system of the heart.

by nerve impulses. For example vagal nerve impulses slow the heart and decrease its force of muscular contraction, whereas impulses from the sympathetic nervous system accelerate the heart rate and increase the force of contraction. If the sinu-atrial node is greatly slowed, a lower part in the conducting pathway may take over the pacemaking role. This may be the atrio-ventricular node, in which case the resting heart rate is about 50/min instead of 70/min.

In the condition known as *heart block*, conduction of the impulse between the atria and the ventricles is impaired. This condition is caused by disease of the atrio-ventricular bundle. Sometimes the effects are mild and there may be delay in the passage of the impulse from the atria to the ventricles, or some atrial contractions will not be followed by a ventricular contraction. Occasionally passage of the impulse is completely blocked and the ventricles contract at their own rate, which is 30–40/min.

The pulse

Every time the left ventricle contracts, blood is forced out into the aorta, and blood from the arterial tree is forced into the capillaries. The arteries, however, have elastic walls which relax with each increase in pressure caused by left ventricular systole, and recoil as the pressure within the artery falls during ventricular diastole. Thus each heart beat can be felt in a peripheral artery as an expansion, followed by a contraction, of the vessel wall, and this is called *the pulse*. By feeling the pulse it is possible to count the heart rate, to notice irregularities in the heart rhythm, to form some idea of the pressure within the artery, and also to note the characteristics of the vessel wall, which may be harder than normal in patients with atherosclerosis.

The electrocardiograph

Each impulse generated by the sinu-atrial node and passing through the conducting system of the heart causes a series of electrical changes in the heart muscle which precede the associated mechanical events. These electrical changes can be recorded by placing electrodes on the surface of the body. By varying the position of the electrodes, information about the passage of the impulse through various parts of the heart can be obtained. The P wave represents the passage of the impulse through the atria; the QRS complex, the impulse passing through the ventricular muscle; and the T wave, ventricular recovery.

Electrocardiography plays an important part in the investigation of the heart. It gives information about the site of the pacemaker and is useful in diagnosing disorders of cardiac rhythm. For example, in the condition known as atrial fibrillation the atria beat rapidly, ineffectively and in a disorganized fashion. The electrocardiograph, instead of showing P waves shows a large number of irregular, lower amplitude fibrillation waves. In atrio-ventricular conduction disorders the interval between the atrial and ventricular impulses on the tracing is increased.

The electrocardiograph also gives information about the size and health of ventricular muscle.

The blood pressure

The blood pressure is the pressure exerted by the blood on the walls of the arteries. The pressure is higher during left ventricular systole, around 120 mm Hg (millimetres of mercury) in healthy young adults. This is known as the *systolic blood pressure*. When left ventricular systole ceases the aortic valve closes and the elastic arterial walls recoil. Although the blood pressure falls, it does not reach zero because left ventricular systole occurs again. Thus with a normal heart rate, there is a constantly varying pressure within the arteries. The lowest blood pressure is usually about 80 mm Hg. This is known as the *diastolic blood pressure.*

Blood pressure is measured with an instrument known as a *sphygomano-meter.* This consists of an inflatable bladder which is wrapped around the upper arm so that the bladder lies over the brachial artery. The bladder is inflated with air until the brachial pulse at the elbow, or the radial pulse at the wrist, can no longer be felt. The pressure required to obliterate either arterial pulse is the approximate systolic blood pressure. The systolic and diastolic blood pressures are then recorded more accurately by listening over the brachial artery at the elbow with a stethoscope.

The blood pressure depends on *cardiac output* and *peripheral resistance.*

The cardiac output is stroke volume × heart rate, the stroke volume being the volume of blood ejected into the aorta per ventricular systole. Cardiac output also depends upon venous return and blood volume. In order to maintain cardiac output sufficient blood must return to the right side of the heart from the periphery.

The return of blood depends upon:

1 *Gravity*. When sitting or standing, blood returns to the heart from the head and neck by gravity.

2 *Muscular contraction*. Contraction of muscles causes pressure on the veins and forces blood towards the heart. This is particularly important in causing venous return from the legs when standing. The valves in the veins prevent backflow of blood.

3 *Intrathoracic pressure*. During inspiration, movements of the ribs and diaphragm tend to create a negative pressure within the thorax to draw air into the lungs. This negative pressure also draws blood into the thorax, towards the heart, from the inferior and superior venae cavae. Furthermore, during inspiration intra-abdominal pressure rises due to descent of the diaphragm, and this causes the veins of the abdominal cavity to empty, favouring venous return to the thorax.

4 *Transmitted pressure*. Transmitted pressure from the arterial tree, via the capillaries, tends to force blood onward in the venous system toward the right heart.

Peripheral resistance. The peripheral resistance depends mainly upon the smooth muscle of the arterioles and the pre-capillary sphincters. Contraction of the arteriolar wall smooth muscle narrows the lumen of the arteriole which decreases the amount of blood reaching the capillaries of that organ or tissue, slows the onward passage of blood and increases the pressure on the arterial side of the circulation. Stimulation of the sympathetic nervous system causes arteriolar constriction and a rise in blood pressure. The circulation to the skin and viscera is particularly affected by this vasoconstriction, being decreased by both adrenaline and noradrenaline. However, the circulation to skeletal muscle is decreased only by noradrenaline and is actually increased by adrenaline. The increase in skeletal muscle blood flow is one of the many aspects of the 'flight or fight' reaction. Angiotensin II (see Chapter oo) is also a powerful constrictor of small blood vessels and has an important role in maintaining blood pressure in potentially hypotensive conditions.

Blood volume

Blood volume, which is about 75−80 ml/Kg or 5 l in an average adult, is kept remarkably constant in health. Fluid is lost from the body in urine, sweat and faeces and there is insensible loss through the lungs as water vapour, and the skin. Fluid intake, which balances these losses, is controlled by thirst. In addition, increased loss via one route is usually offset by decreased loss via another. For example, excessive sweating, caused by heat exposure, is associated with a marked decrease in urine output.

Water depletion is prevented by the secretion of *antidiuretic hormone* (ADH) by the posterior pituitary gland. ADH reduces fluid loss by the kidneys by causing the excretion of a concentrated urine. (see Chapters 13 and 17). Conservation of sodium by the body is also important in the maintenance of blood volume. The *renin—angiotension—aldosterone* system (Chapter 13) is the means by which the body regulates sodium loss by the kidney. The balances of water and sodium are closely linked. The secretion of ADH is stimulated by a rise in the osmolality of the blood and this is sensed by osmoreceptors in the hypothalamus. The renin—angiotensin—aldosterone system is activated by a fall in the pressure of blood perfusing the kidney or a fall in the sodium concentration of the blood (*hyponatraemia*).

Blood volume is increased in disease states which are associated with an increased secretion of aldosterone. These include congestive cardiac failure, cirrhosis of the liver and the nephrotic syndrome, a renal disorder in which there is a massive loss of albumin in the urine with a consequent fall in plasma osmolality. All of thses conditions are characterized by swelling of the extremities (*oedema* or *dropsy*).

Thirst is caused by dehydration either of the extracellular fluid or of the intracellular fluid. This *true thirst* must be distinguished from the thirst which is not due to dehydration, but is solely due to local factors which cause dryness of the tongue such as smoking, talking and eating 'hot' curries.

Blood flow through tissues and organs is dependent upon total blood volume, arterial blood pressure, the state of arteriolar vasoconstriction within the tissue or organ and blood viscosity. When the total blood volume, or arterial blood pressure, falls, the blood flow to certain organs, such as the brain and the kidneys (which have no glycogen stores), is favoured at the expense of the skin and viscera. Vasoconstriction reduces the blood flow to tissues but there is evidence that most organs can auto-regulate their own blood flow, i.e. maintain a stable perfusion when blood pressure is varying. The mechanism for auto-regulation is probably the production of chemical substances, caused by tissue metabolism, which lead to dilatation of arterioles by direct stimulation.

Viscosity refers to that property of a fluid which describes its stickiness. This is affected by the *haematocrit* (Packed Cell Volume). When this rises above 45% the blood viscosity is significantly increased. This predisposes to thrombosis, or clotting of the blood within the vessels, particularly when blood flow is slow.

The pulmonary circulation

The main pulmonary artery

The main pulmonary artery arises from the right ventricle and it is about 5 cm in length. It runs upwards and backwards, lying at first in front of the ascending aorta and then on its left side. Beneath the arch of the aorta it divides into the *right* and *left pulmonary arteries*.

The right pulmonary artery

The right pulmonary artery, which is slightly longer than the left, runs horizontally to the right behind the ascending aorta and the superior vena cava. The distal portion is crossed by the upper right pulmonary vein and behind it lie the oesophagus and the right main bronchus. At the hilum of the right lung the right pulmonary artery divides into an *upper branch* which supplies the upper lobe and a *lower branch* which supplies the middle and lower lobes.

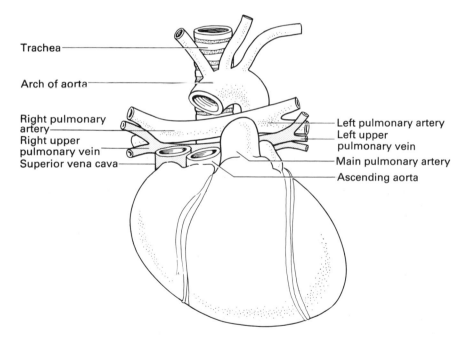

Fig. 11.16. View of the pulmonary arteries from the front. The ascending aorta and the superior vena cava have been removed.

The left pulmonary artery

The left pulmonary artery, which is shorter than the right, runs horizontally to the left in front of the left main bronchus and the descending thoracic aorta. At the hilum of the left lung it divides into an *upper branch* which supplies the upper lobe and a *lower branch* which supplies the lower lobe.

The pulmonary veins

There are four pulmonary veins, an upper and a lower from each lung.

The upper right pulmonary vein is formed by two branches, one from the upper lobe of the right lung and one from the middle lobe of the right lung. It runs horizontally behind the superior vena cava to enter the left atrium. **The lower right pulmonary** vein drains the lower lobe of the right lung and it runs behind the right atrium to enter the left atrium.

The left upper and **lower pulmonary veins** drain the upper and lower lobes of the left lung respectively. They run horizontally in front of the descending aorta to enter the left atrium.

The systemic circulation

The aorta

The aorta commences at the upper border of the left ventricle. It runs upwards and to the right for a short distance and then turns through 180° to run downwards close to the vertebral column to end at the level of the 4th lumbar vertebra by dividing into the right and left common iliac arteries.

The aorta is divided into the following parts:

> *the ascending aorta*
> *the arch of the aorta*   ⎫
> *the descending thoracic aorta* ⎬   *the thoracic aorta*
> *the abdominal aorta*   ⎭

The ascending aorta

The ascending aorta is about 5 cm in length. It starts at the upper border of the left ventricle at the level of the lower border of the 3rd left costal cartilage behind the left half of the sternum. It runs upwards, forwards and to the right and ends at the level of the upper border of the 2nd right costal cartilage, where it is continuous with the arch of the aorta. It gives off two branches: the right and left coronary arteries. These arteries arise from the ascending aorta just above the aortic valve and supply the muscle of the heart.

The right coronary artery runs downwards in the groove between the right atrium and the right ventricle and then runs across the base of the heart to communicate with the left coronary artery.

The left coronary artery gives off a branch, called the interventricular branch, which runs downwards in the groove between the two ventricles on the sternocostal surface of the heart, and then runs on to the posterior surface of the heart and communicates with the right coronary artery.

The arch of the aorta

The arch of the aorta connects the ascending and the descending thoracic aorta. It starts as a direct continuation of the ascending aorta and arches to the left and backwards across the superior mediastinum, in front of the trachea and oesophagus, to become continuous with the descending thoracic aorta on the left side of the lower border of the 4th thoracic vertebra. The lower surface of the arch of the aorta lies on the upper surface of the

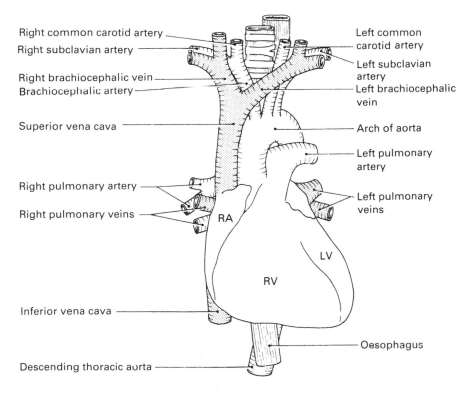

Right common carotid artery

Right subclavian artery

Right brachiocephalic vein

Brachiocephalic artery

Superior vena cava

Right pulmonary artery

Right pulmonary veins

RA

Inferior vena cava

Descending thoracic aorta

Left common carotid artery

Left subclavian artery

Left brachiocephalic vein

Arch of aorta

Left pulmonary artery

Left pulmonary veins

LV

RV

Oesophagus

Fig. 11.17. The heart and the great vessels.

division of the main pulmonary artery into the right and left pulmonary arteries.

Three branches arise from the arch of the aorta:

The brachiocephalic (innominate) artery is the first branch of the arch of the aorta. It runs upwards, backwards and to the right for 4−5 cm and ends behind the right sternoclavicular joint by dividing into the right common carotid artery and the right subclavian artery.

The left common carotid artery is the second branch of the arch of the aorta. It arises from the arch in front of the trachea, and then passes upwards and to the left of the trachea to reach the base of the neck.

The left subclavian artery is the third branch of the arch of the aorta. It arises behind, and slightly to the left of, the left common carotid artery and runs upwards to the base of the neck.

The arteries of the head and neck

The common carotid arteries are the main arteries of supply to the head and neck.

The right common carotid artery arises from the brachiocephalic artery.

The left common carotid artery arises directly from the arch of the aorta.

The course of the common carotid arteries from the base of the neck is the same. Both pass obliquely upwards in the neck, one on either side of the oesophagus and trachea. The common carotid artery is enclosed in a sheet of fascia together with the internal jugular vein, which lies lateral to the artery, and the vagus nerve which lies behind the artery and vein. At

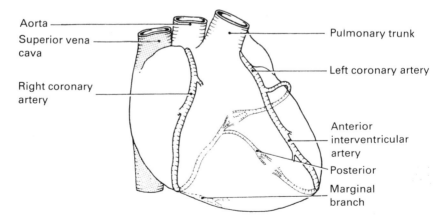

Fig. 11.18. The coronary arteries.

the level of the upper border of the thyroid cartilage the common carotid artery divides into two branches: the *external carotid artery*, and the *internal carotid artery.*

The external carotid artery runs upwards in the line of the common carotid artery to end in the parotid gland, behind the neck of the mandible, by dividing into the superficial temporal artery and the maxillary artery.

The external carotid arteries supply the superficial tissues of the head and neck. Each has the following branches:

The superior thyroid artery arises from the external carotid artery just below the tip of the hyoid bone, and supplies the apex of the lobe of the thyroid and other adjacent structures.

The ascending pharyngeal artery is the smallest branch of the external carotid artery. It passes to the pharynx and also has several small branches which supply part of the middle ear.

The lingual artery arises from the external carotid artery in the neck, and passes upwards to supply the tongue and structures of the floor of the mouth.

The facial artery arises from the external carotid artery in the neck, curves round the lower border of the mandible, and supplies the face.

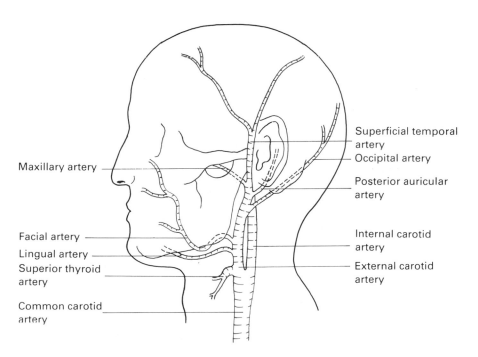

Fig. 11.19. The superficial arteries of the head.

The occipital artery passes backwards from the external carotid artery to supply the posterior part of the scalp.

The posterior auricular artery is a small branch which runs upwards behind the ear and gives branches to the side of the scalp and the ear.

The superficial temporal artery is the smaller of the two terminal branches of the external carotid artery. It runs upwards from the substance of the parotid gland, in front of the ear, and supplies the upper part of the face and the side of the scalp.

The maxillary artery is the larger of the two terminal branches of the external carotid artery. It runs forwards behind the mandible, enters the pterygo-palatine fossa, passes into the orbit and finally reappears on the face through the infraorbital foramen as the *infraorbital artery*. It supplies the upper and lower jaws, the muscles of mastication, the palate and nose, and part of the cranial dura mater. The cranial dura mater is supplied by a branch of the maxillary artery called the *middle meningeal artery* which enters the cranial cavity through the foramen spinosum.

The internal carotid artery runs upwards in the neck close to the external carotid artery and enters the carotid canal in the base of the skull. It gives off no branches in the neck. Its course within the cranial cavity will be described on p. 301.

The arteries of the upper limb

The main arteries of supply to the upper limb are the subclavian arteries.

The subclavian arteries

The right subclavian artery arises from the brachiocephalic artery.

The left subclavian artery arises from the arch of the aorta.

From the root of the neck the course of the two arteries is similar. Each curves laterally over the first rib and is continuous with the axillary artery at the outer border of the first rib. The subclavian artery has four branches:

The vertebral artery runs upwards through the transverse foramina of the 6th to the 1st cervical vertebrae and enters the cranial cavity through the foramen magnum. Its distribution in the cranial cavity will be described on p. 301.

The internal thoracic (mammary) artery runs downwards into the thoracic cavity and lies on the internal surface of the upper costal cartilages just lateral to the sternum. It supplies several structures in the thoracic cavity and anastomoses with the upper intercostal arteries.

(An *anastomosis* is a communication between different arteries or branches of arteries. These communications can be important as an alternative source of blood supply if one of the main supplying arteries to a region becomes blocked.)

The thyrocervical trunk supplies the lower part of the thyroid gland by a branch called the inferior thyroid artery. The other branches are distributed to the muscles of the neck and the scapular region, and the lower part of the pharynx.

The costocervical trunk supplies the muscles of the upper two intercostal spaces and the deep muscles of the neck.

The axillary artery is the direct continuation of the subclavian artery at the outer border of the first rib. It runs downwards along the lateral wall of the axilla close to the medial side of the upper end of the humerus. At the lower margin of the axilla it is directly continuous with the brachial artery. The axillary artery gives branches to the muscles in the shoulder region, to the shoulder joint and it forms anastomoses with other arteries in the region.

The brachial artery is the direct continuation of the axillary artery at the lower border of the axilla. It runs down the upper arm, at first on the medial side of the humerus and then in front of the humerus. It divides in front of the elbow joint into the radial artery and the ulnar artery. The brachial artery has branches which supply the muscles of the upper arm.

The radial artery is the smaller of the two terminal branches of the brachial artery. It passes down the lateral side of the forearm to the wrist. The radial artery is usually felt at the wrist as the pulse. At the wrist it curves backwards to the dorsal surface of the space between the 1st and 2nd metacarpal bones and then runs forwards through this space to complete the deep palmar arch by uniting with a branch of the ulnar artery. At the wrist it gives off a branch which completes the superficial palmar arch by uniting with the ulnar artery.

The ulnar artery is the larger of the two terminal branches of the brachial artery. It runs downwards on the medial side of the forearm to the wrist and then passes on to the palm of the hand. It gives off a branch which completes the deep palmar arch by uniting with the radial artery and then curves laterally across the palm of the hand to form the superficial palmar arch by uniting with a branch of the radial artery.

The palmar arches. The superficial palmar arch is formed by the ulnar artery and a branch of the radial artery.

The deep palmar arch is formed by the radial artery and a branch of the ulnar artery.

The descending thoracic aorta

The descending thoracic aorta commences at the left side of the lower border of the 4th thoracic vertebra and runs downwards in the posterior mediastinum to enter the abdominal cavity through the aortic opening of the diaphragm at the level of the 12th thoracic vertebra.

The descending thoracic aorta is related posteriorly to the thoracic

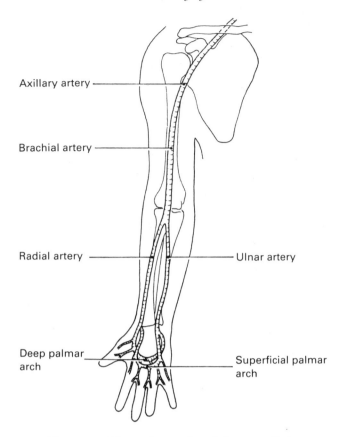

Axillary artery

Brachial artery

Radial artery

Ulnar artery

Deep palmar arch

Superficial palmar arch

Fig. 11.20. The main arteries of the upper limb.

vertebrae, anteriorly to the hilum of the left lung above, and to the heart and pericardium below. The oesophagus initially lies on its right side but in the lower part of the thorax it lies anteriorly. The left side of the descending thoracic aorta is related to the left lung and pleura. The right side of the descending thoracic aorta is related to the thoracic duct, in its upper part to the oesophagus and in its lower part to the right lung and pleura. The descending thoracic aorta has the following branches:

The intercostal arteries, nine on each side of the descending aorta, run laterally to supply the structures of the lower nine intercostal spaces.

Branches arise from the descending thoracic aorta to supply the bronchi and lungs, the pericardium, the lymph nodes of the mediastinum, the oesophagus and the diaphragm.

The abdominal aorta

The abdominal aorta is the direct continuation of the descending thoracic aorta. It begins at the level of the lower border of the 12th thoracic vertebra and runs downwards on the left side of the anterior surface of the lumbar vertebrae to end at the level of the 4th lumbar vertebra, where it divides into the left and right common iliac arteries. The inferior vena cava lies on the right side of the abdominal aorta, and the pancreas, the duodenum and the peritoneum lie in front of it. The branches of the abdominal aorta are:

The inferior phrenic arteries which arise from the sides of the upper part of the abdominal aorta and supply the diaphragm.

The coeliac axis is a wide artery which arises from the front of the abdominal aorta just below the diaphragm. It runs horizontally forwards and to the right for about 1.5 cm and then divides into three branches:

The hepatic artery runs to the right to supply the liver. It gives off right and left branches to the right and left lobes of the liver and also branches to the stomach, the gall bladder, the duodenum and the pancreas.

The left gastric artery is the smallest branch of the coeliac axis. It runs upwards and to the left to supply the cardiac end of the stomach. It also gives three small branches to the lower end of the oesophagus.

The splenic artery is the largest branch of the coeliac axis. It passes horizontally to the left to supply the spleen. It also gives small branches to the stomach and pancreas.

The superior mesenteric artery arises from the front of the abdominal aorta about 1 cm below the coeliac axis. It passes into the mesentery and supplies the whole of the small intestine, apart from the upper part of the duodenum, and the large bowel round to the distal third of the transverse colon.

The right and left suprarenal arteries arise from the side of the abdominal aorta at the level of the superior mesenteric artery and pass laterally to supply the right and left suprarenal glands.

The right and left renal arteries arise from the side of the abdominal aorta and pass laterally to supply the right and left kidneys. Since the abdominal aorta lies on the left side of the anterior surface of the lumbar vertebrae, the right renal artery is longer than the left, and it has to pass behind the inferior vena cava to reach the right kidney.

The testicular arteries (the **ovarian arteries** in the female) arise from the front of the abdominal aorta just below the renal arteries and pass downwards on the posterior abdominal wall into the pelvis to supply the testes (the ovaries in the female).

The inferior mesenteric artery arises from the front of the abdominal aorta about 3—4 cm above its division into the right and left common iliac

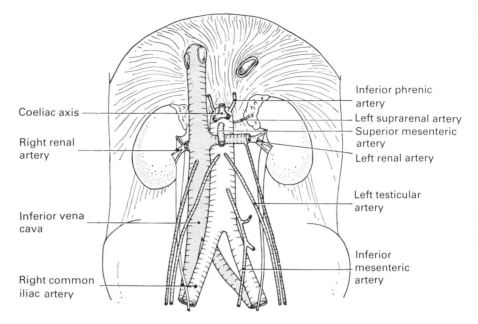

Coeliac axis

Right renal
artery

Inferior vena
cava

Right common
iliac artery

Inferior phrenic
artery

Left suprarenal artery

Superior mesenteric
artery

Left renal artery

Left testicular
artery

Inferior
mesenteric
artery

Fig. 11.21. The abdominal aorta and its main branches.

arteries. The inferior mesenteric artery supplies the large bowel from the
distal third of the transverse colon to the rectum.

The lumbar arteries, four on each side, arise from the posterolateral
aspect of the abdominal aorta. They are equivalent to the intercostal
arteries and they supply branches to the muscles of the back and the
anterior abdominal walls.

The common iliac arteries are the terminal branches of the abdominal
aorta. Each artery passes downwards and laterally from the anterior surface
of the 4th lumbar vertebra and divides opposite the upper border of the
sacroiliac joint into the external and the internal iliac arteries.

The internal iliac artery runs medially into the pelvis and supplies the
reproductive organs, the bladder, the anus, the muscles of the pelvic floor,
the muscles of the gluteal region and the hamstring muscles.

The external iliac artery runs downwards along the medial border of the
psoas muscle and enters the thigh beneath the inguinal ligament, midway
between the anterior superior iliac spine and the symphysis pubis. From
this point it is continued as the femoral artery.

The arteries of the lower limb

The femoral artery is the main artery of supply to the lower limb. It
commences as the direct continuation of the external iliac artery, beneath

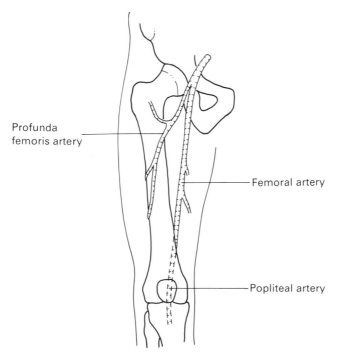

Profunda
femoris artery

Femoral artery

Popliteal artery

Fig. 11.22. The femoral artery.

the inguinal ligament, runs down the front and then the medial side of the thigh, and ends at the junction of the upper two-thirds and the lower third of the thigh, as it passes through a gap in the adductor magnus muscle, to become the popliteal artery. It distributes branches to the skin of the lower part of the abdominal wall and the external genital organs, and to the muscles of the thigh. About 3.5 cm below the inguinal ligament it gives rise to a branch called the *profunda femoris artery* which runs at first lateral to, and then behind, the femoral artery and gives off numerous branches to the muscles of the thigh.

The popliteal artery is the direct continuation of the femoral artery. It runs downwards through the popliteal space behind the knee joint and ends at the lower border of this space by dividing into the anterior and posterior tibial arteries. The popliteal artery gives off branches to the knee joint and branches to the muscles in the region.

The anterior tibial artery arises as one of the terminal branches of the popliteal artery. It runs downwards for a short distance at the back of the leg and then turns forwards, pierces the interosseous membrane between the tibia and fibula, and runs downwards on the anterior surface of this membrane. In the lower part of the leg it comes to lie on the anterior surface of the tibia, and then crosses the ankle joint, midway between the two malleoli, to become the dorsalis pedis artery.

The posterior tibial artery is one of the terminal branches of the popliteal artery. It runs downwards and medially on the posterior aspect of the leg, deep to the gastrocnemius and soleus muscles. It passes behind the medial malleolus and curves forwards to enter the sole of the foot, where it divides into the lateral and medial plantar arteries. The posterior tibial artery gives off a branch called the *peroneal artery* about 2 cm below its origin. The peroneal artery supplies the muscles of the lateral side of the leg.

The dorsalis pedis artery is the direct continuation of the anterior tibial artery. It runs along the medial side of the foot to the base of the 1st metatarsal bone and then dips downwards between the 1st and 2nd metatarsal bones to join the plantar arch. Before leaving the dorsum of the foot it

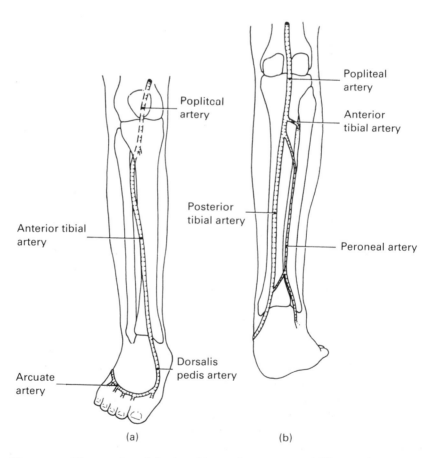

Fig. 11.23. The arteries of the leg: (a) anterior aspect, and (b) posterior aspect.

gives off a branch which supplies the big toe and the medial side of the 2nd toe, and a branch called the *arcuate artery* which curves laterally across the dorsum of the foot, and gives branches to the 2nd to the 5th toes.

The medial plantar artery is one of the terminal branches of the posterior tibial artery. It runs along the medial side of the sole of the foot and gives off a large branch to the big toe and smaller branches to the 2nd, 3rd and 4th toes.

The lateral plantar artery is one of the terminal branches of the posterior tibial artery. It runs obliquely across the sole of the foot to the base of the 5th metatarsal bone and then curves medially across the metatarsal bones to join the dorsalis pedis artery to form the plantar arch. Branches arise from the plantar arch to supply the toes.

The arteries of the brain

Four large arteries enter the cranial cavity to supply the brain. They are:

The right and **left internal carotid arteries** which enter the cranial cavity through the upper part of the foramen lacerum.

The right and **left vertebral arteries** which enter the cranial cavity through the foramen magnum.

The internal carotid arteries supply a large part of the cerebral hemispheres, the eye and its associated structures, and they send branches to the skin of the forehead and the anterior part of the scalp. The *ophthalmic artery* arises from the internal carotid artery close to the anterior clinoid process and runs forwards to enter the orbit, together with the optic nerve, through the optic canal. It supplies the eyeball and other structures in the orbit and gives off the *supraorbital artery* which leaves the orbit through the supraorbital foramen or notch. The internal carotid artery ends by dividing into the *anterior* and *middle cerebral arteries* and the *anterior communicating artery*.

The vertebral arteries enter the cranial cavity through the foramen magnum. Soon after their entrance into the cranial cavity the two vertebral arteries unite in the midline to form the *basilar artery*. The basilar artery passes upwards in the midline on the basal part of the occipital bone and at the upper border of the pons it divides into the two *posterior cerebral arteries*. The posterior cerebral arteries each give off a branch called the *posterior communicating artery* which communicates with the anterior cerebral arteries.

Before uniting, the vertebral arteries give off branches to the spine, the cerebellum and the medulla oblongata.

The basilar artery gives off branches to the cerebellum and the midbrain.

The Circle of Willis. The two internal carotid arteries, the basilar artery and the anterior and posterior communicating arteries form an arterial circle at the base of the brain. This arterial circle is called the *Circle of Willis*.

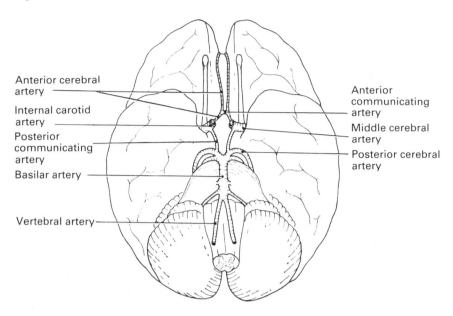

Anterior cerebral
artery

Internal carotid
artery

Posterior
communicating
artery

Basilar artery

Vertebral artery

Anterior
communicating
artery

Middle cerebral
artery

Posterior cerebral
artery

Fig. 11.24. The base of the brain, showing the Circle of Willis.

The following branches arise from the Circle of Willis on each side:

The anterior cerebral artery which passes upwards and backwards in the midline over the corpus callosum (p. 419). It supplies the anterior and medial parts of the cerebral hemisphere.

The middle cerebral artery which is the other terminal branch of the internal carotid. It runs laterally and then upwards in the lateral cerebral sulcus and supplies a large part of the lateral surface of the cerebral hemisphere.

The posterior cerebral artery is one of the terminal branches of the vertebral artery. It runs laterally, gives off the posterior communicating artery and then runs on to the inferior surface of the cerebral hemisphere. It supplies the temporal and occipital lobes of the cerebral hemisphere.

The veins

The veins return the blood from the tissues to the heart. They start as small vessels, or venules, in the periphery and these gradually unite to form large trunks, until finally two main vessels, called the superior and inferior venae cavae open into the right atrium of the heart.

The veins are divided into two main groups:

The superficial veins which run close to the surface of the skin, and are thus frequently visible.

The deep veins many of which accompany the arteries and have the same name as the arteries.

The superior vena cava

The superior vena cava is formed by the junction of the right and left brachiocephalic veins and opens into the superior part of the right atrium. It drains blood from the head and neck, the upper limbs, and most of the thoracic cavity.

The venous return from the head and neck

The superficial veins of the scalp accompany the branches of the external carotid artery which supply the scalp and they receive the same names. They all unite just behind the angle of the mandible to form the external jugular vein.

The external jugular vein runs downwards in front of the sternomastoid muscle to enter the subclavian vein behind the clavicle.

The superficial veins of the face, however, drain into a vein called the **facial vein**. The facial vein starts close to the inner angle of the eye, runs downwards to cross the mandible in front of the angle, and enters the internal jugular vein.

The venous sinuses of the cranial cavity. The venous blood from the brain drains into special channels called the venous sinuses. These are sinuses which are formed between the layers of the dura mater, which is the outer membrane lining the brain. The sinuses are lined by endothelial tissue. The most important venous sinuses are: the *superior sagittal sinus*, the *inferior sagittal sinus*, the *straight sinus*, the *cavernous sinuses*, the *transverse sinuses* and the *sigmoid sinuses*.

The superior sagittal sinus lies in the upper margin of the falx cerebri, a fold of dura mater which lies between the right and left cerebral hemispheres. The superior sagittal sinus commences above the crista galli of the ethmoid bone and passes backwards in the midline of the skull vault to reach the internal occipital protuberance, where it opens into one of the transverse sinuses. It drains the blood from the superior part of the brain.

The inferior sagittal sinus lies in the lower margin of the falx cerebri and runs backwards to drain into the straight sinus.

The straight sinus runs downwards and backwards in the junction of the falx cerebri with the tentorium cerebelli, which is a fold of dura mater which almost completely roofs in the posterior cranial fossa. The straight sinus opens into one of the transverse sinuses. A vein, the great cerebral vein, opens into the anterior end of the straight sinus.

The cavernous sinuses lie on either side of the sella turcica. They

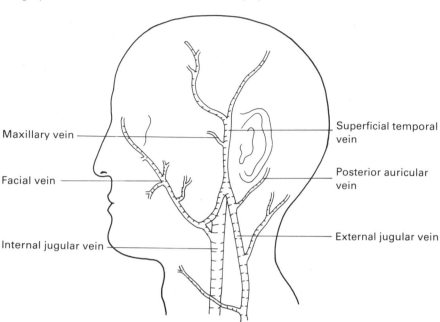

Maxillary vein

Facial vein

Internal jugular vein

Superficial temporal vein

Posterior auricular vein

External jugular vein

Fig. **11.25.** The venous return from the head and neck.

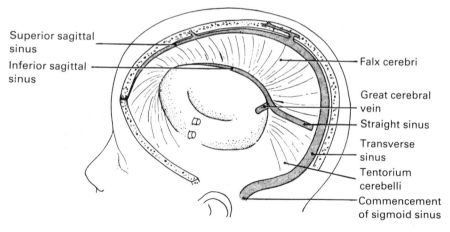

Superior sagittal sinus

Inferior sagittal sinus

Falx cerebri

Great cerebral vein

Straight sinus

Transverse sinus

Tentorium cerebelli

Commencement of sigmoid sinus

Fig. **11.26.** The venous sinuses.

communicate with the sigmoid sinus, the internal jugular vein and the veins of the face by small venous channels.

The transverse sinuses are large sinuses. The right transverse sinus is usually the continuation of the superior sagittal sinus, and the left transverse sinus the continuation of the straight sinus. They commence at the internal occipital protuberance and then run laterally and forwards round the upper margin of the posterior cranial fossa, in the attached margins of the tentorium cerebelli. Near the posterior margin of the lateral ends of the petrous parts of the temporal bones they are continuous with the sigmoid sinuses.

The sigmoid sinuses run downwards and medially, from the ends of the transverse sinuses, to reach the jugular foramen, where they are continuous with the internal jugular vein.

The internal jugular vein runs downwards in the neck from the base of the skull. It passes deep to the sternomastoid muscle and unites with the subclavian vein behind the clavicle to form the brachiocephalic vein. In its course it is joined by the facial vein and small veins from the pharynx and thyroid gland.

The venous return from the upper limb

The veins of the upper limb are divided into a superficial group and a deep group.

The superficial veins of the upper limb commence mainly on the back of the hand. They unite to form three main veins.

The cephalic vein commences on the lateral side of the dorsum of the hand, runs upwards on the dorsal surface of the forearm for a short distance and then winds round the radial side of the forearm and ascends to the elbow. Near the elbow it gives off the *median cubital vein*, which runs upwards and medially to join the basilic vein. The cephalic vein then continues upwards on the lateral side of the biceps brachii muscle and finally crosses medially and passes deep to join the axillary vein.

The basilic vein commences on the medial side of the dorsum of the hand, runs up the dorsal surface of the forearm and then curves forwards, to lie on the anterior surface of the ulnar side of the forearm. It then passes upwards to the elbow, where it is joined by the median cubital vein. The basilic vein then continues upwards on the medial side of the biceps brachii muscle. At about the middle of the upper arm it runs deeply to lie alongside the brachial artery, and is then continuous with the axillary vein at the lower border of the axilla.

The median vein of the forearm drains the blood from the palmar surface of the hand. It runs upwards in the midline of the forearm to end either in the basilic vein, or the cephalic vein.

The deep veins of the arm follow the course of the arteries. They are arranged in pairs, called the venae commitantes, and lie one on each side of the corresponding artery. The venae commitantes of the brachial artery join the axillary vein at the lower border of the axilla.

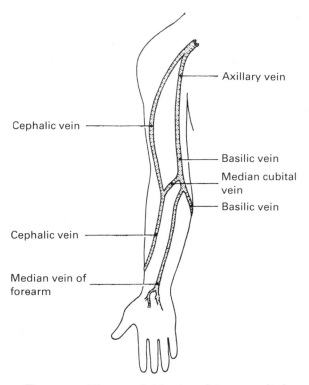

Fig. 11.27. The superficial veins of the upper limb.

The axillary vein accompanies the axillary artery and is continuous with the subclavian vein at the outer border of the first rib.

The subclavian vein accompanies the subclavian artery and unites with the internal jugular vein behind the inner border of the clavicle to form the brachiocephalic vein. In its course it receives the external jugular vein.

The venous return from the thoracic cavity

A large vein, called the **azygos vein**, runs upwards in the posterior mediastinum on the right side of the vertebral column. It receives most of the blood from the thoracic cavity. Its main tributaries are the *posterior intercostal veins*, the *bronchial veins* and the *oesophageal veins*. Also it receives a large tributary called the *hemiazygos vein* which receives the blood from the upper left intercostal spaces. The azygos vein opens into the superior vena cava just above the right atrium.

The right brachiocephalic vein is formed behind the sternal end of the right clavicle by the junction of the right internal jugular vein and the right subclavian vein. It then runs downwards for about 2.5 cm to unite with the left brachiocephalic vein behind the lower border of the 1st right costal cartilage to form the superior vena cava.

The left brachiocephalic vein is formed behind the sternal end of the left clavicle by the junction of the left internal jugular vein and the left subclavian vein. It runs obliquely downwards and to the right for about 6 cm to join the right brachiocephalic vein to form the superior vena cava.

The venous return from the lower limb

The veins of the lower limb are arranged in superficial and deep groups.

The superficial veins from the dorsum of the foot empty into a vein called the **dorsal venous arch** which runs across the dorsum of the foot. It opens, on either side of the foot, into a **lateral** and a **medial marginal vein**.

The superficial veins of the sole of the foot form a plantar venous arch which also opens on either side of the foot into the lateral and medial marginal veins.

At the level of the ankle joint the medial marginal vein is continuous with the long saphenous vein and the lateral marginal vein with the short saphenous vein.

The long saphenous vein is the longest vein in the body. It commences in front of the medial malleolus as a continuation of the medial marginal vein. It runs obliquely upwards and medially on the medial side of the leg. At the knee it passes posteriorly to the medial tibial and femoral condyles and then runs up the medial side of the thigh. A short distance below the inguinal ligament it runs deeply to join the femoral vein.

The short saphenous vein commences behind the lateral malleolus as a continuation of the lateral marginal vein. It runs along the lateral side of the Tendo Achilles and then upwards along the back of the leg to the lower part of the popliteal fossa where it passes deeply to join the popliteal vein.

The deep veins accompany the arteries and are named accordingly. The femoral vein is continuous with the external iliac vein beneath the inguinal ligament.

The venous return from the abdomen and pelvis

The external iliac vein is the direct continuation of the femoral vein. It runs upwards, on the medial side of the external iliac artery, and then passes behind it, to be joined by the internal iliac vein to form the common iliac vein.

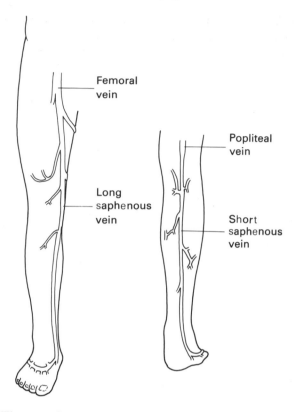

Fig. 11.28. The superficial veins of the lower limb: (a) anterior aspect, and (b) posterior aspect.

The **internal iliac vein** ascends from the pelvis, receiving tributaries in its course from the pelvic organs, and joins the external iliac vein to form the common iliac vein.

The **common iliac veins** commence at the upper borders of the sacroiliac joints. The right common iliac vein runs almost directly upwards to the right side of the anterior surface of the 5th lumbar vertebra where it unites with the left common iliac vein to form the inferior vena cava. The left common iliac vein runs obliquely upwards and to the right and is longer than the right common iliac vein. It crosses behind the right common iliac artery to join the right common iliac vein.

The inferior vena cava

The inferior vena cava is formed by the union of the right and left common iliac veins on the right side of the body of the 5th lumbar vertebra. It runs upwards through the abdomen on the anterior surfaces of the bodies of the lumbar vertebrae on the right side of the abdominal aorta. In the upper part of the abdomen it lies in a deep groove on the posterior surface of the liver from which it emerges to pass through the central tendon of the diaphragm at the level of the upper border of the 9th thoracic vertebra. It then runs a very short course in the thorax before entering the right atrium of the heart. During its course through the abdomen it receives the following tributaries:

The lumbar veins, four on each side, drain the blood from the walls of the abdomen and receive tributaries from the vertebral plexus of veins. The left lumbar veins pass behind the abdominal aorta before entering the inferior vena cava.

The right testicular (ovarian) vein runs upwards from the pelvis to enter the side of the inferior vena cava just below the right renal vein. *The left testicular (ovarian) vein* runs a similar course but drains into the left renal vein and not the inferior vena cava.

The right and left renal veins enter the sides of the inferior vena cava at the level of the upper border of the 2nd lumbar vertebra. They lie in front of the renal arteries and the left renal vein is longer than the right as it

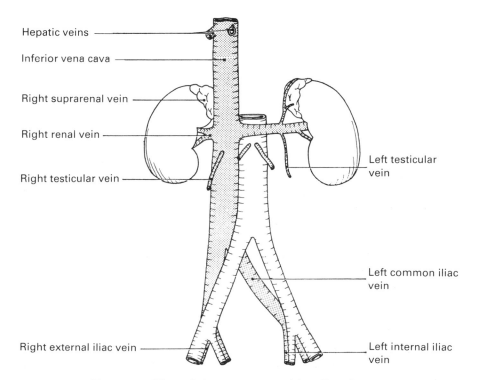

Hepatic veins

Inferior vena cava

Right suprarenal vein

Right renal vein

Right testicular vein

Left testicular vein

Left common iliac vein

Right external iliac vein

Left internal iliac vein

Fig. 11.29. The inferior vena cava and its tributaries.

passes in front of the abdominal aorta to reach the inferior vena cava. The left renal vein receives the left testicular (ovarian) vein and the left suprarenal vein before entering the inferior vena cava.

The right suprarenal vein drains into the side of the inferior vena cava just above the right renal vein. *The left suprarenal vein* enters the left renal vein.

The hepatic veins open directly into the front of the inferior vena cava as it lies in the groove on the posterior surface of the liver.

The portal circulation

The veins from the stomach, the intestine, the spleen, the pancreas and the gall bladder unite to form a single vein called the *portal vein*, which enters the liver. In the liver the portal vein divides into smaller and smaller tributaries which finally drain into the liver sinusoids. The blood from the sinusoids then drains into small veins which unite to form the hepatic vein which in turn joins the inferior vena cava. Thus the venous blood from the intestine carrying foodstuffs, which have been absorbed following the process of digestion, passes first to the liver and not directly back to the heart.

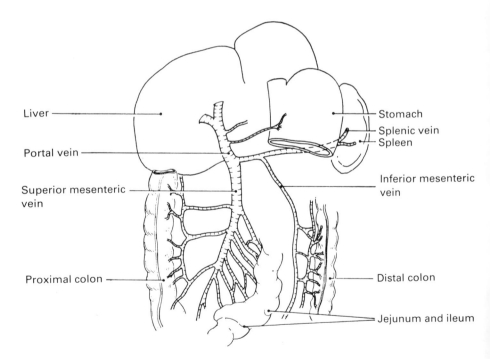

Fig. 11.30. Diagrammatic illustration of the formation of the portal vein.

The portal vein is formed at the level of the 2nd lumbar vertebra by the junction of the superior mesenteric vein and the splenic vein. The portal vein runs upwards and to the right for about 8 cm and then divides, just before entering the liver, into right and left branches which pass to the right and left lobes of the liver respectively. The right branch of the portal vein is joined by the cystic vein from the gall bladder. The main portal vein receives several tributaries from the stomach.

The superior mesenteric vein accompanies the superior mesenteric artery and drains the blood from the small intestine and the colon round to the distal third of the transverse colon. It receives several tributaries from the stomach, the pancreas and the duodenum.

The splenic vein runs alongside the splenic artery and receives small tributaries from the stomach and the pancreas. A short distance before it joins the superior mesenteric vein it is joined by the inferior mesenteric vein.

The inferior mesenteric vein drains the blood from the distal third of the transverse colon, the splenic flexure, the descending colon, the sigmoid colon and the rectum.

CHAPTER 12

The lymphatic system

The lymphatic system consists of:

1 Plexuses of minute vessels, the *lymph capillaries*, which start with closed ends in the tissues and which contain a fluid called *lymph*. Their wall is composed of a single layer of endothelial cells.

2 *Lymph vessels* which are formed by union of the lymph capillaries, and which drain the lymph into regional lymph nodes and ultimately into the venous system.

3 *Lymph nodes* which consist of solid masses of lymphoid tissue and through which lymph must usually pass before reaching the venous system.

4 Collections of *lymphoid tissue* in the walls of the alimentary canal and the respiratory tract as well as in the spleen and the thymus.

5 *The circulating lymphocytes.*

The lymph capillaries form a network in the tissues throughout the body, except for the central nervous system, the splenic pulp, the bone marrow, and certain avascular tissues, such as cartilage, the nails and the hair. In the lungs the network does not reach out as far as the alveoli (see Chapter 13).

Lymph

The thin walls of the lymph capillaries are permeable to larger molecules than are the blood capillaries and are therefore able to absorb proteins, debris of damaged cells and micro-organisms. They are important for returning tissue fluid and larger molecules to the blood stream. If the lymph vessels become obstructed, the tissues drained by those vessels become swollen, or oedematous, with a tissue fluid which has a high protein content. Large molecules, cell debris and micro-organisms reach the lumen of the lymph capillaries either through the gaps between the cells which make up the single layer of the wall, or by micropinocytosis, that is to say they are taken into the substance of the cells of the wall in vacuoles and are then extruded into the lumen of the lymph capillary. Although the lymph draining from most of the tissues is clear, the lymph draining from the small intestine, after a fatty meal, is milky due to the presence of fats in the form of minute droplets, called chylomicrons. This lymph with its high fat content is called *chyle*, and the lymph vessels of the small intestine are called *lacteals*.

The lymphocyte

Lymphocytes develop from stem-cells in the bone marrow. They are the second most common white cell in the blood, constituting 20−30% of the total leucocyte count. Unlike the other leucocytes, however, they can also be developed in large numbers outside the bone marrow in the lymph nodes and the spleen. They are commonly classified as small or large lymphocytes. The *small lymphocytes* are usually 5−8 μm in diameter and the *large lymphocytes* are 10−15 μm in diameter but there is a gradation in size between the two types. Small lymphocytes have a relatively large, densely staining nucleus with a thin surrounding rim of cytoplasm which contains no granules. Large lymphocytes are either immature lymphocytes or they are small lymphocytes which have been stimulated by an antigen to become functionally active.

The lymphocytes are known to be of two further types which cannot be distinguished by microscopic examination. These two types are called T-lymphocytes and B-lymphocytes.

The *T-lymphocytes*, which form about 80% of the total blood lympho-cytes, depend on a factor produced by the thymus to become antigenically active or competent. The T-cells are involved in cell-mediated immunity and they produce factors which mobilize macrophages to the site of infection or the site of reaction with an antigen.

The *B-lymphocytes* form 20% of the total blood lymphocytes and they depend upon a gut-produced factor to become immunologically competent. In birds this gut factor is in a structure called the *bursa of Fabricus* and this is the reason for the title B-cell as opposed to the T-cells which are thymus dependent. It is not clear where this gut factor originates in man but it may well be the lymphoid tissue of the gut. B-cells, when antigenically stimulated, can proliferate to produce plasma cells which can in turn produce antibodies.

A third category of lymphocyte is also described called *K-cells*, or killer cells, which, although they are not specifically sensitized, can destroy cells which are coated with antibody.

The roles of the lymphocytes and the macrophage system in the immune response are described in more detail in the section on disorders of immunity (see Chapter 3).

Lymph vessels

The wall of the lymph capillaries consists of a single layer of endothelial cells, but, unlike the blood capillaries, the cells are not attached to a basement membrane. The precise microscopic structure of lymph vessels varies slightly from one tissue to another. The smaller lymph vessels, which are formed by union of the capillaries, have an outer layer of connective tissue which surrounds and supports the endothelial layer. The walls of the larger lymph vessels, which are the conducting trunks, consist of three layers:

1   an *inner layer* of endothelial cells supported by a thin layer of connective tissue;

2   a *middle layer* of smooth muscle cells, which run in a circular fashion around the vessel, and interspersed with these are some yellow elastic connective tissue fibres; and

3   an *outer layer* of fibrous tissue which contains a few smooth muscle cells.

Lymph vessels contain many semilunar valves and in this respect differ from the small veins. The valves are usually in pairs and are formed by a double fold of the endothelial lining, supported by connective tissue. The lymph vessel is slightly expanded on the proximal (downstream) side of the valve, producing a beaded appearance. The purpose of the valves is to prevent backflow of lymph towards the tissues.

The flow of lymph in the vessels is due to:

1   the pressure of fluid in the tissues;

2   the squeezing of the lymph vessels by muscles during movement; the valves in the lymph vessels ensure that this flow of lymph is away from the tissues;

3   the pulsation of the arteries also squeezes the lymph vessels;

4   respiratory movements and consequent negative pressure in the brachiocephalic veins; and

5   contraction of smooth muscle in the wall of the larger lymph vessels. This contraction is under the control of the sympathetic nervous system.

Lymph flow is slow, and about 2−4 l a day pass through the thoracic duct into the venous system whereas the cardiac output is about 5 l/min. When venous obstruction occurs, however, the flow in the lymph vessels can increase to five or ten times the normal rate of flow.

Usually lymph has to pass through a series of regional lymph nodes before reaching the blood stream. However, lymph from the thyroid gland, the oesophagus and parts of the liver passes directly into the thoracic duct and thus directly into the blood stream. The larger lymph vessels in the skin run alongside the superficial veins and are near the deep fascia. Generally the superficial lymph vessels have few connections with the deep lymph system. Virtually all the lymph from the body enters the blood stream via the *thoracic duct* and the *right lymphatic duct* which enter the left brachiocephalic and the right brachiocephalic veins respectively.

The larger lymph vessels are surrounded by a network of fine blood vessels. If the lymph vessels become infected (*lymphangitis*) then this plexus of blood vessels becomes engorged and the position of the superficial lymph vessels in the skin becomes visible as red lines which run longitudinally up the limb.

The lymph vessels form a pathway by which infected material and malignant cells can spread. They are, however, a means by which infective and foreign materials (antigens) are brought into contact with macrophages and lymphocytes (see below). Such spread, apart from causing lymphangitis in the case of infections, usually causes enlargement of the regional lymph nodes, which then become palpable (i.e. they can be felt whereas normally they cannot). In extreme cases they may actually become visible, but this usually only occurs with the spread of a malignant tumour.

The lymph nodes

The lymph nodes are small oval-shaped structures which are 0.1−2.5 cm in length. They are indented on one side, forming the hilum of the gland, and blood vessels enter and leave the node at the hilum. A single efferent lymph vessel leaves the node also at the hilum, but the afferent lymph vessels enter it at different points around its periphery. The node is subdivided into an outer cortex, which is absent at the hilum, and a darker medulla, or central portion, which reaches to the surface of the node at the hilum. There is a relatively indefinite line of division between the cortex and the medulla.

The lymph nodes consist of:

1 *the outer capsule* and *trabeculae*;
2 *the reticular framework*;
3 *the lymphatic channels*;
4 *the lymphatic blood system*; and
5 *the lymphocytes* and *macrophages* which are entangled in the reticulum.

The capsule and the trabeculae consist of collagen fibres, fibroblasts, some yellow elastic fibres and a few smooth muscle cells. The capsule envelops the outer surface of the lymph node and projects inwards to form the trabeculae, which incompletely partition the node. **The reticulum** is a meshwork of reticulin fibres found within the spaces formed by the capsule and the trabeculae, and this meshwork supports the contained cells.

The lymphatic channels (sinuses) within the lymph node are fed by the afferent lymph vessels and they extend throughout the lymph node allowing the lymph to come into contact with the macrophages and lymphocytes which are entrapped in the reticular framework. The lymphatic

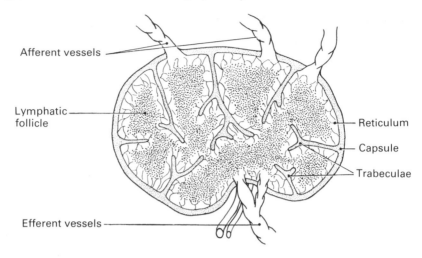

Afferent vessels

Lymphatic follicle

Reticulum

Capsule

Trabeculae

Efferent vessels

Fig. 12.1. The structure of a lymph gland.

channels join together in the medulla of the node to form the efferent vessel which leaves the node at the hilum.

The lymphatic blood system consists of an artery and a vein, which respectively enter and leave the node at the hilum, and arcades of arterioles and capillaries which supply the inter-trabecular spaces. **The lymphatic blood supply** has no direct communication with the lymph sinuses, although movement of lymphocytes, in both directions, between the blood stream and the substance of the lymph node is possible. When the lymph node is antigenically stimulated (see below) to produce more lymphocytes the vascular pattern of the lymph node is altered and the vascularity is considerably increased.

The partitions of the interior of the lymph node each contain a lymphatic follicle which consists of densely packed cells within the lymphatic sinuses. The outer layer of the lymphatic follicle, which is relatively loosely packed, consists of small lymphocytes, macrophages and some plasma cells. The inner layers are more closely packed and contain the germinal centres, which are composed of large lymphoblasts, some of which are undergoing mitosis. Generally the lymphocytes are formed by subdivision of the lymphoblasts in the centre of the follicle and, as they pass outwards towards the periphery of the follicle, they become smaller, eventually entering the lymphatic sinuses. This scheme is, however, complicated by the flow of lymphocytes into the follicle from the lymphatic system and from the blood stream.

Generally the B-lymphocytes are found in the outer, or cortical, parts of the follicle and the T-lymphocytes are found in the region between the germinal centres and the medulla of the node.

The functions of the lymph nodes

The lymph nodes have the following functions:
1 the filtration of lymph;
2 the trapping of foreign material (antigens) by phagocytes;
3 the production of lymphocytes;
4 the production of humoral antibodies; and
5 the provision of a pathway by which lymphocytes in the blood stream can re-enter the lymphatic system.

The lymphatic drainage of the body

The thoracic duct

The thoracic duct is the main lymph vessel of the body. It drains the lymph from the whole of the body, apart from the right side of the head and neck, the right arm and the right side of the thorax. It begins at the level of the lower border of the 12th thoracic vertebra at the upper end of the *cisterna chyli*, which is a sac-like dilatation of the vessels which drain the abdomen, the pelvis and the lower limbs, and passes through the aortic opening in the diaphragm. Although it is 38–45 cm in length in adults, its diameter is only about 5 mm at its origin and it tapers as it passes upwards. There is usually a slight increase in its diameter just proximal to its termination.

In the thorax the thoracic duct ascends in the posterior mediastinum on the right side of the ascending aorta, lying behind the oesophagus. At the level of the 5th thoracic vertebra it runs towards the left and enters the superior mediastinum where it passes upwards along the left side of the oesophagus. At the thoracic inlet, at the level of the transverse process of the 7th cervical vertebra, it arches laterally, behind the left common carotid artery, the left vagus nerve, the left internal jugular vein and the left subclavian vein and ends by draining into the angle of the junction of the left internal jugular and the left subclavian veins as they unite to form the left brachiocephalic vein. Just before its termination the thoracic duct is usually joined by the *left jugular* and *left subclavian lymph trunks* which drain the lymph from the left side of the head and neck, and the left arm respectively. These lymph trunks, however, may open independently into the left internal jugular and left subclavian veins respectively. There are other variations in the anatomy of the thoracic duct which are occasionally encountered: it may branch into a number of tributaries close to its termination, or it may split into a network of interconnecting vessels in the middle of its course through the thorax.

The thoracic duct has a number of valves, including one at its junction with the venous system which prevents backflow of blood from the left brachiocephalic vein.

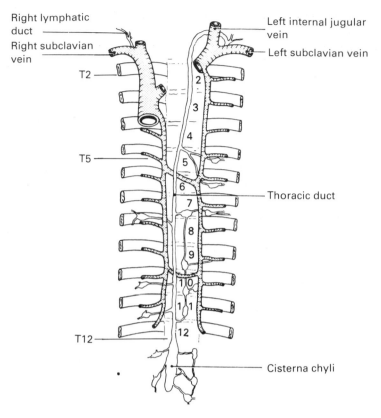

Fig. 12.2. The thoracic duct, the cisterna chyli and the right lymphatic duct.

The cisterna chyli

The cisterna chyli is a sac-like dilatation of the lymph vessels which drain the abdomen, the pelvis and the lower limbs. It is about 5−7 cm in length and it lies on the anterior surfaces of the bodies of the 1st and 2nd lumbar vertebrae, to the right of the abdominal aorta. It is covered anteriorly by the medial side of the right crus of the diaphragm. It is formed by the junction of the *right* and *left lumbar* and the *intestinal lymph trunks*. Frequently there is no cisterna chyli present, the lumbar and intestinal lymph trunks draining directly into the lower end of the thoracic duct. Occasionally two, or even three, cisternae are present.

The right lymphatic duct

The right lymphatic duct is about 1 cm long and drains into the angle of the junction of the right internal jugular and the right subclavian veins as they unite to form the right brachiocephalic vein. The right lymphatic duct is formed by the junction of:

1 *the right subclavian trunk* which drains the right arm;

2 *the right bronchomediastinal trunk* which drains the right side of the thorax, the right heart and part of the surface of the liver; and

3 *the right jugular trunk* which drains the right side of the head and neck.

The anatomy of the right lymphatic duct is also variable and in most cases the three trunks open separately into the area of the junction of the veins which form the right brachiocephalic vein.

Obstruction of the thoracic duct, or of the right lymphatic duct, due to malignant disease often causes no symptoms because of other communications between the lymphatic system and the venous system. Damage to the thoracic duct in the root of the neck, either during surgery or due to a stab wound, is a rare event but it does lead to loss of lymph. If widespread lymphatic obstruction occurs, fluid collections may be found in the pleural cavity (a *pleural effusion*) or the peritoneal cavity (*ascites*). Because chyle contains fat these fluid collections are opaque and cream-coloured.

The other lymph vessels of the body are too numerous, and too variable in their anatomy, for them all to be mentioned or to be described in detail, but the main directions of the lymph flow and the nodes through which the lymph passes will be outlined. A knowledge of the nodes and of the principles of lymph drainage is important when considering the spread both of malignant tumours and of infection.

The lymphatic drainage of the head and neck

The lymph draining from the head and neck ultimately reaches the deep cervical lymph nodes which consist of two groups:

1 *The superior deep cervical lymph nodes* which lie alongside the internal jugular vein and which are usually deep to the sternomastoid muscle. Efferent vessels pass from this group either to the inferior deep cervical group or directly to the jugular trunk.

2 *The inferior deep cervical lymph nodes* which lie mainly under the lower end of sternomastoid muscle close to the subclavian artery and vein and to the brachial plexus. Efferent vessels drain into the jugular trunk.

During its passage to the deep cervical lymph nodes the lymph passes through secondary lymph nodes and these are divided into superficial and deep groups.

The regional superficial lymph nodes in the head

The lymph draining the superficial tissues of the head and neck passes to one of eight regional groups of nodes, although some of the lymph vessels

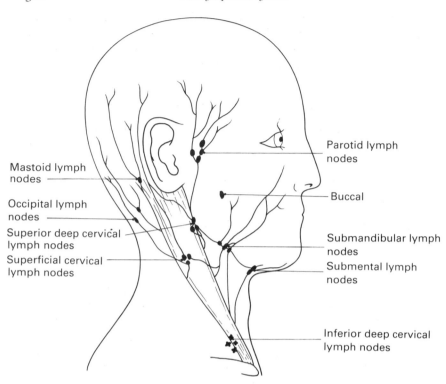

Mastoid lymph
nodes

Occipital lymph
nodes

Superior deep cervical
lymph nodes

Superficial cervical
lymph nodes

Parotid lymph
nodes

Buccal

Submandibular lymph
nodes

Submental lymph
nodes

Inferior deep cervical
lymph nodes

Fig. 12.3. The superficial lymph nodes of the head and neck.

from the superficial structures may drain directly into the deep cervical lymph nodes.

1   *The occipital lymph nodes* lie on the surface of the trapezius muscle, close to its attachment to the occipital bone. They drain the lymph from the occipital part of the scalp. Efferent vessels from this group pass to the inferior deep cervical lymph nodes.

2   *The retro-auricular (mastoid) lymph nodes* lie behind the ear on the surface of the attachment of the sternomastoid muscle. They drain the lymph from an area of scalp behind the ear, from the posterior surface of the pinna and from the posterior wall of the external acoustic meatus. Efferent vessels pass to the superior group of the deep cervical lymph nodes.

3   *The parotid lymph nodes* lie on the superficial surface of the parotid salivary gland. They drain lymph from the greater part of the forehead, the temporal part of the scalp, the anterior surface of the pinna and the anterior wall of the external acoustic meatus. The efferent vessels pass to the superior deep cervical lymph nodes.

4 *The buccal lymph nodes* lie alongside the facial vein just anterior to the masseter muscle. They drain the lymph from the surface of the nose, the cheeks, the upper lip and the lateral part of the lower lip. The efferent vessels pass to the submandibular lymph nodes (see below).

The regional superficial lymph nodes in the neck

1 *The submandibular lymph nodes* lie under the inferior border of the mandible around, or sometimes within, the submandibular salivary gland. They drain the lymph from a wide area and afferent vessels pass to them from the submental, the buccal and the lingual lymph nodes. The efferent vessels pass from them to both the superior and the inferior deep cervical lymph nodes.

2 *The submental lymph nodes* lie on the surface of the muscles below the chin. They drain the lymph from the central part of the lower lip, the tip of the tongue and the floor of the mouth. The efferent vessels pass to the submandibular and the inferior deep cervical lymph nodes.

3 *The anterior cervical lymph nodes* lie along the anterior jugular veins close to the anterior border of the sternomastoid muscle. They drain the lymph from the anterior part of the neck, below the hyoid bone. The efferent vessels pass to the inferior group of the deep cervical lymph nodes.

4 *The superficial cervical lymph nodes* lie on the upper part of the sternomastoid muscle alongside the external jugular vein. They drain the lymph from the lobe of the ear, the floor of the external acoustic meatus and the skin around the angle of the mandible. Efferent vessels pass to the superior and inferior deep cervical lymph nodes.

The lymphatic drainage of the deep structures of the head and neck

The lymph vessels from the deep structures of the head and neck either pass directly to both groups of the deep cervical lymph nodes or to one of four groups of secondary lymph nodes.

1 *The retropharyngeal lymph nodes* lie between the posterior wall of the pharynx and the fascia covering the pre-vertebral muscles of the cervical spine. They drain the lymph from the nose, the sinuses and the nasopharynx. Efferent vessels pass to the superior group of the deep cervical lymph nodes.

2 *The paratracheal lymph nodes* lie on either side of the cervical parts of the trachea and the oesophagus. They drain the lymph from the upper oesophagus and trachea and from part of the thyroid gland. The efferent vessels pass to both groups of the deep cervical lymph nodes (superior and inferior).

3 *The pre-tracheal, pre-laryngeal and infrahyoid lymph nodes* lie along the course of the recurrent laryngeal nerves. They drain the lymph from the trachea, the larynx and part of the thyroid gland. The efferent vessels pass to both groups of the deep cervical lymph nodes.

4 *The lingual lymph nodes* are small and lie deep in the floor of the mouth. Frequently they are absent.

The lymphatic drainage of the tongue

The lymphatic plexuses in the mucous membrane and in the muscular part of the tongue are continuous. The lymph vessels from the tip of the tongue drain into the submental lymph nodes.

The anterior two-thirds of the tongue, which is that part of the tongue in front of the vallate papillae, drains into the submental and the submandibular lymph nodes and also directly into the deep cervical lymph nodes. The lymph vessels from the posterior two-thirds of the tongue drain directly into the superior and inferior deep cervical lymph nodes.

The lymphatic drainage of the upper limb

All the lymph from the upper limb ultimately passes through the *axillary lymph nodes*. The lymph vessels are divided into a *superficial group* which tend to follow the superficial veins and a *deep group* which follow the course of the deep veins and nerves.

The axillary lymph nodes

There are 20 to 30 axillary lymph nodes and they are divided into five groups.

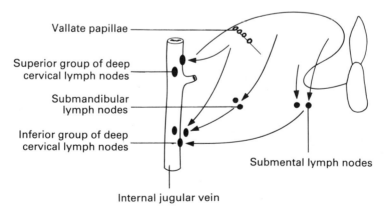

Vallate papillae

Superior group of deep cervical lymph nodes

Submandibular lymph nodes

Inferior group of deep cervical lymph nodes

Submental lymph nodes

Internal jugular vein

Fig. 12.4. The lymphatic drainage of the tongue.

1 *The lateral group of axillary lymph nodes* which drain virtually the whole limb lie on the lateral wall of the axilla, close to the axillary vein. Their efferent vessels pass to the central and apical groups of the axillary lymph nodes and also to the inferior deep cervical lymph nodes.

2 *The anterior (or pectoral) group of axillary lymph nodes* lie on the lower border of the pectoralis minor muscle and they drain the lymph from the skin and muscles of the anterior and lateral walls of the chest and of the abdomen down to the level of the umbilicus. This group also drains the lateral part of the breast. The efferent vessels pass to the central and apical groups of the axillary lymph nodes.

3 *The posterior group of axillary lymph nodes* lie along the lower edge of the posterior wall of the axilla and they drain lymph from the skin and muscles of the lower part of the neck and of the back down to the level of the iliac crest. The efferent vessels pass to the central and apical groups of the axillary lymph nodes.

4 *The central group of the axillary lymph nodes* lie in the axillary fat and they drain lymph from the lateral, anterior and posterior groups. The efferent vessels pass to the apical group of the axillary lymph nodes.

5 *The apical group of the axillary lymph nodes* lie in the upper part of the axilla close to the axillary vein. They receive afferent vessels from all the other groups of the axillary lymph nodes and also afferent vessels from the upper, outer part of the breast. The lymph vessels accompanying the cephalic vein also drain directly into this group. The efferent vessels from this group join together to form the subclavian trunk. This drains into the right lymphatic duct on the right side and the thoracic duct on the left side.

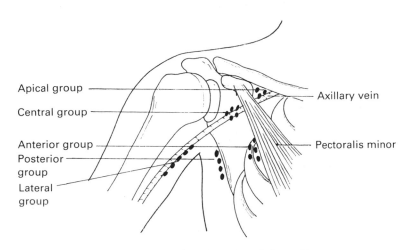

Fig. 12.5. The lymph nodes of the axilla.

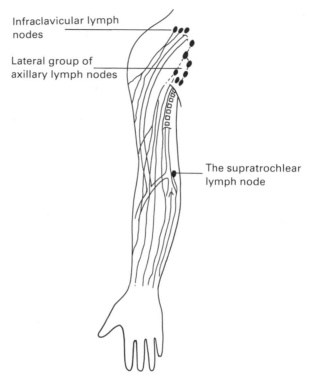

Infraclavicular lymph nodes

Lateral group of axillary lymph nodes

The supratrochlear lymph node

Fig. **12.6.** The lymphatic drainage of the upper limb.

The subclavian trunks, however, may drain directly and independently into the subclavian vein (see above).

There are two groups of lymph nodes in the upper limb through which the lymph may pass before reaching the axillary lymph nodes.

1 *The supratrochlear lymph nodes*, of which there are usually one or two, lie superficially above the medial epicondyle of the humerus, close to the basilic vein. Some lymph vessels from the medial side of the forearm pass through these nodes. The efferent vessels drain into the axillary lymph nodes.

2 *The infraclavicular lymph nodes* lie below the clavicle between the pectoralis major and the deltoid muscles, close to the cephalic vein. Some of the lymph vessels accompanying the cephalic vein drain into this group of lymph nodes. The efferent vessels drain into the apical group of the axillary lymph nodes.

The lymphatic drainage of the breast

The lymph vessels from the skin over the breast, the nipple and the areola drain into the subareolar plexus of lymph vessels. This communicates with a plexus of lymph vessels on the surface of the pectoralis major muscle deep to the breast tissue. From these plexuses efferent vessels pass to the anterior (or pectoral) group of the axillary lymph nodes. From the upper part of the breast some efferent vessels pass directly to the apical group of the axillary lymph nodes. Seventy-five per cent of the lymph from the breast drains into the axillary lymph node system. The remaining 25% drains into the parasternal lymph nodes which lie alongside the internal thoracic artery.

The lymphatic drainage of the lower limb

Most of the lymph vessels from the lower limb drain directly, or indirectly, into the inguinal lymph nodes at the groin.

The superficial lymph vessels of the lower limb leave the foot in two groups, a *medial group*, which is by far the most numerous group and which follow the course of the great saphenous vein, and a *lateral group*, which follow the course of the small saphenous vein. At the knee some of the lateral group of superficial lymph vessels join the medial group while others pass to *the popliteal lymph nodes* which are the only group of secondary lymph nodes in the lower limb. Apart from draining the lateral surface of the dorsum of the foot these popliteal lymph nodes also drain the knee

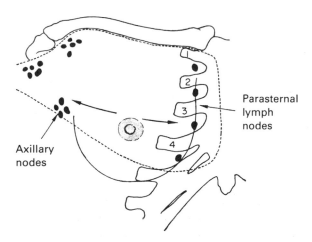

Axillary nodes

Parasternal lymph nodes

Fig. 12.7. The principal directions of the lymphatic drainage of the breast.

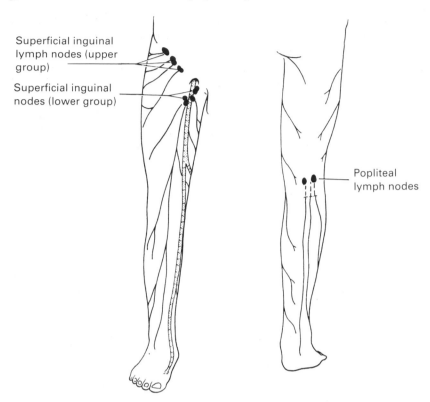

Superficial inguinal
lymph nodes (upper
group)

Superficial inguinal
nodes (lower group)

Popliteal
lymph nodes

Fig. 12.8. The lymphatic drainage of the lower limb.

joint. Efferent vessels from the popliteal lymph nodes pass to the deep inguinal lymph nodes.

The deep lymphatic vessels of the lower limb accompany the main vessels (i.e. the anterior tibial, the posterior tibial, the peroneal, the popliteal and the femoral vessels). The deep lymph vessels from the foot and leg drain into the popliteal lymph nodes, but the deep vessels of the thigh pass to the inguinal lymph nodes.

The inguinal lymph nodes

The inguinal lymph nodes consist of a superficial and a deep group.

1 *The superficial inguinal lymph nodes* are divided into an upper and a lower group. The *upper group* lie just below the inguinal ligament. They drain lymph from the gluteal region, the skin and muscles of the anterior abdominal wall below the level of the umbilicus, the external genitalia and the perianal region together with the lower part of the anal canal. In the female

they also drain the lower part of the vagina and receive afferent vessels from part of the uterus. The *lower group* drain lymph from the whole of the lower limb except for the lymph which passes to the popliteal lymph nodes. The efferent vessels from the superficial inguinal lymph nodes pass to the external iliac lymph nodes. Many lymph vessels connect the individual inguinal lymph nodes.

2 *The deep inguinal lymph nodes* lie on the medial side of the femoral vein close to its junction with the great saphenous vein. They receive afferent lymph vessels from the superficial inguinal lymph nodes, the vessels from the deep lymph system which accompany the femoral artery and they also drain the glans penis (the glans clitoris in the female). The efferent vessels pass to the external iliac lymph nodes.

The lymphatic drainage of the abdomen and pelvis

The lymph vessels from the abdominal viscera and the pelvis drain ultimately into the cisterna chyli and the thoracic duct. The lymph vessels run with the arteries which supply the various viscera and they drain into one of three main groups of lymph nodes which lie close to the abdominal aorta. The outlying intermediate groups of lymph nodes, like the lymph vessels, lie close to the arteries.

In the small intestine each villus (p. 376) contains a small lymph vessel, called a *lacteal*. In other parts of the alimentary canal a submucous plexus of lymph vessels is formed. The submucous plexus and the lacteals join small lymph vessels in the muscular layers of the alimentary canal. These vessels pierce the muscle coat of the gut and join larger vessels which follow the same course as the arteries of supply, and which, in the case of the small intestine and the colon, run in the mesentery.

The pre-aortic lymph nodes

The pre-aortic lymph nodes lie on the anterior surface of the abdominal aorta close to the origins of the arteries which arise from the front of the aorta. They are thus divided into the coeliac, the superior mesenteric and the inferior mesenteric groups of lymph nodes. The efferent vessels from these groups of lymph nodes unite to form the intestinal trunk which enters the cisterna chyli.

1 **The coeliac lymph nodes** lie close to the origin of the coeliac axis. Their secondary lymph nodes lie along the arterial branches of the coeliac axis.

a) *The gastric lymph nodes* lie along the greater and lesser curvatures of the stomach, around the lower oesophagus and in the angle between the

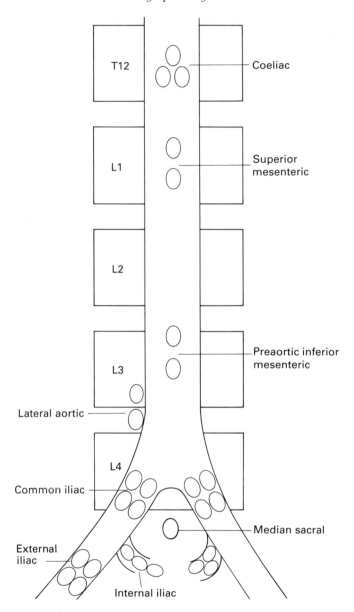

Fig. 12.9. The main lymph nodes of the abdomen and pelvis.

pylorus and the duodenum. They receive afferent vessels from the abdominal part of the oesophagus, the stomach and the first part of the duodenum. Their efferent vessels pass to the coeliac lymph nodes.

b) *The hepatic lymph nodes* lie along the course of the hepatic artery and its branches and along the course of the bile duct. They receive afferent vessels from the stomach, the duodenum, the liver, the gall bladder, the bile ducts and the pancreas. Their efferent vessels pass to the coeliac lymph nodes.

c) *The pancreatico-splenic lymph nodes* lie along the course of the splenic artery, which runs along the superior border of the pancreas. They receive afferent vessels from the stomach, the pancreas and the spleen. Their efferent vessels pass to the coeliac lymph nodes.

2 The superior and inferior mesenteric lymph nodes lie on the anterior surface of the abdominal aorta close to the origins of these two arteries. They receive afferent lymph vessels from lymph nodes which lie along the courses of the two arteries. They therefore drain the lymph from the whole of the intestine from the duodenal-jejunal flexure to the upper part of the anal canal. The largest group of secondary lymph nodes numerically are the lymph nodes of the mesentery and there are usually more than 100 of these. These nodes can enlarge markedly if involved by malignant tumours or certain infections, such as typhoid fever or tuberculosis, and they can then on occasions be felt through the anterior abdominal wall.

The lateral aortic lymph nodes

The lateral aortic lymph nodes lie on either side of the abdominal aorta and are related to the medial border of the psoas major muscle. These lymph nodes are associated with the larger paired branches of the abdominal aorta and they drain lymph from the structures and organs which these arteries supply. The efferent lymph vessels from these lymph nodes join on each side to form the right and left lumbar trunks which, with the intestinal trunk, unite to form the cisterna chyli. There are usually connections across the midline between the right and left lumbar trunks and between the right and left lateral aortic lymph nodes.

The lateral aortic lymph nodes drain lymph from the kidneys, the adrenal glands, the abdominal parts of the ureters, the gonads (ovaries or testes), the uterine tubes, the upper part of the uterus and the posterior abdominal wall.

The main groups of secondary lymph nodes are:

1 The common iliac lymph nodes which lie alongside the common iliac arteries and drain the lymph from the external and internal iliac groups of lymph nodes. The efferent vessels pass to the lateral aortic lymph nodes.

2 The external iliac lymph nodes which lie alongside the external iliac arteries. They drain lymph from the deeper layers of the anterior abdominal wall below the level of the umbilicus, the medial side of the thigh, the glans penis in the male or the glans clitori in the female, part of the urethra, the prostate in the male, the fundus of the urinary bladder and the upper part of the vagina and part of the cervix uteri in the female. They also drain lymph from the lower limb which reaches them via the inguinal lymph nodes. The efferent vessels pass to the common iliac lymph nodes.

3 The internal iliac lymph nodes which lie along the internal iliac arteries. They receive afferent vessels from all the pelvic organs, the muscles of the pelvic floor, the gluteal muscles and the muscles of the back of the thigh. The efferent vessels pass to the common iliac lymph nodes.

The lymphatic drainage of the uterus

The lymph vessels from the upper part of the body of the uterus and the uterine tubes drain into the lateral aortic and pre-aortic lymph nodes. The lymph vessels from the lower part of the uterus pass mainly to the external iliac lymph nodes. There is a small area of the uterus, close to the point of entry of the uterine tubes, which drains into the superficial inguinal lymph nodes. The lymphatic system of the uterus enlarges considerably during pregnancy.

The lymph vessels from the cervix uteri pass in three directions:

1 laterally to the external iliac lymph nodes;
2 posterolaterally to the internal iliac lymph nodes;
3 posteriorly to the sacral lymph nodes which are outlying nodes of the internal iliac group of nodes.

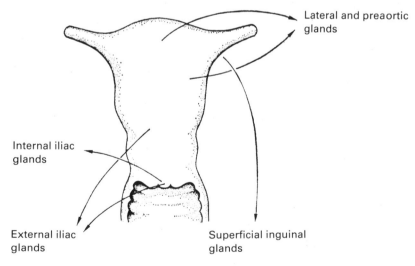

Fig. 12.10. The lymphatic drainage of the uterus.

The lymphatic drainage of the testes and ovaries

The lymph vessels from the ovaries and testes accompany the course of the arteries of supply and drain into the lateral aortic and pre-aortic lymph nodes.

The lymphatic drainage of the thorax

The lymphatic drainage of the walls of the thoracic cavity

The lymph vessels from the superficial tissues of the walls of the thoracic cavity unite to form a number of lymph trunks which drain into the *axillary lymph nodes*. Some of the superficial lymph vessels near the lateral border of the sternum, however, pass to the *parasternal lymph nodes*. Other lymph vessels from the upper part of the pectoral region pass to the *inferior deep cervical lymph nodes*.

The lymph vessels of the deep tissues of the thoracic walls drain into three main groups of lymph nodes:

1 **The parasternal (internal thoracic) lymph nodes** lie alongside the internal mammary artery, just lateral to the sternum. They drain lymph from the deeper parts of the anterior abdominal wall above the umbilicus, the upper surface of the liver, part of the breast and the deeper parts of the anterior walls of the thorax. Their efferent vessels join with the efferent vessels from the tracheobronchial and brachiocephalic lymph nodes to form the *bronchomediastinal trunk*.

2 **The intercostal lymph nodes** lie in the posterior ends of the intercostal spaces and receive lymph vessels from the posterolateral thoracic walls and some lymph vessels from the breast. The efferent vessels in the lower four or five intercostal spaces unite to form a trunk which enters the cisterna chyli or the lower part of the thoracic duct. The efferent vessels from the upper intercostal spaces on the left side enter the thoracic duct, and on the right side they enter the right lymphatic duct.

3 **The diaphragmatic lymph nodes** lie on the upper surface of the diaphragm and they receive lymph vessels from the diaphragm and from the upper surface of the liver. The efferent lymph vessels pass to the posterior mediastinal, parasternal and brachiocephalic lymph nodes.

The lymphatic drainage of the thoracic viscera

The lymph draining from the thoracic viscera passes through one of three groups of lymph nodes:

1 **The brachiocephalic lymph nodes** lie in the anterior part of the superior mediastinum, in front of the brachiocephalic veins. They drain

lymph from the pericardium, the thyroid and thymus glands and the outer parts of the diaphragm. The efferent vessels join with the efferent vessels of the parasternal and tracheobronchial lymph nodes to form the broncho-mediastinal trunk.

2   The posterior mediastinal lymph nodes lie behind the pericardium, alongside the oesophagus and the descending thoracic aorta. They drain lymph from the oesophagus, the posterior part of the pericardium and from the diaphragmatic lymph nodes. They sometimes also drain the left lobe of the liver. Their efferent vessels pass mainly to the thoracic duct but a few vessels pass to the tracheobronchial lymph nodes.

3   The tracheobronchial lymph nodes are large nodes which lie around the trachea and the bronchi. They are divided into five groups:

a)   *the pulmonary lymph nodes* which lie alongside the larger bronchi in the substance of the lung;

b)   *the bronchopulmonary lymph nodes*, which are not sharply divided from the pulmonary lymph nodes, and which lie around the main bronchi at the hilum of each lung. In clinical medicine these lymph nodes are referred to as the 'hilar lymph nodes', and they may become visibly enlarged on plain radiographs of the chest in malignant or infective disease of the lung;

c)   *the inferior tracheobronchial lymph nodes* lie at the carina, below the division of the trachea into the left and right main bronchi;

d)   *the superior tracheobronchial lymph nodes* lie above each main bronchus close to the tracheal bifurcation; and

e)   *the paratracheal lymph nodes* lie alongside the thoracic part of the trachea.

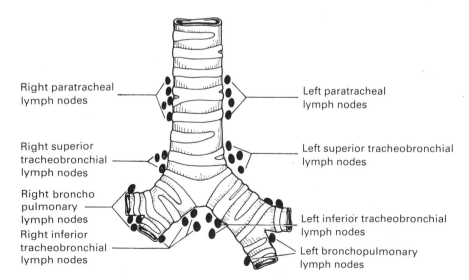

Fig. 12.11. The lymph nodes of the trachea and bronchi.

These lymph nodes drain the lungs, the bronchi, the thoracic part of the trachea and the heart. They are joined by a few efferent vessels from the posterior mediastinal lymph nodes. Their efferent lymph vessels pass upwards and join with the efferent vessels from the parasternal and brachiocephalic lymph nodes to form the *right* and *left bronchomediastinal trunks*.

The variability of the anatomy of the lymphatic system

Like many other organs and systems of the body, the anatomy of the lymphatic system is inconstant and variable. Some of the variations have been mentioned briefly above, for example the absence of the cisterna chyli and the differing terminations of the thoracic duct and the right lymphatic duct. A number of lymph vessels bypass regional lymph nodes and pass directly to the next set of nodes and, where vessels lie close to the midline, there are frequent interconnections between the right and left sides.

The spleen

The spleen lies in the left hypochondriac region of the abdomen, between the diaphragm and the fundus of the stomach. It is a soft organ of dark purple colour and it is about 12 cm long, 7 cm wide and 3−4 cm thick.

The spleen has a diaphragmatic surface, a visceral surface and superior, inferior, posterior and anterior borders. *The superior border* usually has one or two notches close to its anterior end. On the lower medial part of the visceral surface of the spleen is the *hilum*, through which the vessels and nerves enter and leave the spleen.

The diaphragmatic surface of the spleen faces upwards, backwards and to the left and it is related to the lower surface of the diaphragm which separates it from the left lung, the pleura and the 9th, 10th and 11th ribs.

The visceral surface faces downwards and to the right, into the abdominal cavity and it has gastric, renal, colic and pancreatic impressions. The size of the colic impression determines the overall shape of the spleen. The spleen is related anteriorly to the fundus of the stomach, medially to the left kidney and inferiorly to the splenic flexure of the colon. The lateral end of the hilum of the spleen is related to the tail of the pancreas.

The structure of the spleen

The spleen has an outer serous coat, formed by the peritoneum which is adherent to the fibro-elastic coat. The *fibro-elastic coat* surrounds the whole organ, except at the hilum, and projects into the interior of the organ as *trabeculae*. The largest trabeculae run inwards from the hilum of the spleen and the splenic vessels run in these trabeculae. In man there are only a few muscle cells in the fibro-elastic capsule or trabeculae and variations in the

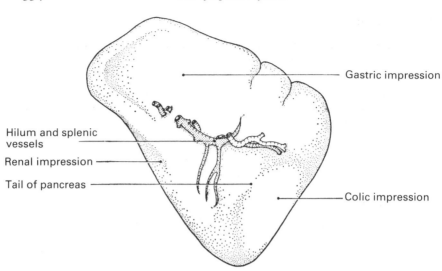

Gastric impression

Hilum and splenic
vessels

Renal impression

Tail of pancreas

Colic impression

Fig. 12.12. The spleen.

size of the spleen are, in general, related to changes in the blood flow within it and thus to the amount of blood it contains. Arising from the subdivisions of the trabeculae is a fine reticular network which contains the *splenic pulp*.

The splenic pulp is of two types, white and red. *The white pulp* consists of sheaths of lymphatic tissue surrounding the arterioles, and of splenic lymphatic follicles, which contain masses of lymphoctes and which are enlargements of these sheaths. The lymph vessels of the spleen begin in the white pulp and run alongside the arterioles and arterial branches to emerge at the hilum with the splenic vessels.

The red pulp consists of a labyrinth of venous sinusoids divided by splenic cords, which are a network of macrophages attached to the reticulum.

The cut surface of the spleen is dark red due to the red splenic pulp, but the splenic lymphatic follicles are visible as white dots which are 0.25−1 mm in diameter.

The blood supply of the spleen

The splenic artery, which is the largest branch of the coeliac axis, divides into five or more branches before it reaches the hilum. It then permeates the spleen in the trabeculae. The venous drainage of the spleen is via the splenic vein which unites with the superior mesenteric vein to enter the liver as the portal vein.

The functions of the spleen

1 *Phagocytosis*. The macrophages in the red pulp of the spleen phagocytose old erythrocytes, leucocytes and platelets as well as any circulating micro-organisms. These cells also engulf foreign material (antigens).

The haemoglobin from the old red blood cells is broken down to bile pigments and iron which then pass to the liver. In the liver the iron is stored and the pigments are excreted in the bile.

2 *Erythropoiesis*. In the human fetus the spleen is a site of red blood cell formation from the fourth month of intrauterine life. The red pulp is capable of resuming this function in adult life in myeloid leukaemia and certain other disorders.

3 *Lymphopoiesis*. The white pulp of the spleen is an important source of lymphocytes and large mononuclear cells (see below).

4 *Immune response*. When the body's immune system is responding to the presence of an antigen the macrophages in the spleen increase in number and the lymphoid tissue in the spleen also enlarges. In certain chronic infections (bacterial and protozoal) the spleen may become permanently enlarged. T-lymphocytes, which are found mainly in the periarteriolar sheaths, and B-lymphocytes, which are found mainly in the lymphatic follicles of the white pulp, migrate into the splenic cords where they become immunologically active.

5 *Storage of red blood cells*. In many mammals red blood cells are separated from plasma and are stored in the venous sinusoids or in the red pulp. In conditions of hypoxia the capsular muscle fibres contract and force the stored red blood cells into the circulation, thus increasing the oxygen-carrying capacity of the blood. This is not an important function of the spleen in man, although the spleen does contain about 1% of the total blood volume.

Splenectomy

In children, splenectomy, which may be necessary following abdominal trauma or in certain blood disorders, is usually followed by an increased susceptibility to infection. In adults splenectomy is followed by a transient increase in the numbers of circulating leucocytes, lymphocytes, platelets and red cells lasting a few weeks. Most of the functions of the spleen, however, are taken over by other parts of the macrophage system.

The thymus

The thymus is now known to be an organ which is important in controlling the activity of the T-lymphocytes. The thymus lies in the anterior and superior mediastinum behind the sternum. It is related posteriorly to the pericardium, the heart, the aortic arch, the great vessels and the trachea. It has right and left lobes, the upper parts of which extend into the neck. It grows in size from birth until puberty and then diminishes in size during adult life, although cellular reduction starts at about the age of five. In the newborn child the thymus is important for the proper development of the lymphoid system but by adolescence the lymphoid system is fully mature and the relative size of the thymus decreases. Nevertheless, in the adult, thymectomy results in a reduced capability to respond to new antigens.

Lymphoid tissue

Masses of lymphoid tissue, which are not encapsulated like the spleen or the lymph nodes, are found in various sites in the body, particularly associated with the alimentary tract, and the respiratory tract. They are found particularly in the following sites:

1   **Around the oropharynx** where the tonsils form a circle, known as *Waldeyer's ring*, around the openings into the alimentary and respiratory tracts. There are four groups of tonsils:

a)   the *palatine tonsils*, which lie on either side of the pharynx at the back of the mouth and are known as the 'tonsils' to the layman;

b)   the *pharyngeal tonsil*, which lies in the roof and posterior wall of the nasopharynx and is known as the 'adenoids';

c)   the *tubal tonsils* which are small and lie on either side of the nasopharynx just behind the opening of the auditory tube; and

d)   the *lingual tonsils* which are numerous small masses of lymphoid tissue and which lie in the posterior surface of the tongue.

2   **In the appendix, the colon and the rectum**, and in the small intestine where they are known as Peyer's patches.

3   **In the trachea and bronchi**.

Although it is not encapsulated, the structure of the mucosa associated lymphoid tissue resembles the structure of the lymph nodes and contains T- and B-lymphocytes as well as lymph follicles with germinal centres. Similar tissue is found in the lacrimal and salivary glands, in the breasts and in the genito-urinary system. The function of these centres probably is to provide defence against infective organisms by secreting an immunoglobulin (IgA—see Chapter 3).

The macrophage system

The macrophage system (mononuclear phagocyte system) consists of a widely scattered system of phagocytic cells and it was previously known as the reticulo-endothelial system. The latter term has now been abandoned because many of the cells in the system are neither reticular nor endothelial. The cells originate in the bone marrow as monocytes and then migrate to a number of different sites and they remain fixed in these sites until they are activated by the presence of invading micro-organisms, or other antigens, when they become mobile and phagocytic. The cells of the macrophage system are found in the following sites:

1   in *connective tissue;*
2   in *serosal cavities* (the pleura and the peritoneum);
3   in *the blood* as monocytes;
4   lining *the blood sinuses of the liver*, where they are known as Kupffer cells;
5   in *the reticular tissue of the spleen, the lymph nodes, the mucosa associated lymphoid tissue and the bone marrow;*
6   in *the meninges*;
7   in *the alveoli of the lung*;
8   in *areas of infection*; and
9   in *resorbing bone*, where they are called osteoclasts.

This system forms another important defence against infection and tumour growth. To be fully effective in dealing with foreign antigens the macrophages are dependent on normal lymphocyte function and vice versa.

Hodgkin's disease

Hodgkin's disease is the commonest of the lymphomas which are a group of diseases due to a malignant process of the lymphoid system. The lymph nodes are invariably involved, but the liver, the spleen and the bone marrow may also be affected. The prognosis depends upon how widespread are the areas of lymph node involvement and what are the microscopic appearances of the involved nodes. The better prognosis is associated with a single area of lymph node involvement and the presence of many lymphocytes on microscopic examination of an involved node.

   The clinical presentation is with smooth, rubbery, painless enlargement of lymph nodes. Although any group of lymph nodes may be affected, the commonest initial site is in the cervical group. Lymph node enlargement in the mediastinum may produce symptoms due to pressure on adjacent structures (see below). Fever is another common presenting symptom. In order to distinguish Hodgkin's disease from other causes of lymph node enlargement, such as infection, lymphatic leukaemia or secondary carcinoma, microscopic examination of an excised involved lymph node, or other involved tissue, is necessary. Treatment consists of radiotherapy or chemotherapy and occasionally both treatments are used. In patients with

Hodgkin's disease localized to one region of lymph nodes, who have no other symptoms, 90% may live for five years or more from the time that the diagnosis is made.

Another group of lymphomas is the **non-Hodgkin's lymphoma**, a condition in which there is uncontrolled growth of one clone (group) of lymphoid cells, usually the B-cells. The prognosis is generally worse than in Hodgkin's disease.

Multiple myeloma (myelomatosis)

Multiple myeloma is a neoplastic process involving plasma cells. Normally plasma cells produce antibodies and in this condition the malignant clone of plasma cells produces large quantities of an immunoglobulin which can be identified in the blood or urine. The bone marrow is usually involved. The osteoclasts, which are thought by some to be macrophages, are stimulated to cause resorption of bone and in some areas, particularly in the vault of the skull, the erosion of bone by plasma cell masses produces punched-out areas of bone destruction which are visible on plain radiographs.

Mediastinal obstruction

Enlargement of mediastinal lymph nodes due to Hodgkin's disease or local carcinomas, such as the bronchus or the breast, can produce symptoms due to pressure on adjacent structures in the mediastinum. A similar picture may occur with tumours of the thymus, mediastinal cysts, which are developmental abnormalities, and aneurysms of the ascending aorta.

Pressure on the following structures may occur:

1 the trachea, causing cough, breathlessness and stridor, which is a crowing sound occurring particularly during inspiration;

2 the bronchus, causing breathlessness and later collapse of a lobe of the lung or the whole lung;

3 the oesophagus, producing difficulty in swallowing and narrowing or displacement of the oesophagus which is visible on a barium swallow examination;

4 the phrenic nerve, causing paralysis of the diaphragm;

5 the left recurrent laryngeal nerve, causing paralysis of the left vocal cord and consequent hoarseness;

6 the superior vena cava, causing dilatation of the jugular veins and swelling and cyanosis of the head and neck; and

7 the sympathetic trunk, causing constriction of the pupil and drooping of the upper lid.

CHAPTER 13

The respiratory system

Every cell requires oxygen for combustion to provide the energy for its activities. The oxygen is conveyed to the cells in the blood, which also removes the carbon dioxide produced as a waste-product of combustion. The interchange of oxygen and carbon dioxide between the blood and the tissues is called *internal respiration*.

The function of the respiratory system is to replenish the blood with oxygen, and to remove the carbon dioxide from the blood. The interchange of gases which occurs in the respiratory system between the blood and the atmospheric air is called *external respiration*. The organs composing the respiratory system are: the *nose*, the *pharynx*, the *larynx*, the *trachea*, the *bronchi* and the *lungs*.

The nose

The bony structure of the nasal cavity has already been described (p. 106). The nasal cavity is lined by ciliated epithelium containing goblet cells, which secrete mucus. The mucus traps dust particles and bacteria, which are present in the air passing over the surface of the epithelium, and the cilia sweep these towards the exterior. Opening into the nasal cavity are the accessory nasal sinuses, which are similarly lined by ciliated epithelium. In the upper part of the nose are special cells which are concerned with the sense of smell (p. 502).

The pharynx

The pharynx is a muscular tube which is in part shared by the respiratory system and the digestive system. It lies behind the nasal cavities, the mouth and the larynx. It is approximately 12–14 cm long, extending from the base of the skull to the level of the 6th cervical vertebra, where it opens into the oesophagus. It is widest in its uppermost part and gradually narrows throughout its length. It opens anteriorly into the nasal cavities, the mouth and the larynx. Posteriorly it is separated by loose areolar tissue from the upper six cervical vertebrae.

The pharynx is divided for descriptive purposes into three parts: the *nasopharynx*, the *oropharynx* and the *laryngopharynx*.

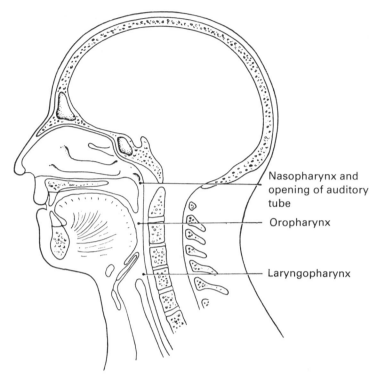

Nasopharynx and
opening of auditory
tube

Oropharynx

Laryngopharynx

Fig. 13.1. A sagittal section through the head and neck to show the subdivisions of the pharynx.

The nasopharynx is the portion behind the nasal cavity and above the level of the soft palate. It communicates anteriorly with the nasal cavity through the posterior nares. Between the soft palate and the posterior wall of the pharynx is an opening called the *nasopharyngeal isthmus*, which connects the nasopharynx and the oropharynx. This opening becomes obliterated during the act of swallowing. Opening into the lateral wall of the nasopharynx is the auditory tube (p. 499). In the roof, and extending onto the posterior wall, of the nasopharynx there is a collection of lymphoid tissue called the *pharyngeal tonsil* (the adenoids). This lymphoid tissue starts to atrophy at puberty, and has virtually disappeared by early adult life.

The oropharynx extends from the soft palate to the level of the tip of the epiglottis. It opens anteriorly, through the *oropharyngeal isthmus*, into the cavity of the mouth. The oropharyngeal isthmus is bounded above by the soft palate, below by the tongue and on either side by muscular folds, covered with mucous membrane, which are called the *pillars of the fauces*. In between the folds of either side lie the *palatine tonsils*, which are collections of lymphoid tissue.

The laryngopharynx extends from the tip of the epiglottis to the level of the 6th cervical vertebra, behind the cricoid cartilage of the larynx, where it is continuous with the oesophagus. Anteriorly it faces the opening of the larynx above and the posterior aspect of the larynx below. On either side of the larynx the laryngopharynx forms two pouches called the *pyriform fossae*.

The structure of the pharynx

The walls of the pharynx are composed of three coats:
1 *The mucous coat*. The pharynx is lined by mucous membrane. The mucous membrane of the nasopharynx possesses an inner lining of *ciliated* epithelium, but the oropharynx and the laryngopharynx possess *stratified* epithelium.
2 *The fibrous coat* lies between the inner mucous layer and the outer muscular layer. The fibrous coat is thickest above, where there are deficiencies in the muscular coat. It is firmly attached to the base of the skull, and is thickened posteriorly, in the midline, to give attachment to the muscles of the pharynx.
3 *The muscular coat*. The muscles of the pharynx are called the constrictor muscles of which there are three, the superior constrictor muscle, the middle constrictor muscle and the inferior constrictor muscle. The constrictor muscles arise from the base of the skull, the mandible, the hyoid bone and the larynx. They sweep round the pharynx to be attached to the posterior thickened portion of the fibrous coat.

The blood supply to the pharynx

The arteries of the pharynx are derived mainly from branches of the external carotid artery.

The veins of the pharynx empty into the internal jugular and the facial veins.

The nerve supply to the pharynx

The nerve supply to the pharynx is derived from the 9th cranial nerve, the 10th cranial nerve and the sympathetic nervous system.

The lymphatic drainage of the pharynx

The lymph vessels from the pharynx drain either directly into the deep cervical lymph nodes, or into the retropharyngeal and paratracheal lymph nodes.

The larynx

The larynx is a box-like structure which extends from the root of the tongue to the level of the upper end of the trachea. In the adult male it lies opposite the 3rd to the 6th cervical vertebrae but it is slightly higher in position in children and adult females. It acts as the organ for the production of the voice and also as a valve to prevent the passage of food into the trachea during the act of swallowing. It opens above into the laryngopharynx and below into the trachea.

The structure of the larynx

The larynx is composed of a number of cartilages which are attached to each other by ligaments and membranes. The cartilages are: the *thyroid cartilage*, the *cricoid cartilage*, the *epiglottis*, the two *arytenoid cartilages* and, in addition, there are two small *corniculate cartilages* and two small *cuneiform cartilages*.

The thyroid cartilage is the largest cartilage of the larynx. It is composed of two quadrilateral plates, or laminae, which are fused anteriorly, in the midline, to form the *laryngeal prominence* (the Adam's apple). Above the laryngeal prominence is a notch, called the *thyroid notch*. The two laminae are widely separated posteriorly. The upper and lower posterior angles of the laminae bear slender processes called the *superior* and *inferior horns*. Each inferior horn has a rounded facet on its inner surface which articulates with the cricoid cartilage.

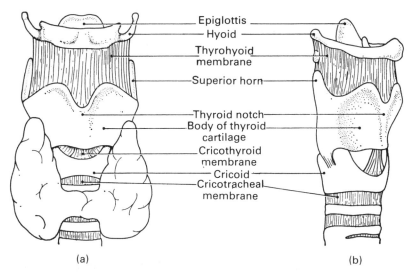

Epiglottis
Hyoid
Thyrohyoid membrane
Superior horn
Thyroid notch
Body of thyroid cartilage
Cricothyroid membrane
Cricoid
Cricotracheal membrane

(a)   (b)

Fig. 13.2. External views of the larynx: (a) anterior aspect, and (b) lateral aspect.

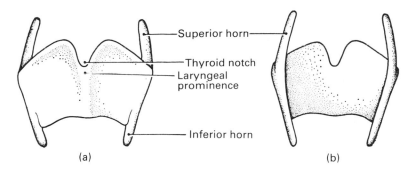

Fig. 13.3. The thyroid cartilage: (a) anterior aspect, and (b) posterior aspect.

The cricoid cartilage is shaped like a signet-ring, consisting of a quadrilateral lamina, which lies posteriorly, and a narrow arch, which lies anteriorly. The lateral surfaces of the lamina each bear two articular facets; one above for the arytenoid cartilage and one below for the inferior horn of the thyroid cartilage.

The arytenoid cartilages are pyramidal in shape, having an apex, a base and three surfaces, posterior, anterolateral and anteromedial. The

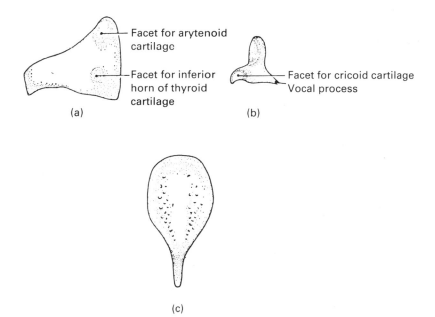

Fig. 13.4. (a) The cricoid cartilage, lateral aspect, (b) the arytenoid cartilage, lateral aspect, and (c) the epiglottis posterior aspect.

anterior angle of the arytenoid cartilage is called the *vocal process*, and gives attachment to the vocal ligament. The lateral angle gives attachment to muscles. The base of the arytenoid cartilage is concave and articulates with the facet on the upper part of the lateral surface of the lamina of the cricoid cartilage. The apex of the arytenoid cartilage articulates with the corniculate cartilage.

The epiglottis is a leaf-shaped cartilage. Its narrow lower end is attached, by the thyroepiglottic ligament, to the posterior surface of the thyroid cartilage, just below the thyroid notch. The upper end of the epiglottis projects upwards behind the hyoid bone and the base of the tongue. The lower part of the anterior surface is attached to the hyoid bone by the hyoepiglottic ligament.

The ligaments and membranes of the larynx
1   The thyrohyoid membrane joins the superior border of the thyroid cartilage to the body and greater horns of the hyoid bone.
2   The cricotracheal ligament joins the lower border of the cricoid cartilage to the 1st cartilaginous ring of the trachea.
3   The hyoepiglottic ligament joins the anterior surface of the lower part of the epiglottis to the posterior surface of the body of the hyoid bone.
4   The cricothyroid ligament is composed of a thick anterior portion, part of which joins the thyroid and cricoid cartilages, and a thin lateral portion, called the *cricovocal membrane*, which is attached, posteriorly, to the cricoid cartilage below and the vocal process of the arytenoid above. The upper edge of the lateral part of the cricothyroid ligament is free and forms the *vocal ligament* which is the framework of the *vocal cord*.
5   The cricoarytenoid ligaments are small ligaments which attach the arytenoid cartilages to the cricoid cartilage.

The cavity of the larynx
The larynx opens above into the pharynx and below into the trachea.

The cavity of the larynx is lined by mucous membrane, which is continuous above with the mucous membrane of the tongue and pharynx. The mucous membrane projects as an upper and lower pair of folds which divide the cavity of the larynx into three parts. The upper folds are called the *vestibular folds*, and the lower folds are called the *vocal folds*. The vocal folds cover the *vocal ligament* thus forming the *vocal cords*. The opening of the larynx into the pharynx faces backwards and slightly upwards and the anterior wall of the opening is formed by the upper part of the epiglottis. The opening is bounded on each side by a fold of mucous membrane

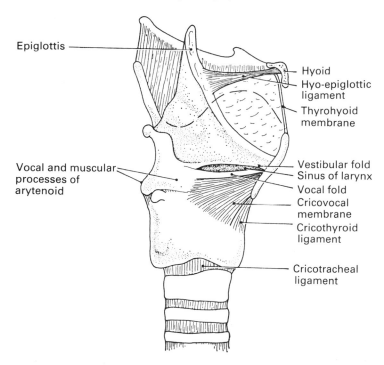

Epiglottis

Hyoid

Hyo-epiglottic ligament

Thyrohyoid membrane

Vocal and muscular processes of arytenoid

Vestibular fold

Sinus of larynx

Vocal fold

Cricovocal membrane

Cricothyroid ligament

Cricotracheal ligament

Fig. 13.5. The cartilages and ligaments of the larynx, lateral aspect.

called the *aryepiglottic fold*, which extends from the side of the epiglottis to the apex of the arytenoid cartilage.

The blood supply of the larynx

The arteries supplying the larynx are branches of the superior and inferior thyroid arteries. The veins from the larynx drain into the thyroid veins and, from there, into the internal jugular vein.

The nerve supply to the larynx

The nerve supply to the larynx is from the laryngeal and recurrent laryngeal nerves.

The lymphatic drainage of the larynx

The lymphatic vessels from the portion of the larynx above the vocal cords drain into the superior deep cervical lymph nodes. The lymphatic vessels from the portion of the larynx below the vocal cords drain into the inferior deep cervical lymph nodes and the pretracheal lymph nodes.

The functions of the larynx

The functions of the larynx are:
1   to allow air to pass from the pharynx to the trachea;
2   to prevent food entering the trachea during swallowing; and
3   to produce the voice.

During swallowing the larynx is drawn upwards by muscular action and the epiglottis becomes more horizontal, covering the opening of the larynx. This ensures that food passes from the pharynx into the oesophagus.

The production of the voice is dependent on the vocal cords, which extend from the midline of the thyroid cartilage anteriorly to the vocal processes of the two arytenoid cartilages posteriorly. During quiet breathing the vocal cords are separated posteriorly (abducted).

During phonation small muscles of the larynx rotate the arytenoid cartilages so that the vocal cords lie close together. Passage of air between the cords then produces the voice.

The volume, or loudness, of the sound depends on the force with which the air is forced between the cords.

The pitch, or note, of the sound depends on the tension of the cords. The larynx only produces sound. Speech is produced by modifications to the sound by the lips, the tongue and the teeth.

The trachea

The trachea is a tube about 10 cm long which extends downwards from the lower part of the larynx. It commences at the level of the 6th cervical vertebra, and ends at the level of the upper border of the 5th thoracic vertebra, by dividing into the right and left main bronchi. The portion of the trachea in the neck lies in the midline but in the thoracic cavity it is displaced slightly to the right by the arch of the aorta.

The relations of the trachea

In the neck. Anteriorly the trachea is crossed by the isthmus of the thyroid gland which covers the 2nd to the 4th cartilaginous rings.

Laterally the trachea is related to the lobes of the thyroid gland which separate it from the common carotid artery.

Posteriorly the trachea is related to the oesophagus, which separates it from the vertebral column.

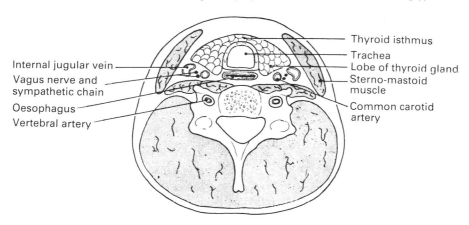

Fig. 13.6. Transverse section of the neck at the level of the lower part of the 6th cervical vertebra to show the relations of the trachea.

In the thoracic cavity. Anteriorly the trachea is related from above downwards to the remnants of the thymus gland, the brachiocephalic artery, the left common carotid artery and the arch of the aorta.

On the right side the trachea is related to the right lung and pleura.

On the left side the trachea is related to the left common carotid artery, the left subclavian artery and, below, to the arch of the aorta.

Posteriorly the trachea is related to the oesophagus.

The structure of the trachea

The trachea is composed of 16 to 20 C-shaped cartilaginous rings, deficient posteriorly, which are joined by a membrane composed of fibrous tissue, which contains unstriped muscle fibres. Posteriorly, where the cartilaginous rings are deficient, the wall is flattened and is composed of fibrous tissue containing elastic fibres and unstriped muscle fibres. The fibrous membrane of the rest of the walls is divided into two layers which split to enclose the cartilaginous rings. The trachea is lined by mucous membrane which possesses a lining of ciliated and columnar epithelium with large numbers of goblet cells. Beneath the mucous membrane is a loose layer of connective tissue which contains vessels and nerves.

The blood supply to the trachea

The arteries supplying the trachea are branches of the inferior thyroid artery. The veins from the trachea empty into the brachiocephalic veins.

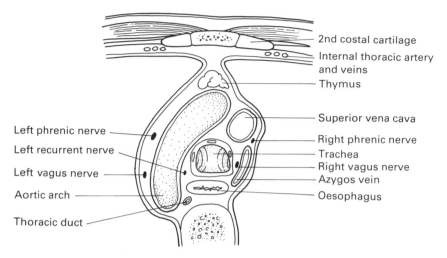

Left phrenic nerve

Left recurrent nerve

Left vagus nerve

Aortic arch

Thoracic duct

2nd costal cartilage

Internal thoracic artery and veins

Thymus

Superior vena cava

Right phrenic nerve

Trachea

Right vagus nerve

Azygos vein

Oesophagus

Fig. 13.7. Transverse section through the 4th thoracic vertebra to show the relations of the trachea.

The nerve supply to the trachea

The nerve supply to the trachea is from branches of the vagus nerve, the recurrent laryngeal nerves and the sympathetic nervous system.

The bronchi

The right and left main bronchi commence at the bifurcation of the trachea, at the level of the upper border of the 5th thoracic vertebra.

The right main bronchus is wider and shorter than the left, and runs more vertically downwards. It is about 2.5 cm in length, and enters the right lung opposite the 5th thoracic vertebra.

The left main bronchus is narrower than the right and it runs more horizontally. It is about 5 cm in length, and it enters the left lung at the level of the 6th thoracic vertebra. In its course it passes behind the arch of the aorta and in front of the oesophagus and the descending aorta. The left pulmonary artery lies first above and then in front of it.

Within the lung the two main bronchi divide and subdivide to form smaller and smaller branches. The principal divisions of the two main bronchi are shown in Fig. 13.8.

The structure of the bronchi

The main bronchi and the larger bronchi within the lungs are of similar structure to the trachea, but the cartilaginous rings are more irregularly placed.

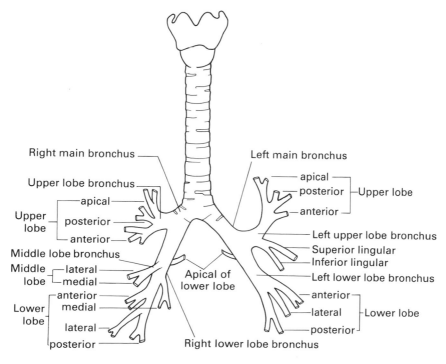

Fig. 13.8. The trachea and the main bronchi, anterior aspect.

The outer coat is composed of fibrous tissue in which lie the cartilaginous plates.

The middle layer is composed of circularly arranged unstriped muscle fibres.

The inner layer is the mucous membrane which contains mucous and serous glands and is lined by columnar ciliated epithelium.

The pleura

Each lung is covered by a closed sac of serous membrane called the pleura. The pleura consists of an outer layer called the *parietal pleura* and an inner layer called the *visceral pleura*.

The parietal pleura is attached to the chest wall, the mediastinum, and the upper surface of the diaphragm and it extends upwards into the base of the neck.

The visceral layer is attached to the surface of the lungs and extends into the fissures, which divide the lungs into lobes.

The two layers of the pleura are continuous around the hilum, or root, of each lung and the pleura extends downwards below each hilum as the *pulmonary ligament*. The two layers of the pleura are normally in close contact, being separated only by a thin film of serous fluid. This fluid holds the two layers of the pleura together by suction but, if air is allowed to leak into the pleural cavity, the two layers separate and the lung collapses. This condition is called a *pneumothorax*.

The lungs

The right and left lungs lie freely on either side of the mediastinum in the right and left pleural cavities, attached only at the hila, where the bronchi and pulmonary vessels enter and leave the lungs.

Each lung is roughly conical in shape, having an apex which lies superiorly, a base, or diaphragmatic surface, a costal surface and a mediastinal surface. The right lung is slightly larger than the left. At birth the lungs are pink in colour but with advancing years they become increasingly grey or black, due to deposits of inhaled particles of carbon from the atmosphere.

The apices of the lungs extend upwards into the base of the neck for 2−3 cm above the anterior margin of the 1st rib. Due to obliquity of the 1st rib, however, the apex of the lung is level with the neck of the 1st rib posteriorly.

The base of each lung is concave and rests on the diaphragm. The right dome of the diaphragm lies higher than the left dome and the right lung is therefore shorter than the left. The diaphragm separates the base of the right lung from the right lobe of the liver, and the base of the left lung from the left lobe of the liver, the fundus of the stomach and the spleen.

The costal surface of each lung is related to the chest wall.

The medial surface of each lung is related posteriorly to the thoracic vertebrae and anteriorly to the mediastinum. The lower part of the mediastinal surface of each lung is concave where it is related to the heart. The concavity of the right lung is less deep than that of the left lung and is related to the right atrium and part of the right ventricle. The concavity of the left lung is deep and is related to the left ventricle and the left auricle.

Above the concavity is the hilum of the lung where the pulmonary vessels, the bronchi, lymphatic vessels and nerves enter and leave the lung. Above the hilum the mediastinal relations of the two lungs differ.

The right lung (Fig. 13.9) is related to the superior vena cava, the lower end of the right brachiocephalic vein, the azygos vein, the right vagus nerve, and the oesophagus.

The left lung (Fig. 13.10) is related to the arch of the aorta, the left subclavian artery, the phrenic nerve and the oesophagus.

The inferior border of each lung is sharp where it lies in the recess between the chest wall and the diaphragm.

The anterior border of each lung is also sharp, and overlies the front of the pericardium. The anterior border of the left lung has a notch in its lower part, called the *cardiac notch*.

The lobes of the lung

The right lung is divided into three lobes by an *oblique fissure* and a *transverse fissure*. The three lobes are the upper lobe, the middle lobe and the lower lobe. The lower lobe lies below and behind the upper and middle lobes.

The oblique fissure of the right lung begins posteriorly at the level of the neck of the 5th rib and runs downwards and forwards, roughly following the line of the 5th rib and ends at the inferior border of the lung close to the 6th costal cartilage. The horizontal fissure runs horizontally and laterally from the hilum at the level of the 4th costal cartilage to meet the oblique fissures at about the midline of the axilla.

The left lung is divided into two lobes by an *oblique fissure*.

The two lobes are called the upper and the lower lobes, the lower lobe lying below and behind the upper lobe. The lower anterior portion of the upper lobe is called the *lingula*.

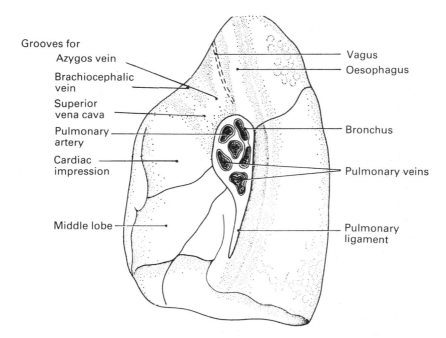

Fig. 13.9. The mediastinal aspect of the right lung.

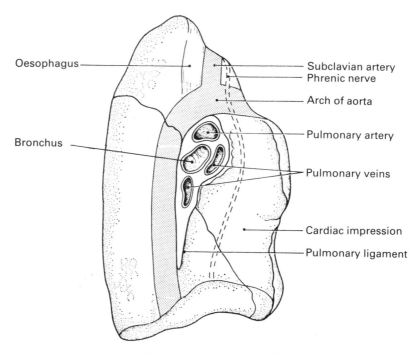

Oesophagus

Bronchus

Subclavian artery
Phrenic nerve

Arch of aorta

Pulmonary artery

Pulmonary veins

Cardiac impression
Pulmonary ligament

Fig. 13.10. The mediastinal aspect of the left lung.

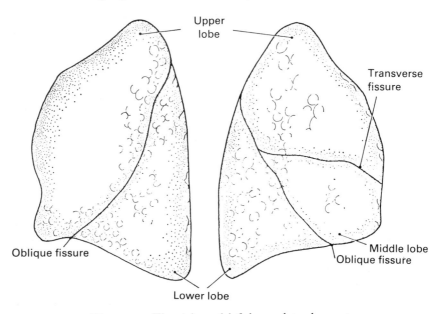

Upper
lobe

Transverse
fissure

Oblique fissure

Middle lobe
Oblique fissure

Lower lobe

Fig. 13.11. The right and left lungs, lateral aspect.

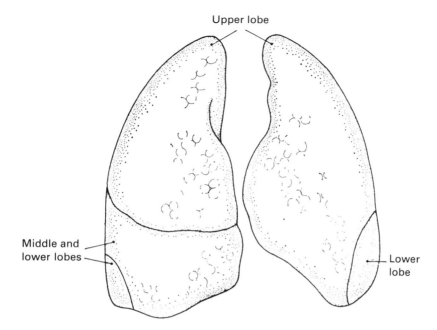

Fig. 13.12. The right and left lungs, anterior aspect.

The oblique fissure of the left lung begins posteriorly at about the level of the neck of the 4th rib, but it may start above or below this level. It then follows the same course as the oblique fissure of the right lung.

The hila of the lungs

The hilum of each lung transmits the bronchus, the pulmonary artery, the two pulmonary veins, the bronchial arteries, lymphatic vessels and nerves. The arrangement of these structures at the hilum is shown in Figs. 13.9 and 13.10.

The structure of the lungs

The lungs are composed of very large numbers of lobules which are closely connected by areolar tissue. Each lobule is composed of a terminal bron-chiole, air sacs, small branches of the pulmonary and bronchial arteries, lymphatic vessels and nerves.

The bronchi to each lobe of the lung divide and subdivide to form smaller and smaller branches. The terminal branches of the bronchi are called *bronchioles*. The bronchioles themselves divide to form terminal branches, called *respiratory bronchioles*, which are about 0.2 mm in diameter. The respiratory bronchioles contain no cartilage in their walls and they are lined by cuboidal epithelium which is not ciliated.

Each respiratory bronchiole divides into a number of *alveolar ducts* which in turn divide into a number of wide structures called *atria*. Each of these atria in turn divide into *air saccules* (Fig. 13.13). The thin walls of the air saccules, the atria and the alveolar ducts are covered by thin-walled structures, called *pulmonary alveoli*. The walls of the pulmonary alveoli are formed by a layer of flat epithelium, and it is here that the exchange of gases takes place between the alveoli and the pulmonary capillaries.

The blood supply of the lungs

The right and left pulmonary arteries carry deoxygenated blood from the right ventricle to the right and left lungs. Each pulmonary artery divides into branches which follow the divisions of the bronchi and bronchioles. Finally a dense network of capillaries is formed around the air sacs and alveoli.

The pulmonary veins, two from each lung, are formed by the union of veins which return from the pulmonary capillaries. The two pulmonary veins from each lung open into the left atrium.

The lungs are supplied with oxygenated blood by the bronchial arteries, which arise from the descending aorta. The bronchial veins into the azygos vein on the right, and the hemiazygos vein on the left.

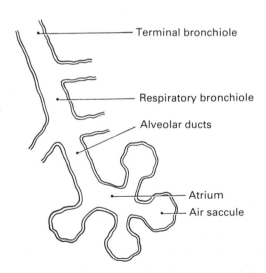

Fig. 13.13. The mode of termination of the bronchioles in the lungs.

The nerve supply to the lungs

The lungs are supplied by branches of the vagus nerve and the sympathetic nervous system.

The lymphatic drainage of the lungs

The lymphatic vessels of the lungs are arranged in a superficial plexus which drains the visceral pleura, and a deep plexus, which drains the bronchi. Both of these plexuses empty into the bronchopulmonary lymph nodes, which lie at the site of division of the main bronchi. From these nodes efferent vessels pass into the tracheobronchial lymph nodes.

The physiology of respiration

The mechanism of respiration

Air is drawn into the lungs during inspiration and expelled during expiration. This is achieved by movements of the ribs and diaphragm.

During *inspiration* the scalene muscles contract to prevent the first rib moving downwards and, at the same time, the external intercostal muscles contract to draw the anterior ends of the remaining ribs upwards. The ribs are articulated so that when their anterior ends are drawn upwards their lateral margins move outwards and the diameter of the thorax is increased in both the antero-posterior and lateral directions. At the same time as the ribs are moving outwards the muscular fibres of the diaphragm contract and pull the central tendon of the diaphragm downwards. Both sets of movements produce an increase in the volume of the thorax. The intercostal muscles are innervated by the motor neurons of T_1-T_{12}, and the diaphragm by the motor neurons of C_3, 4 and 5.

The lungs move with the chest wall and the diaphragm due to the suction exerted between the visceral and parietal layers of the pleura. Thus an increase in the volume of the thorax produces an increase in the volume of the lungs and air is therefore drawn into the lungs through the trachea and bronchi eventually reaching the alveoli. During *quiet respiration* the majority of the increase in volume required is obtained by movement of the diaphragm. In *forced inspiration* the external intercostal muscles and diaphragm contract fully and the 1st rib is actually moved upwards with the manubrium sterni by the scalene and sternomastoid muscles.

During quiet breathing *expiration* is produced by relaxation of the scalene and external intercostal muscles and the diaphragm. The ribs move downwards again by elastic recoil and the diaphragm is pushed upwards by the pressure of the intra-abdominal organs. During *forced expiration* the abdominal muscles contract and raise the intra-abdominal pressure as well as drawing the lower ribs downwards and medially. The muscles of the shoulder girdle can also be used to assist the internal intercostal muscles to pull the ribs downwards.

During quiet breathing there is usually a short pause between inspiration and expiration.

Lung volumes and capacities

During quiet breathing about 500 ml of air are taken into and expelled from the lungs with each breath. This is called the *tidal volume*. Of this 500 ml only about 360 ml actually reach the alveoli. The remainder fills the air passages and this 140 ml is known as the dead space since it does not take part in gas exchange (see below).

During forced breathing the maximum amount of air that can be forced out after maximum inspiration is about 350 ml. This volume is called the *vital capacity*. The *residual volume* is the volume of air which remains in the lungs after maximum expiration and is about 1200 ml.

The control of respiration

A resting adult takes a breath about 16 to 18 times per minute. The rate of breathing in children is higher. During exercise more oxygen is required by the tissues and more carbon dioxide is produced. In order to deal with this, ventilation is increased. This increase is brought about partly by nervous factors and partly by chemical factors. The rate and depth of respiration is controlled by a group of interconnected nerve cells in the medulla oblongata, pons and lower midbrain known as the *respiratory centre*. Fibres pass from the respiratory centre to the phrenic nerve, which supplies the diaphragm, and to the intercostal nerves, which supply the intercostal muscles. Rhythmic impulses pass down these nerves to produce the movements of respiration. Breathing normally takes place without conscious control, but it is possible to control the rate and depth of breathing voluntarily.

The respiratory centre is stimulated by a rise in the concentration of carbon dioxide in the blood. During exercise the amount of carbon dioxide in the blood increases and the respiratory centre responds by increasing the rate and depth of respiration, thus increasing the rate of excretion of carbon dioxide. The respiratory centre is also stimulated by a rise in the acidity of the blood (increased hydrogen ion concentration).

Although a rise in carbon dioxide concentration is the most important chemical stimulus to the respiratory centre, a fall in arterial oxygen concentration also stimulates ventilation of the lungs. Oxygen lack is an important stimulus to respiration at altitudes where the partial pressure of oxygen is low, and also in chronic lung diseases. Whereas a rise in the carbon dioxide concentration in the blood stimulates chemical receptors in the medulla oblongata, a fall in oxygen concentration stimulates receptors in the aortic and carotid bodies. The carotid bodies are specialized structures which lie at the bifurcation of the common carotid arteries into the internal and external carotid arteries, while the aortic bodies are found close to the aortic arch and right subclavian artery. The afferent pathways from these receptors are branches of the vagus (Xth) nerve.

Gas exchange in the lungs

The constituents of inspired and expired air are shown in table 13.1.

In the lungs oxygen is removed from the air and carbon dioxide is added. The blood flowing through the pulmonary capillaries surrounding the alveoli has a high tension of carbon dioxide and a low tension of oxygen. Therefore carbon dioxide passes into the alveoli from the blood and oxygen passes from the alveoli into the blood where it combines with haemoglobin to form oxyhaemoglobin.

Gas transport in the blood

Oxygen in the blood is carried mainly in the red cells as oxyhaemoglobin. In the tissues, where oxygen tension is low, oxyhaemoglobin dissociates releasing oxygen for use in tissue respiration. When a tissue is active, such as exercising muscle, the oxygen extraction may be virtually complete. Carbon dioxide, which is more soluble than oxygen, is carried in the blood partly in solution, partly as bicarbonate ions and partly combined with haemoglobin as carbaminohaemoglobin. Reduced (non-oxygenated) haemoglobin more readily takes up carbon dioxide to form carbaminohaemoglobin.

Thus, in the tissues where oxygen tension is low, oxyhaemoglobin dissociates to yield oxygen, and carbon dioxide is taken up by the blood as bicarbonate ions in solution or by haemoglobin as carbaminohaemoglobin.

Table 13.1. The constituents of inspired and expired air.

	Inspired air	Expired air
Nitrogen	79%	79%
Oxygen	20%	16%
Carbon dioxide	0.04%	4%
Water vapour	+	++

In the lungs the mechanisms are reversed, the low concentrations of carbon dioxide in the alveoli and its great solubility causing passage of the gas from the blood into the alveolus. The high oxygen concentration in the alveolus and the presence of reduced haemoglobin favour the passage of oxygen from the alveolus into the blood.

CHAPTER 14

The digestive system

The digestive system is composed of the organs which are concerned with the digestion and absorption of the food which we eat, and with the elimination of the undigested and unabsorbed residue. Digestion involves the mechanical and chemical breakdown of food into forms in which it can be absorbed into the body, and used for growth, repair and the provision of energy.

The digestive system is composed of the alimentary canal, which extends from the mouth to the anus, and of certain accessory organs.

The alimentary canal, which is about 9 m in length, is composed of: the *mouth*, the *pharynx*, the *oesophagus*, the *stomach*, the *small intestine* and the *large intestine*.

The accessory organs are: the *three pairs of salivary glands*, the *pancreas*, the *liver* and the *biliary apparatus*.

The mouth

The mouth, or oral cavity, is composed of an outer portion called the vestibule, and an inner portion, which is the mouth cavity proper.

The vestibule is bounded externally by the cheeks and lips, and internally by the teeth and gums. The ducts of the parotid salivary glands open into the vestibule of the mouth.

The mouth cavity proper is bounded anteriorly and laterally by the teeth and gums. It communicates, posteriorly, with the pharynx through the *oropharyngeal isthmus*. The lateral margins of the oropharyngeal isthmus are formed by two folds of mucous membrane, which are called the *pillars of the fauces*. Between these two folds, on each side, lies a mass of lymphoid tissue called the *palatine tonsil*. The roof of the mouth is formed anteriorly by the *bony palate* and posteriorly by the *soft palate*. A small conical projection, called the uvula, hangs downwards from the midline of the posterior border of the soft palate. The floor of the mouth is formed by the anterior part of the tongue and by the mylohyoid muscles which formed a muscular diaphragm across the arch of the mandible. Each mylohyoid muscle arises from the mylohyoid line of the mandible and is inserted into the hyoid bone and into a fibrous cord, which extends between the two muscles from the symphysis menti to the hyoid bone.

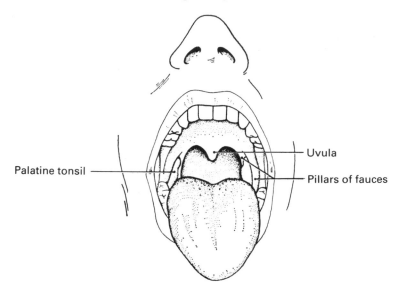

Fig. 14.1. View of the mouth with the tongue depressed.

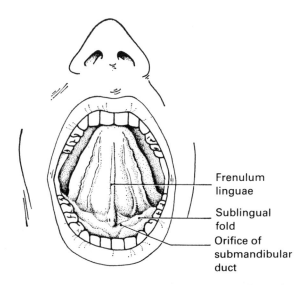

Fig. 14.2. View of the mouth with the tongue elevated.

When the tongue is elevated a fold of mucous membrane, called the *frenulum linguae*, can be seen extending from the midline of the under surface of the tongue to the floor of the mouth. On each side of the lower end of the frenulum linguae is a small papilla, called the *submandibular papilla*, on the surface of which the duct of the submandibular salivary gland opens into the mouth. Extending backwards and laterally from the submandibular papilla is the *sublingual fold*, beneath which lies the sublingual salivary gland. The ducts of the sublingual gland open into the mouth through a series of minute openings along the edge of this fold.

The lips

The lips are two fleshy folds which surround the anterior opening of the mouth. They are formed by the orbicularis oris muscle which is covered externally by skin and internally by mucous membrane.

The palate

The palate consists of the hard palate anteriorly and the soft palate posteriorly.

The hard, or bony, palate is formed, in its anterior two-thirds, by the palatine processes of the maxillae and, in its posterior one-third, by the horizontal plates of the palatine bones.

The soft palate is a moveable fold composed of fibrous tissue, which contains some muscle fibres.

The cheeks

The cheeks form the lateral margins of the vestibule of the mouth. They are continuous anteriorly with the margins of the lips. The main muscle which forms each cheek is the buccinator muscle.

The tongue

The tongue is a muscular organ which lies partly in the mouth and partly in the pharynx. It is concerned with the sense of taste, with speech and with the act of swallowing, or deglutition. It is attached by muscles to the hyoid bone, to the mandible and to the styloid process of the temporal bone.

The upper surface, or dorsum, of the tongue is divided into an anterior, oral two-thirds, which faces upwards, and a posterior, pharyngeal third, which faces backwards. The junction of the two parts is marked by a V-shaped groove, called the *sulcus terminalis*. The posterior part of the dorsum of the tongue forms the anterior wall of the oropharynx. The anterior part of the tongue lies in the floor of the cavity of the mouth and its anterior and lateral margins are free.

The tongue is composed of striped muscle fibres which run longitudinally, transversely and vertically. Intermingling with these fibres are fibres of the muscles by which the tongue is attached to the mandible, the hyoid bone and the styloid process of the temporal bone.

The tongue is covered by mucous membrane which has an outer layer of stratified squamous epithelium. The mucous membrane of the oral part of the dorsum of the tongue bears numerous small projections, called *papillae*. These papillae are concerned with the sense of taste (p. 503) and are of three types.

1 **Vallate papillae** are large papillae found in a V-shaped row in front of the sulcus terminalis. They vary in number from eight to twelve. They are about 2 mm in diameter and are easily recognizable.

2 **Fungiform papillae** are more numerous and smaller than the vallate papillae. They are found mainly at the sides and tips of the tongue, but a few lie on the dorsum of the tongue. They are deep red in colour.

3 **Filiform papillae** are small conical projections which cover the anterior two-thirds of the dorsum of the tongue.

The blood supply to the tongue

The tongue is supplied with arterial blood from the lingual artery, which is a branch of the external carotid artery. The veins from the tongue drain into the internal jugular vein.

The nerve supply to the tongue

The muscles of the tongue are supplied by the hypoglossal nerve.

General sensations from the anterior two-thirds of the tongue pass to the brain in the mandibular division of the trigeminal nerve. Sensations of taste from the anterior two-thirds of the tongue, excluding the vallate papillae, are carried to the brain in a branch of the facial nerve.

General sensations and sensations of taste are carried to the brain from the posterior third of the tongue by the glossopharyngeal nerve.

The lymphatic drainage of the tongue

See Chapter 12.

The mucous membrane of the mouth

The mucous membrane of the mouth is continuous *anteriorly* with the skin of the external surface of the lips and posteriorly with the mucous membrane of the pharynx. It is lined with stratified squamous epithelium, which does not have an outer horny layer. The mucous membrane is reflected from the inner surface of the cheeks and lips on to the gums. From the inner surface

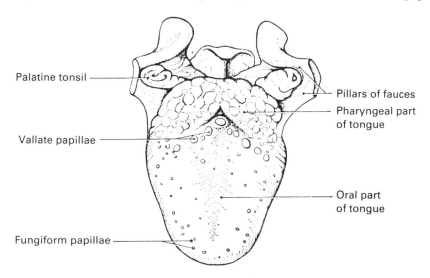

Palatine tonsil

Vallate papillae

Fungiform papillae

Pillars of fauces

Pharyngeal part of tongue

Oral part of tongue

Fig. 14.3. The dorsum of the tongue.

of the gums it is reflected above on to the palate and below on to the floor of the mouth and the tongue.

The teeth

The teeth are embedded in the sockets of the alveolar processes of the maxillae and the mandible. Two sets of teeth are developed during the lifetime of each individual. The first set, called the *deciduous teeth*, erupt through the gums during the 1st and 2nd years of life. The second set, called the *permanent teeth*, start to replace the deciduous teeth after the 6th year of life, and have usually all erupted by the end of the 25th year of life.

1   **The deciduous teeth.** There are 20 deciduous teeth, five in each half of each jaw. They are: two incisors, one canine, two molars.

2   **The permanent teeth.** There are usually 32 permanent teeth, eight in each half of each jaw. They are: two incisors, one canine, two premolars, three molars.

The teeth are usually referred to by the dental formula, in which the first incisor of each half of each jaw receives the number 1 and the third molar the number 8. A full set of teeth would thus be written:

$$\begin{array}{c|c} 87654321 & 12345678 \\ \hline 87654321 & 12345678 \end{array}$$

The right upper first molar would thus be written 6⌋ and the left lower canine ⌈3.

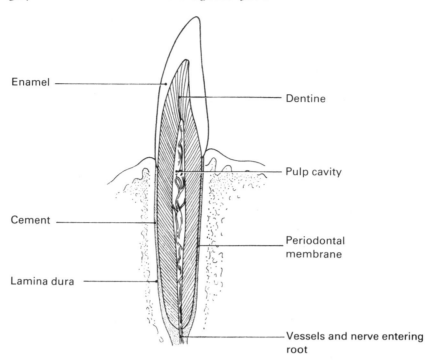

Enamel

Dentine

Pulp cavity

Cement

Periodontal membrane

Lamina dura

Vessels and nerve entering root

Fig. 14.4. Diagrammatic longitudinal section of a tooth.

The structure of the teeth

The part of a tooth which projects above the gum is called the *crown*, the part which is embedded in the alveolar process is called the *root* and the slight constriction between the crown and the root is called the *neck*.

When a vertical section is made through a tooth, a cavity, called the *pulp cavity*, is seen in the centre of the crown, the neck and each root. Within the pulp cavity is the *dental pulp*, which consists of loose connective tissue which contains blood vessels, nerves and lymphatic vessels. The vessels and nerves enter the pulp cavity at the apex of each root.

The bulk of each tooth is composed of *dentine*, which is a hard, yellowish-white, avascular substance. The dentine resembles dense bone, but, instead of Haversian canals, it contains minute canals which open into the pulp cavity. The dentine of the crown of each tooth is covered by a layer of a dense white avascular substance, called *enamel*. The enamel, which is chiefly composed of calcium phosphate, extends downwards to the neck of the tooth.

The roots of each tooth are covered by a layer of coarse bone, called *cement*, which fixes the root in its socket. The cement is fixed to the periosteum of the socket, and the periosteum is referred to as the *periodontal membrane*. The cortex of the bone of the socket is called the *lamina dura*.

The shape of individual teeth (Fig. 14.5)

The incisor teeth have a sharp chisel-shaped crown, which is adapted for biting food. The outer surface of the crown is convex, the inner surface is concave and there is one root.

The canine teeth have a blunt pointed crown, and a long single root.

The premolar teeth bear, on the crown, two rounded projections, or cusps. One cusp lies on the outer side, and one on the inner side, of the crown. There is usually one root, but this bears a longitudinal groove.

The molar teeth are the largest teeth and they are adapted for grinding the food. The crown of each tooth is roughly cuboidal in shape, and the upper surface bears three or four cusps. Each upper molar has three roots, and each lower molar has two roots.

The salivary glands

There are three pairs of salivary glands, the parotid glands, the submandibular glands and the sublingual glands. They secrete the saliva which passes through their ducts into the mouth.

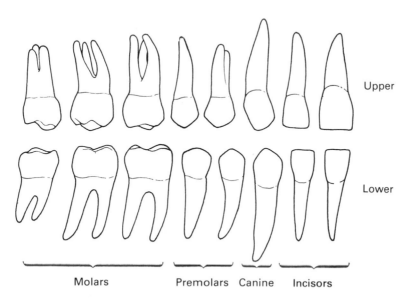

Upper

Lower

Molars   Premolars Canine   Incisors

Fig. 14.5. The various types of permanent teeth.

The parotid glands

The parotid glands are the largest of the salivary glands. Each gland is a lobulated mass which lies below the external acoustic meatus, between the mandible and the muscle. It is roughly pyramidal in shape and has superficial, anteromedial and posteromedial surfaces.

The superficial surface of the gland is covered by the skin and the subcutaneous tissue. It extends upwards to the external acoustic meatus, and downwards to the angle of the mandible. Its anterior margin extends on to the masseter muscle, and its posterior margin on to the sternocleidomastoid muscle.

The posteromedial surface of the gland lies against the floor of the external acoustic meatus, the mastoid process, the styloid process and the sternocleidomastoid muscle. The external carotid artery enters the lower part of the posteromedial surface and divides, within the substance of the gland, into the maxillary artery and the superficial temporal artery. The facial nerve enters the upper part of the posteromedial surface of the gland and divides, within the substance of the gland, into its terminal branches, which pass out through the anteromedial surface.

The anteromedial surface of the gland is grooved by the posterior border of the ramus of the mandible and lies partly on the surface of the masseter muscle.

The *parotid duct* (Stenson's duct) is about 5 cm in length. It emerges from the anterior border of the gland and runs forwards over the outer surface of the masseter muscle. At the anterior end of this muscle it turns inwards to pierce the buccinator muscle, and it opens on the surface of a

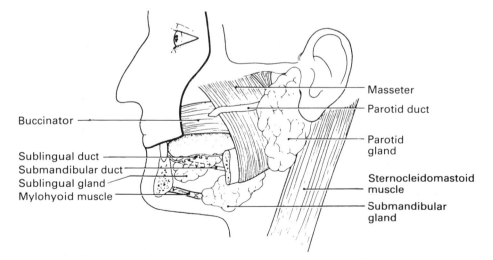

Fig. 14.6. The salivary glands: a portion of the mandible has been removed.

small papilla on the inner surface of the cheek, opposite the second upper molar tooth.

The submandibular gland

The submandibular gland lies between the angle of the mandible and the posterior part of the mylohyoid muscle. Its lower surface is superficial, and is covered only by the skin and subcutaneous tissue. Its lateral surface is related to the submandibular fossa of the inner surface of the body of the mandible. Its medial surface is related to the mylohyoid muscle. A portion of the gland, called the deep portion, extends upwards on to the upper surface of the mylohyoid muscle.

The *submandibular duct* (Wharton's duct) is about 5 cm in length. It emerges from the deep surface of the gland and runs forwards, beneath the mucous membrane of the floor of the mouth, to open on the surface of the small papilla at the base of the frenulum linguae.

The sublingual glands

The sublingual glands are the smallest of the salivary glands. Each lies beneath the mucous membrane of the floor of the mouth, in contact with the sublingual fossa of the inner surface of the anterior part of the body of the mandible. It lies on the surface of the mylohyoid muscle, and is in contact, in the midline, with the gland of the opposite side. It opens into the mouth by a series of small ducts which emerge on the surface of the sublingual fold.

The structure of the salivary glands

The salivary glands are compound racemose glands, and consist of large numbers of lobules, which are interconnected by areolar tissue containing vessels and nerves. The lobules are composed of secretory alveoli which are of two types:
1   **mucous alveoli**, which are lined by columnar epithelium and secrete mucus; and
2   **serous alveoli**, which are lined by cuboidal epithelium and secrete a watery fluid which contains an enzyme called *ptyalin*.

The parotid gland contains only serous alveoli.

The pharynx

The pharynx is divided into three portions, the nasopharynx, the oropharynx and the laryngopharynx. Only the oropharynx and the laryngopharynx are concerned with the digestive system. The anatomy of the pharynx has already been described (p. 339).

The oesophagus

The oesophagus is about 25 cm in length and extends from the pharynx to the stomach.

It commences in the neck at the level of the 6th cervical vertebra, where it is continuous with the lower end of the laryngopharynx. It runs downwards, in front of the vertebral column, through the superior and posterior parts of the mediastinum and passes through the diaphragm, slightly to the left of the midline, at the level of the 10th thoracic vertebra. At the level of the 11th thoracic vertebra it opens into the cardiac orifice of the stomach.

At its commencement the oesophagus lies in the midline, but it curves slightly to the left down to the level of the thoracic inlet and then curves back again to reach the midline at the level of the 5th thoracic vertebra. Below this it again curves slightly to the left, before passing through the diaphragm.

The relations of the oesophagus

In the neck. *Anteriorly* the oesophagus is related to the trachea, *posteriorly* to the cervical vertebrae, and *laterally* to the common carotid artery and the lower part of the lobe of the thyroid gland. The thoracic duct is related to the left side of the oesophagus.

In the thorax. *Anteriorly* the oesophagus is related, from above downwards, to the trachea, the left main bronchus, the right pulmonary artery and the pericardium and the left atrium of the heart. *Posteriorly* the oesophagus is related to the thoracic vertebrae, the thoracic duct, and, just above the diaphragm, to the descending thoracic aorta. *On the left side* the oesophagus is related, in the superior mediastinum, to the left subclavian artery, the thoracic duct, the aortic arch and the left lung and pleura. In the posterior mediastinum it is related to the descending thoracic aorta and the left lung and pleura. *On the right side* the oesophagus is related to the right lung and pleura.

The vagus nerves form a plexus around the lower end of the oesophagus. The *abdominal part* of the oesophagus is about 2.5 cm in length. After piercing the diaphragm it lies on the posterior surface of the left lobe of the liver. The right border of the oesophagus is smoothly continuous with the lesser curvature of the stomach, but the left border meets the fundus of the stomach at a sharp angle which is called the cardiac notch.

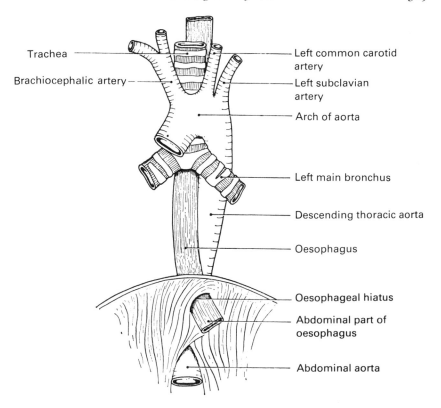

Trachea

Brachiocephalic artery

Left common carotid artery

Left subclavian artery

Arch of aorta

Left main bronchus

Descending thoracic aorta

Oesophagus

Oesophageal hiatus

Abdominal part of oesophagus

Abdominal aorta

Fig. **14.7.** The oesophagus and its anterior relations.

The lumen of the oesophagus is not uniform, but is slightly narrowed at three sites:

1 at its commencement;
2 at the level of the left main bronchus; and
3 at the level at which it passes through the diaphragm.

The structure of the oesophagus

The wall of the oesophagus is composed of four coats:

1 *an outer coat* of areolar tissue, which contains elastic fibres;
2 *a muscular coat*, which is composed of an outer layer of longitudinal muscle fibres and an inner layer of circular muscle fibres;
3 *a submucous coat* of loose areolar tissue which contains vessels and nerves; and

4 *an inner mucous coat* which is lined by stratified squamous epithelium: the mucous membrane of the oesophagus has a deep layer, called the *muscularis mucosae*, which contains longitudinal smooth muscle fibres. The mucous coat lies in a series of longitudinal folds when the oesophagus is empty, but these are smoothed out when the oesophagus is distended with food.

The blood supply to the oesophagus

The oesophagus is supplied with arterial blood by branches of the inferior thyroid arteries, branches of the descending thoracic aorta and branches of the left gastric artery. The veins from the oesophagus mainly drain into the azygos veins and into the left gastric vein.

The nerve supply to the oesophagus

The oesophagus receives its nerve supply from branches of the vagus nerve and from the sympathetic nervous system.

The lymphatic drainage of the oesophagus

The lymphatic vessels from the cervical part of the oesophagus pass to the deep cervical lymph glands.

The lymphatic vessels from the thoracic part of the oesophagus pass to the posterior mediastinal lymph glands.

The lymphatic vessels from the abdominal part of the oesophagus pass to the lymph glands which lie alongside the left gastric artery.

The stomach

The stomach is the most dilated portion of the alimentary canal. It lies in the left hypochondrium, the epigastrium and the umbilical regions of the abdomen. It communicates, above, with the oesophagus at the cardiac orifice and, below, with the duodenum at the pyloric orifice. The pyloric orifice is surrounded by a thickened band of smooth muscle called the pyloric sphincter. There is no sphincter at the cardiac orifice, but it is thought that the fibres of the right crus of the diaphragm compress the oesophagus, and prevent regurgitation of the gastric contents.

The volume of the stomach in an average adult is about 1000−1500 cc. The exact shape of the stomach varies both with the volume of its contents and with the build of the individual. When the stomach is studied radiologically, during barium meals, the commonest type of stomach is seen to be J-shaped. Less commonly the stomach lies more transversely and is called

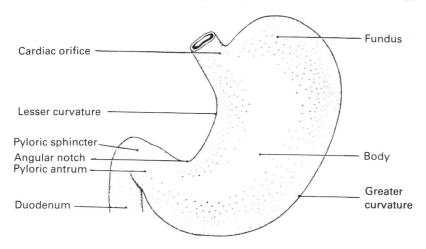

Cardiac orifice —

Fundus

Lesser curvature —

Pyloric sphincter
Angular notch
Pyloric antrum —

Body

Greater
curvature

Duodenum —

Fig. 14.8. The stomach.

a 'steerhorn' stomach, or it tends to sag downwards when it is called a 'fish-hook' stomach (Fig. 14.9).

The stomach has anterior and posterior surfaces and upper right and lower left borders, called the lesser and greater curvatures respectively. The lesser curvature usually has, in its dependent portion, a sharp notch, called the *angular notch*, or the incisura.

The stomach is divided, for descriptive purposes, into three parts:

1 **The fundus** lies above, and to the left of, the cardiac orifice, below the left dome of the diaphragm.

2 **The body** lies below the fundus and above the more narrowed portion of the stomach, which is called the pyloric antrum. The junction between the pyloric antrum and the body of the stomach is indicated by the angular notch.

3 **The pyloric antrum** lies between the body of the stomach and the pyloric orifice.

The relations of the stomach

Anteriorly the stomach is related to the diaphragm, the left costal margin, the left lobe of the liver and the anterior abdominal wall.

Posteriorly the stomach is related to the diaphragm, the left suprarenal gland, the left kidney, the pancreas, the splenic artery, the transverse mesocolon and the left half of the transverse colon, the splenic flexure of the colon and the spleen. These structures are frequently referred to as the 'stomach-bed' (Fig. 14.10).

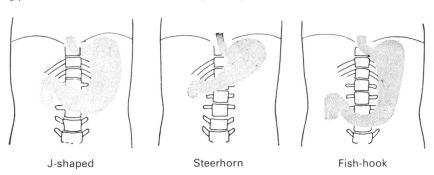

J-shaped   Steerhorn   Fish-hook

Fig. 14.9. Types of stomach.

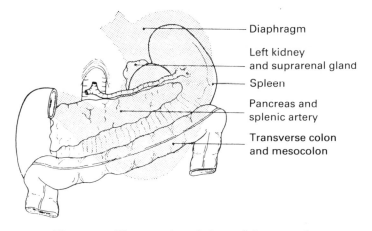

— Diaphragm

— Left kidney
and suprarenal gland

— Spleen

Pancreas and
splenic artery

Transverse colon
and mesocolon

Fig. 14.10. The posterior relations of the stomach.

The *lesser omentum* is attached to the lesser curvature of the stomach and the *greater omentum* to the greater curvature.

The structure of the stomach

The wall of the stomach is composed of four coats:

1 An *outer serous coat*, which is provided by the peritoneum.

2 A *muscular coat*, which is composed of an outer layer of longitudinal smooth muscle fibres, a middle layer of circular smooth muscle fibres and an inner, incomplete layer, of obliquely arranged smooth muscle fibres; the circular muscle is thickened at the pyloric orifice to form the *pyloric sphincter*.

3 A *submucous coat* of loose areolar tissue which contains vessels and nerves.

4 An *inner thick layer of mucous membrane*. The mucous membrane lies in a series of longitudinal folds when the stomach is empty. The mucous membrane has an inner lining of columnar epithelium and it contains large numbers of gastric glands. These glands are of three types:

a)   *the cardiac glands*, which lie near the cardiac orifice, and mainly secrete mucus;

b)   the *main gastric glands*, which lie in the fundus and body of the stomach, and secrete the gastric juice which is responsible for the digestion of food in the stomach; and

c)   *the pyloric glands*, which lie in the pyloric antrum and mainly secrete mucus.

The deep layer of the mucous membrane contains smooth muscle fibres and is called the *muscularis mucosae*.

The blood supply to the stomach

The stomach is supplied with arterial blood by the left gastric artery and by branches of the hepatic and splenic arteries. The veins from the stomach drain into the portal circulation.

The nerve supply to the stomach

The stomach receives its nerve supply from branches of the vagus nerve and from branches of the sympathetic nervous system.

The lymphatic drainage of the stomach

The lymphatic vessels from the stomach follow the course of the arteries of supply, and drain, ultimately, into the coeliac lymph nodes which lie on the front of the abdominal aorta.

The small intestine

The small intestine is about 6 m in length, and it extends from the pyloric orifice of the stomach to the ileocaecal valve, where it is continuous with the large intestine. It is divided into three parts: the *duodenum*, the *jejunum* and the *ileum*.

The duodenum

The duodenum is about 25 cm in length, and it is the shortest and widest part of the small intestine. It is relatively fixed in position, and it describes a

C-shaped curve around the head of the pancreas. It is divided, for descriptive purposes, into four parts:

1 The first part of the duodenum runs upwards and to the right for about 2.5 cm. It is entirely covered with peritoneum, whereas the remaining three parts of the duodenum are retroperitoneal. At the under surface of the liver the first part of the duodenum turns sharply downwards to become continuous with the second part.

2 The second part of the duodenum runs downwards for about 7.5 cm describing a slight curve around the head of the pancreas. The common bile duct and the pancreatic duct open into the medial aspect of this part, about midway along its length. The opening is called the *duodenal papilla* and it is surrounded by a sphincter called the *sphincter of Oddi*.

3 The third part of the duodenum is 10 cm in length, and it runs horizontally to the left, across the inferior vena cava and the abdominal aorta.

4 The fourth part of the duodenum runs upwards and to the left for 2.5 cm, and then turns forwards to join the jejunum at the *duodenojejunal flexure*.

The relations of the duodenum

The first part of the duodenum is related above and anteriorly to the liver and the gall bladder. Posteriorly it is related to the portal vein and the common bile duct. Below it is related to the head of the pancreas.

The second part of the duodenum lies on the medial aspect of the right kidney. It is related medially to the head of the pancreas and laterally to the hepatic flexure of the colon. The transverse colon crosses its anterior surface.

The third part of the duodenum is related posteriorly to the right psoas muscle, the inferior vena cava and the abdominal aorta. The root of the mesentery and the superior mesenteric vessels cross the anterior surface of the third part about midway along its length. It is related above to the head of the pancreas.

The fourth part of the duodenum is related anteriorly to the transverse colon and the transverse mesocolon. Above it is related to the head of the pancreas. Posteriorly it is related to the left psoas muscle, the left renal vein and the inferior mesenteric vein.

The blood supply to the duodenum

The duodenum receives arterial blood from branches of the hepatic and superior mesenteric arteries. The veins from the duodenum ultimately drain into the portal vein.

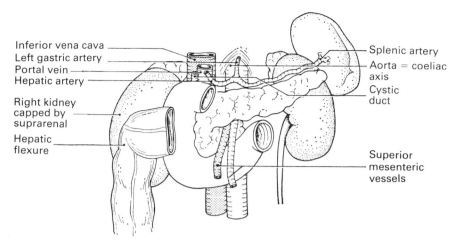

Inferior vena cava
Left gastric artery
Portal vein
Hepatic artery

Right kidney
capped by
suprarenal

Hepatic
flexure

Splenic artery
Aorta = coeliac
axis
Cystic
duct

Superior
mesenteric
vessels

Fig. 14.11. The relations of the duodenum.

The jejunum and the ileum

The jejunum and the ileum extend from the duodenojejunal flexure to the ileocaecal valve. They lie in a complicated series of loops and are completely enclosed in peritoneum. The jejunum, which forms the upper two-fifths of the remaining part of the small intestine, tends to lie in the umbilical region. The ileum tends to lie in the hypogastric region and in the upper part of the pelvic cavity. The last part of the ileum passes upwards to open into the caecum at the *ileocaecal valve*. There is no definite point of demarcation between the jejunum and ileum, but there are structural differences between the upper part of the jejunum and the lower part of the ileum:

1 The mucosa of the jejunum lies in a series of folds called the *valvulae conniventes*,
2 The jejunum is of slightly greater diameter than the ileum.

There is no abrupt transition in these characteristics but there is a gradual change between the upper jejunum and the lower ileum.

The blood supply to the jejunum and ileum

The jejunum and ileum receive arterial blood from the superior mesenteric artery, which runs in the folds of the mesentery. The veins from the jejunum and ileum drain into the superior mesenteric vein, which joins the portal circulation.

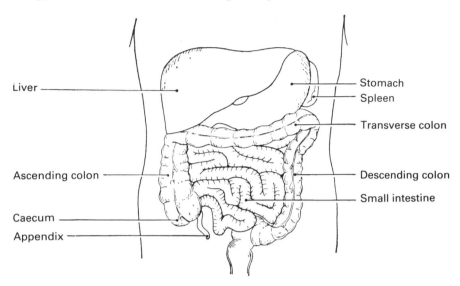

Fig. 14.12. The arrangement of the small and large bowel.

The structure of the small intestine

The wall of the small intestine is composed of four coats:

1 An *outer serous coat* which is provided by the peritoneum.

2 A *muscular coat* which is composed of an outer layer of longitudinal smooth muscle fibres and an inner thicker layer of circular smooth muscle fibres.

3 A *submucous coat*, which is composed of loose areolar tissue containing vessels and nerves.

4 A *mucous coat*, which is thicker in the upper part of the small intestine than in the lower part. The mucous coat has a deep layer of smooth muscle fibres called the muscularis mucosae. The mucous membrane is lined by columnar epithelium, and is characterized by the following structures:

a) **Circular folds**. The mucous membrane from the second part of the duodenum to the upper part of the ileum is thrown into a series of circular folds, called the valvulae conniventes, which are permanent and are not smoothed out when the intestine is distended. These folds are most numerous in the upper part of the jejunum, and gradually disappear as the intestine is traced downwards.

b) **Intestinal villi** are minute, highly vascular processes which project from the surface of the mucous membrane. They are numerous in the duodenum and the jejunum, but are less numerous and smaller in the ileum. Each contains a plexus of blood vessels and a single lymphatic

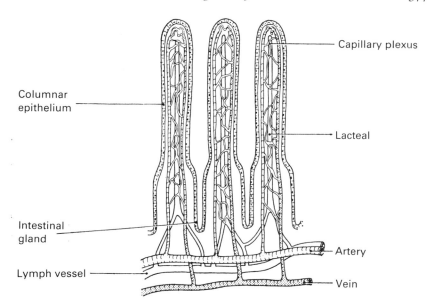

Capillary plexus

Columnar epithelium

Lacteal

Intestinal gland

Artery

Lymph vessel

Vein

Fig. **14.13**. The structure of the villi.

vessel, which is called a *lacteal*. They greatly increase the surface area of the mucosa and thus aid the absorption of digested foodstuffs.

c) **Intestinal glands** are simple tubular glands which open into the small intestine between the villi. They produce the succus entericus, which is responsible for digestion of food in the small intestine.

d) **Duodenal glands** are small compound tubular glands which are found only in the duodenum.

e) **Solitary lymphatic follicles** are small collections of lymphoid tissue which are found scattered throughout the mucous membrane of the small intestine. They are most numerous in the lower part of the ileum.

f) **Peyer's patches** are aggregated lymphatic follicles which are found in the lower part of the ileum. They may be as large as 10 cm in length.

The large intestine

The large intestine is about 1.5 m in length and it extends from the ileum to the anus. It is of greater diameter than the small intestine and it is so arranged that it describes an arch which almost encircles the small intestine. Its external appearance differs from that of the small intestine in that it has a sacculated, or haustrated, appearance. The large intestine is composed of the following parts: the *caecum*, the *colon*, the *rectum*, and the *anal canal*.

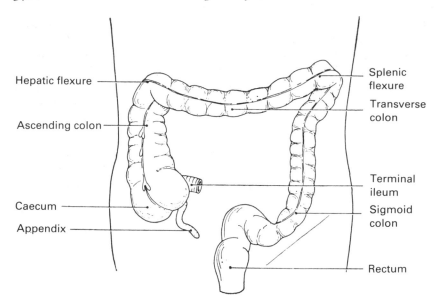

Hepatic flexure

Ascending colon

Caecum

Appendix

Splenic flexure

Transverse colon

Terminal ileum

Sigmoid colon

Rectum

Fig. 14.14. The large intestine.

The caecum

The caecum is the first part of the large intestine and it lies in the right iliac fossa. It is about 6 cm long and 7.5 cm wide. Below it ends blindly, but above it is continuous with the colon. On the posteromedial aspect of the caecum, just below its junction with the colon, is the *ileocaecal orifice* which is guarded by the *ileocaecal valve*. **The vermiform appendix** opens into the caecum about 2 cm below the ileocaecal valve. The appendix varies considerably in length and its position is also variable, but in the majority of people it lies behind the caecum and the ascending colon. The appendix has the same structure as the small intestine, but the submucous coat contains large amounts of lymphoid tissue.

The colon

The colon is divided, for descriptive purposes, into four parts: the *ascending colon*, the *transverse colon*, the *descending colon*, and the *sigmoid colon*.

The ascending colon commences at the upper end of the caecum and passes upwards, on the right side of the abdomen, to the upper surface of the right lobe of the liver, where it turns sharply to the left, forming the *hepatic flexure*. It is about 15 cm in length and is of slightly smaller diameter than the caecum. The hepatic flexure is related, posteriorly, to the anterior surface of the right kidney.

The transverse colon commences at the hepatic flexure, and passes to the left, across the abdomen, to the lower border of the spleen, where it turns downwards forming the splenic flexure. It is about 50 cm in length and describes an arc, which is concave upwards, during its course across the abdomen. It is suspended from the posterior abdominal wall by a fold of peritoneum, called the transverse mesocolon.

The descending colon passes downwards on the left side of the abdomen curving slightly towards the midline in the lower part of its course. It is about 25 cm in length and it is continuous with the sigmoid colon at the level of the pelvic brim.

The sigmoid colon describes an S-shaped loop in the pelvis and is continuous, at its lower end, with the rectum. It is usually about 40 cm in length.

The rectum

The rectum commences at the lower end of the sigmoid colon at the level of the 3rd sacral vertebra. It passes downwards, following the curve of the sacrum and coccyx. Just below the tip of the coccyx it bends backwards and becomes continuous with the anal canal. The upper part of the rectum is of the same diameter as the sigmoid colon, but the lower part of the rectum is dilated, and is known as the *rectal ampulla*. The anterior surface of the upper part of the rectum is covered with peritoneum, but the lower part has no peritoneal covering.

The anal canal

The anal canal is about 4 cm in length. It passes downwards and backwards, from the lower end of the rectum, to end at the anal orifice. The anal canal is surrounded by a thickened band of circular smooth muscle fibres, which constitute the *internal anal sphincter*. Surrounding the internal anal sphincter is a band of voluntary muscle which constitutes the *external anal sphincter*.

The structure of the large intestine

The wall of the large intestine is composed of four coats:
1 *An outer serous coat* formed by the peritoneum.
2 *A muscular coat* composed of an inner layer of circular smooth muscle fibres and an outer layer of longitudinal smooth muscle fibres. The longitudinal fibres are not arranged in a layer of uniform thickness, but are gathered into three bands of increased thickness, called the *taenia coli*.

These bands of longitudinal smooth muscle are somewhat shorter than the other coats of the wall of the large intestine and as a result they produce the puckering, or *haustration*, which gives the colon its characteristic appearance. The taenia coli do not extend on to the rectum, so that the rectal wall is smooth.

3 *A submucous coat* composed of areolar tissue containing vessels and nerves.

4 *A mucous coat* lined by columnar epithelium. The mucous coat of the colon has no villi, but it is thrown into a number of folds which correspond to the haustrations. The mucous coat contains large numbers of mucus-secreting glands.

The blood supply to the large intestine

The caecum, the appendix, the ascending colon and the right two-thirds of the transverse colon are supplied with arterial blood by branches of the superior mesenteric artery.

The left one-third of the transverse colon, the descending colon, the sigmoid colon, the rectum and the upper part of the anal canal are supplied with arterial blood by the inferior mesenteric artery.

The veins from the large intestine drain into the superior and inferior mesenteric veins which in turn drain into the portal circulation.

The lymphatic drainage of the large intestine

The lymphatic vessels from the colon drain into the superior and inferior mesenteric lymph glands, the lymphatic vessels following the course of the corresponding arteries.

The lymphatic vessels from the upper half of the rectum drain into the inferior mesenteric lymph glands.

The lymphatic vessels from the lower half of the rectum and the upper part of the anal canal drain into the internal iliac lymph glands.

The lymphatic vessels from the lower part of the anal canal drain into the superficial inguinal lymph glands.

The pancreas

The pancreas is a greyish-pink gland, about 15 cm in length, which extends transversely across the posterior abdominal wall, from the duodenum to the spleen. It is composed of a *head*, which lies within the curve of the duodenum, a *neck*, a *body*, which passes across the abdomen, behind the stomach, and a *tail*, which is the narrowed extremity of the pancreas which lies in contact with the spleen.

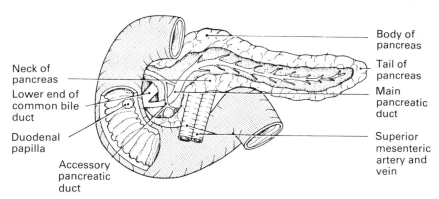

Body of pancreas

Neck of pancreas

Lower end of common bile duct

Duodenal papilla

Accessory pancreatic duct

Tail of pancreas

Main pancreatic duct

Superior mesenteric artery and vein

Fig. 14.15. The pancreas and its ducts.

The common bile duct passes behind the head of the pancreas, in a groove on its posterior surface. The inferior vena cava, the abdominal aorta, the superior mesenteric vessels, and the left kidney lie behind the body of the pancreas. The splenic artery runs along the upper border of the pancreas and the splenic vein runs behind the upper border of the pancreas.

The structure of the pancreas

The pancreas is composed of a large number of lobules, and the whole gland is surrounded by areolar tissue which forms septa between the lobules. Each lobule is composed of a number of tubular *alveoli* lined by columnar epithelial cells, which secrete the pancreatic juice. The alveoli drain into ducts which unite to form a single lobular duct. The duct from each lobule drains into the *main pancreatic duct* which runs the whole length of the gland and opens, together with the common bile duct, into the second part of the duodenum, on the surface of an elevation called the duodenal papilla. The orifice of the common bile duct and the pancreatic duct is surrounded by a sphincter, which is called the sphincter of Oddi. The lobules of the lower part of the head of the pancreas frequently drain into an *accessory pancreatic duct*, which passes in front of the main pancreatic duct and opens into the second part of the duodenum, a short distance above the duodenal papilla.

Interspersed, between the alveoli, are groups of cells called the *Islets of Langerhans*. There are three main types of cell in each islet, A-cells, B-cells and D-cells. The most numerous are the A- and B-cells which secrete glucagon and insulin respectively. The D-cells, which tend to lie peripherally in each islet, secrete the hormone somatostatin. The actions of these hormones will be described in Chapter 17.

The blood supply of the pancreas

The pancreas receives arterial blood from branches of the splenic, right gastric, and superior mesenteric arteries.

The liver

The liver is the largest gland in the body, and it weighs slightly more than 1.5 kg. It lies in the upper part of the abdomen, mainly in the right hypochondrium and the epigastrium, but it also extends into the left hypochondrium. It is approximately wedge-shaped; the base of the wedge lies against the right costal margin and the apex of the wedge extends into the left hypochondrium. The liver thus has anterior, posterior, superior, inferior and right surfaces. The borders between the superior, anterior and right surfaces are rounded, but the anterior and inferior surfaces meet at a sharp edge.

The liver is divided into two main lobes, the right and left lobes. When the under surface of the liver is examined however, two smaller lobes, the quadrate lobe and the caudate lobe, can be seen. These are subdivisions of the right lobe.

The relations of the liver

The superior surface of the liver is related to the under surface of the diaphragm.

The anterior surface of the liver is related to the diaphragm and the anterior abdominal wall.

The right surface of the liver is related to the right dome of the diaphragm and to the costal cartilages.

The posterior surface of the liver is related to the diaphragm, the right suprarenal gland, the inferior vena cava, the abdominal aorta, and the abdominal part of the oesophagus. The inferior vena cava lies in a deep groove on the posterior surface. The hepatic veins, which drain blood from the liver, enter the inferior vena cava in this groove.

The inferior surface of the liver is related to the right kidney, the right suprarenal gland, the duodenum, the hepatic flexure of the colon, the gall bladder, the bile duct and the stomach. The gall bladder lies in a depression on the inferior surface of the liver and the fundus of the gall bladder protrudes anteriorly below the inferior border.

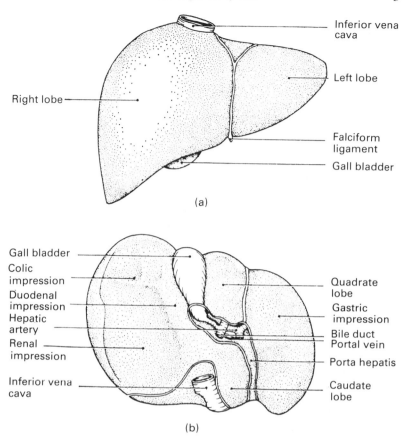

Fig. 14.16. The liver: (a) anterior aspect, and (b) inferior aspect.

On the inferior surface of the liver, between the quadrate lobe anteriorly and the caudate lobe posteriorly, is the *porta hepatis*, at which the portal vein, the hepatic artery and nerves enter the liver and the hepatic ducts, and lymph vessels leave the liver.

Except for a small area on the posterior surface the liver is completely covered by the peritoneum. From the inferior surface of the liver a peritoneal fold, the *lesser omentum*, extends downwards to the lesser curvature of the stomach. Anteriorly another fold of peritoneum, called the *falciform ligament*, extends from the liver to the diaphragm and to the anterior abdominal wall as far down as the umbilicus. Other smaller folds of peritoneum extend from the superior and posterior surfaces of the liver to the diaphragm.

The blood supply to the liver

The liver receives its arterial supply from the hepatic artery which is a branch of the coeliac axis. In addition venous blood, containing absorbed foodstuffs, is brought to the liver from the alimentary canal by the portal vein. The liver thus has a double blood supply. The blood from the liver passes in the hepatic veins to the inferior vena cava.

The structure of the liver

The liver is enclosed in a thin fibrous capsule which lies immediately below the peritoneal covering.

   The liver is composed of a very large number of *lobules*, which, on sections of the liver, are seen to have a hexagonal shape. The lobules are composed of columns of cuboidal cells which contain large amounts of glycogen and fat. Between these columns of cells are *blood sinusoids*. These sinusoids are of a greater diameter than capillaries and they are lined by endothelial cells which form part of the macrophage system (p. 336).

   Running through the centre of each lobule is a vein, called the *central vein*, which is a tributary of the hepatic vein. At the edge of each lobule is the *portal canal*, which contains a branch of the hepatic artery, a branch of the portal vein, and a branch of the hepatic duct. The blood from the branch of the hepatic artery and the branch of the portal vein enters the blood sinusoid and passes between the columns of cells to reach the central

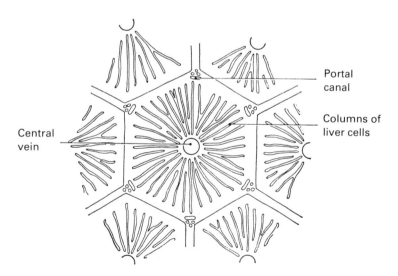

Portal canal

Columns of liver cells

Central vein

Fig. 14.17. The structure of the liver.

vein. Surrounding the liver cells are minute vessels, called bile canaliculi. The bile, which is secreted by the liver cells, passes in these canaliculi to the branch of the hepatic duct which lies in the portal canal.

The biliary apparatus

The bile is drained from the liver by two main ducts, the *right* and *left hepatic ducts*. The right and left hepatic ducts leave the liver at the porta hepatis and unite almost immediately to form the *common hepatic duct*.

The common hepatic duct passes downwards for about 3 cm and is then joined at an acute angle, on its right side, by the cystic duct, which is the duct from the gall bladder.

The cystic duct and the hepatic duct together form the common bile duct.

The common bile duct, which is about 7.5 cm in length, passes downwards behind the first part of the duodenum and then enters a groove on the posterior surface of the head of the pancreas. It then runs downwards in this groove, in front of the inferior vena cava, and finally unites with the pancreatic duct to enter the second part of the duodenum, on the surface of the duodenal papilla. The opening into the duodenum is surrounded by a sphincter muscle, the sphincter of Oddi.

The gall bladder

The gall bladder is a pear-shaped sac which lies on the under surface of the liver, to which it is attached by connective tissue. The gall bladder is divided, for descriptive purposes, into a fundus, a body and a neck.

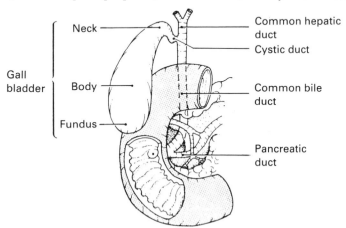

Fig. 14.18. The biliary apparatus.

The fundus is the outer expanded end of the gall bladder. It projects below the inferior surface of the liver and lies in contact with the anterior abdominal wall just below the right 9th costal cartilage at the level of the lateral border of the rectus abdominis muscle.

The body of the gall bladder is directed upwards, backwards and to the left on the under surface of the liver. It is continuous with the neck of the gall bladder close to the porta hepatis.

The neck of the gall bladder is narrow. It passes initially upwards and forwards and then curves sharply downwards to become continuous with the cystic duct.

The cystic duct is about 3 cm in length. It passes downwards and backwards from the neck of the gall bladder to join the right side of the hepatic duct. The mucous membrane lining the cystic duct is thrown into a series of circular folds which are called the *spiral valve*.

The structure of the gall bladder

The wall of the gall bladder is composed of three coats:

1 *an outer serous coat* formed by the peritoneum: the peritoneal covering is absent on the upper surfaces of the body and the neck of the gall bladder;

2 *a middle coat* composed of a thin layer of fibrous tissue and smooth muscle fibres which run mainly in a longitudinal direction; and

3 *an inner layer of mucous membrane* possessing a series of minute folds which give it a honeycombed appearance; the mucous membrane is lined by columnar epithelium.

The blood supply to the gall bladder

The gall bladder is supplied with arterial blood by the cystic artery, which is a branch of the hepatic artery. The veins from the gall bladder drain into the cystic vein which joins the portal vein.

The functions of the alimentary system

Diet

The alimentary system consists of organs which are concerned with the intake of food and its preparation for digestion and absorption into the circulation. Before describing the processes which are involved in digestion and absorption it is important to review the essential constituents of a normal diet.

A normal diet should include: *carbohydrates, fats, proteins, vitamins, mineral salts* and *water.*

Carbohydrates

These are essential to the body as a source of energy. They are present in the food in three forms:

1 *Monosaccharides* are the simplest form of carbohydrate. They are absorbed directly into the blood stream from the jejunum and upper ileum. Glucose and fructose are the main monosaccharides in the diet and are found in fruits and vegetables. Galactose is formed in the intestinal mucosa from lactose. Glucose is the main monosaccharide in the body. Small quantities of other monosaccharides are absorbed from the alimentary canal but are converted to glucose in the liver.

2 *Disaccharides* are composed of two monosaccharide units joined together. The commonest disaccharides present in the diet are sucrose and lactose. Disaccharides are broken down in the alimentary canal into their constituent monosaccharides and are then absorbed.

3 *Polysaccharides*, such as starch, consist of a large number of monosaccharide units. They are found in vegetable foods and in bread. Polysaccharides are broken down mainly to the disaccharide maltose which is subsequently broken down to two glucose units. Celluloses and pectins are not digested by the enzymes of the human alimentary canal and are, therefore, unabsorbable. These substances constitute the dietary fibre which increases the volume (bulk) of the colonic contents and is thought to protect against certain colonic diseases.

Fats

These are an important energy source in the diet and animal fats are the main source of the fat-soluble Vitamins A and D (see below). Weight for weight fat has twice the energy value of starch or sugar and deprivation of fat for more than short periods has serious consequences, particularly when energy expenditure is high, e.g. muscular work or exposure to cold. Fats are broken down in the alimentary canal into fatty acids and glycerol and are almost completely absorbed. The fatty acids are then absorbed via the lacteals into the lymphatic system and pass through the thoracic duct into the blood stream. Glycerol is water-soluble and is absorbed into the villous capillaries. After absorption some of the fatty acids are reconstituted as fat. Fat can be stored in the fat depots as an energy source, or metabolized to yield energy immediately. Fat is also used for the construction of cell membranes and the myelin sheaths of medullated nerve fibres.

Fats are present in the diet in two forms, *animal fat* and *vegetable fat*. Vegetable fats are an equally important source of energy but contain no vitamins, except Vitamin E.

Animal fat is found in butter, milk, cream, cheese, eggs and fatty meat.

Vegetable fat is found in oils (e.g. olive, sunflower or nut oils) and margarine.

Proteins

These are complex substances consisting of the elements carbon, hydrogen, oxygen, nitrogen, phosphorous and sulphur. Proteins are constructed from a number of amino acid units linked together to form polypeptides. Proteins are a type of polypeptide. In the alimentary canal proteins are broken down into peptides and amino acids, which are then available for absorption.

Proteins are necessary to the body for growth and repair, for the maintenance of the osmotic pressure of the blood and tissue fluids, and they can be used as a source of energy. Proteins are also a source of essential amino acids which cannot be sythesized (manufactured) in the body.

Animal protein is present in meat, fish, eggs and dairy produce.

Vegetable protein is present in adequate quantities in peas, beans and lentils as well as in bread and potato.

Vitamins

These are substances which are essential to the body for normal health, are needed in small amounts and cannot be synthesized by the body. They are divided into two groups:

Vitamins A, D, E and *K* are fat-soluble vitamins.

Vitamins B and *C* are water-soluble.

Vitamin A, or retinol, is essential for the maintenance of normal epithelial tissue. It is also necessary for normal night vision since it is concerned with the regeneration of the pigment in the rods of the retina (see p. 492). It is found in milk and milk products, eggs, and in very large amounts in liver fat, especially halibut-liver oil, but also in cod-liver oil. Carrots and green vegetables contain no vitamin A but do contain carotene which is converted into the vitamin.

Vitamin B is not a single vitamin but a complex group of water-soluble vitamins. Most of this group are involved with important enzyme systems and are necessary for normal function of all tissues. The most important are:

1   *Thiamin* (Vitamin B1) is necessary for the normal functioning of the nervous and cardiovascular systems. A deficiency leads to the disease *beri-beri* which is characterized by impaired functioning of the heart and peripheral nervous system. Thiamin is found in cereals, pulses (peas, beans and lentils) and in yeast.

2 *Niacin* (nicotinic acid and nicontinamide) is necessary for normal functioning of epithelial tissue and the nervous system. Deficiency leads to *pellagra* (once common in the southern states of the USA) which is characterized by skin pigmentation, diarrhoea and mental changes. Niacin is found in liver, kidney, yeast, flour and green vegetables.

3 *Riboflavin* deficiency leads to changes in skin, eyes and mucous membranes. It is found in liver, kidney, milk, meat, cheese and eggs. It is also present in wholemeal flour and in beer.

4 *Pyridoxine* (Vitamin B6) is necessary for the proper functioning of skin, liver, blood vessels and nervous tissue, but deficiency of this vitamin in man is rare. It is found in liver, wholegrain cereals and bananas.

5 *Folic acid* is necessary for the normal production of red blood cells. It is present in liver, green vegetables and meat.

6 *Vitamin B12* (cobalamin) is necessary for normal production of red blood cells and for normal functioning of nervous tissue. A deficiency of Vitamin B12 leads to anaemia and abnormalities of the spinal cord. The vitamin is present in animal foodstuffs but is virtually absent from vegetable sources. Absorption of Vitamin B12 from the small intestine depends on the production of intrinsic factor in the gastric juice (p. 393). Deficiency of Vitamin B12 is found in vegans (absolute vegetarians), in patients who lack intrinsic factor and in those who have diseases of the small intestine.

Vitamin C (ascorbic acid) is necessary for maintenance of normal connective tissue. A deficiency of Vitamin C leads to the disease *scurvy*. Scurvy is characterized by bleeding into the tissues, failure of wound healing and swollen, spongy gums. In children the major features are anaemia, painful limbs, due to subperiosteal haematomas, and lassitude. Vitamin C is found in large quantities in fresh fruit and vegetables. The vitamin is destroyed by heat and is therefore often removed by cooking.

Vitamin D is a fat-soluble vitamin which is present in a number of forms and is necessary for normal absorption of calcium from the alimentary canal. The natural form is cholecalciferol but in the body this is converted into a more active form, first in the liver (25-hydroxy-cholecalciferol) and then in the kidney (1,25-hydroxy-cholecalciferol). The latter can be considered a hormone secreted into the blood stream by the kidney and acting on distant organs, namely gut and bone. Deficiency of Vitamin D causes failure of adequate calcium absorption from the alimentary canal and results in defective bone mineralization. In children Vitamin D deficiency causes *rickets* which is characterized by bone softening and deformity of bone. *Osteomalacia* is the adult equivalent of rickets. Vitamin D (cholecalciferol) can be formed in the skin by the action of sunlight, or ultraviolet light, on ergocalciferol. Vitamin D is present in fish oils, dairy produce and eggs.

Vitamin E is a fat-soluble vitamin which is necessary for the protection of cell membranes by preventing oxidation of fatty acids. Deficiency causes visual disturbances in adults and anaemia in premature infants. It is present in vegetable oils, wholegrain cereals and fish.

Vitamin K is necessary for the formation of clotting factors (including prothrombin) in the liver. It is present in vegetables and is also synthesized by bacteria in the colon. Deficiency of Vitamin K causes bruising and also blood loss from the alimentary canal, the renal tract, the nose and the uterus.

In general vitamin deficiency can be caused by dietary lack, failure of absorption or increased demand, e.g. during pregnancy or childhood. Diseases due to vitamin deficiency are still common in many parts of the world. In developed (industrialized) countries vitamin deficiency is commoner among the old, the young, the poor and the immigrant population. Vitamin deficiency is rare among people eating an adequate, mixed diet. Food preparation may influence the amount of vitamin that is available for absorption (see Vitamin C). The taking of vitamin supplements in large doses by those without evidence of vitamin deficiency (as a 'tonic') is unecessary but usually harmless. However some vitamins in excess can cause toxic effects. For example Vitamin A excess leads to loss of appetite, painful bony swelling, hair loss and skin rashes. Vitamin C excess can cause the formation of oxalate stones in the kidney, and hypervitaminosis D causes hypercalcaemia.

Minerals

The main minerals which are essential to the body are: *sodium, potassium, magnesium, calcium, phosphorous, iron, sulphur* and *iodine*. Most of the mineral is found in the skeleton. Minerals are ingested in the form of soluble salts which are absorbed into the blood from the alimentary canal.

Sodium is present in the diet mainly in the form of common salt (sodium chloride). Sodium salts are the main mineral salts of the extracellular fluid. The minimum daily requirement is $1-2$ g of sodium chloride.

Potassium is present in many foods. Potassium salts are the main salts in the intracellular fluid.

Magnesium is found mainly in cereals and vegetables. With potassium, it is an important constituent of intracellular fluid. About half the body magnesium is found in the skeleton. Deficiency is not due to dietary lack but is caused by diarrhoea.

Calcium is found in milk, cheese, cereals, vegetables and some breads. It is deposited in the organic matrix of bones and teeth and is necessary for maintaining their hard structure. Calcium is also necessary for the normal functioning of cardiac and skeletal muscle and nerves. The minimum daily requirement is about 600 mg.

Phosphorous is present in all natural foods. Like calcium, it is necessary for bone synthesis. The minimum daily requirement is about 900 mg.

Iron is present in meat and vegetables. It is necessary for the formation of haemoglobin. Iron is absorbed in the duodenum and upper jejunum. The absorption rate is increased if the person is iron-deficient. It is stored in the liver, the spleen, the bone marrow and lymph nodes until needed for haemoglobin synthesis.

Sulphur is present in certain amino acids and is therefore obtained from proteins in the diet. Deficiency of sulphur occurs with protein malnutrition.

Iodine is present in most foods but particularly in seafood. Vegetables and milk are the main dietary sources of iodine for most people. In areas of endemic goitre (see p. 457) iodization of table salt has eradicated goitre. Iodine is necessary for the formation of thyroid hormones.

Digestion and absorption

Digestion is the process by which foodstuffs are altered into forms in which they can be absorbed into the body from the alimentary canal. Chemical and mechanical processes are involved in digestion.

The chemical changes which occur during digestion are due to proteins called *enzymes* (biological catalysts), which are present in the digestive juices. Complex foodstuffs are broken down, by the action of these enzymes, into simpler forms which can be absorbed and used by the body. Enzymes are specific in their action, so that an enzyme which breaks down a protein molecule will not break down a polysaccharide or a fat molecule.

The mechanical processes involved in digestion ensure that the food is broken up and well mixed with the digestive juices, so that adequate contact takes place between the food and the enzymes in the digestive juices. The food is propelled through the alimentary tract by *peristalsis*. Peristalsis is a coordinated process in which a wave of contraction in the smooth muscle of the wall of the alimentary canal moves along behind a bolus of food preceeded by a wave of relaxation in front of the bolus. Normal peristalsis is controlled by the vagus nerve and by local nervous reflexes via nerve plexuses in the gut wall.

Digestion in the mouth

Mastication, or chewing, is the process by which food is broken up into smaller pieces in the mouth and is thoroughly mixed with the saliva. In the

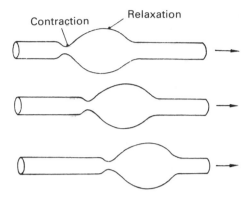

Fig. 14.19. Peristalsis.

mouth the food is moulded into a small rounded mass, or bolus, in preparation for swallowing.

Saliva is secreted by three pairs of salivary glands (parotid, submandibular and sublingual). The thought, sight, taste or smell of food stimulates the secretion of saliva.

Saliva contains: *water, mineral salts* (of sodium and potassium), *mucin,* and *ptyalin*, which is an enzyme.

The **functions of saliva** are as follows:

1 *Mechanical.* It moistens the food and aids mastication and swallowing, as well as speech. It also lowers the risk of a mouth infection.

2 *Chemical.* Ptyalin is an amylase, an enzyme which initiates the breakdown of starches to maltose. The action of ptyalin ceases when the food becomes mixed with the acid contents of the stomach.

Swallowing

Three phases are involved in swallowing: oral, pharyngeal and oesophageal. Only the oral phase is voluntary.

1 The bolus of food is pushed through the oropharyngeal isthmus by an upward and backward movement of the tongue. When the bolus reaches the posterior pharyngeal wall, stimulation of sensory nerves, which are branches of the glossopharyngeal nerve, initiate the complex, coordinated and involuntary phases of swallowing.

2 The soft palate is raised to close off the nasal cavity. At the same time the muscles around the hyoid bone lift the larynx upwards towards the base of the tongue, the epiglottis closes off the larynx and the posterior pillars of the fauces come together to close off the mouth. As a result of these movements the food bolus enters the oesophagus. During this phase respiration is temporarily suspended.

3 When the bolus reaches the oesophagus a wave of peristalsis propels it downwards into the stomach. Gravity plays little or no part in the process of swallowing.

Digestion in the stomach

Food which enters the stomach initially lies in the fundus, close to the oesophageal opening. It is gradually pushed downwards into the body of the stomach as more food boluses enter from the oesophagus. When the stomach is empty the walls lie close together.

The gastric juice. In the stomach the food is brought into contact with the gastric juice which prepares food for intestinal digestion and initiates protein digestion. The wall of the stomach contains three types of glands: *main gastric glands, pyloric glands* and *cardiac tubular glands.*

Main gastric glands are the most numerous of the gastric glands and contain three types of cell: chief, or peptic, cells which secrete *pepsinogen,* parietal, or oxyntic cells, which secrete *hydrochloric acid* and mucous cells which secrete an *alkaline mucus.* The main gastric glands also secrete *intrinsic factor.*

Pyloric glands are found mainly in the region of the pylorus and secrete alkaline mucus.

Cardiac tubular glands are found mainly in the gastric mucosa surrounding the oesophageal opening into the stomach.

The functions of gastric juice

1 *Hydrochloric acid* acidifies the stomach contents and converts pepsinogen into the active enzyme pepsin. The high concentration of acid in the stomach kills most bacteria ingested with the food.

2 *Pepsin* initiates the digestion of proteins, breaking them down into peptones and proteoses, which are smaller, simpler molecules.

3 *Rennin* is an enzyme present in the gastric juice of young animals and is not present in adult human gastric juice. It curdles milk, converting the soluble protein caseinogen into the insoluble casein. Pepsin can also perform this function.

4 *Mucus* forms a surface layer on the resting gastic mucosa. It is alkaline and protects the mucosa from damage by pepsin and hydrochloric acid.

5 *Intrinsic factor* combines with Vitamin B12 in the food and the complex is then absorbed in the ileum.

Control of gastric juice secretion

Both nervous and chemical stimuli act to control the secretion of gastric juice.

1   *Nervous.* The thought, sight, smell or taste of food stimulates gastric secretion by impulses passing to the gastric glands via the vagus nerves.
2   *Chemical.* Both the presence of certain foods (meat and proteins) in the pyloric antrum and vagal stimulation cause the secretion of a hormone, called gastrin, from the pyloric antral mucosa. Gastrin passes in the blood stream to stimulate the parietal cells to secrete hydrochloric acid. Gastric acid secretion is inhibited by fear and nausea, by a high acid concentration in the pyloric antrum or duodenum and by fat in the duodenum.

The movements of the stomach

As food enters the stomach the smooth muscle of the stomach wall relaxes to accomodate the increase in its contents. Soon after the entry of the food, waves of peristalsis start to pass along the body and pyloric antrum of the stomach. Initially the pyloric sphincter remains closed and these waves of peristalsis mix the food thoroughly with the gastric juice. After a variable time the pyloric sphincter relaxes and each wave of peristalsis ejects a small quantity of liquid stomach contents into the duodenum. The stomach is usually empty from four to five hours after eating, the time depending on the nature of the food eaten.

The smooth muscle of the stomach wall derives its nerve supply from both sympathetic and parasympathetic nervous systems. The passage of the peristaltic waves requires the coordination of successive relaxation and contraction of the stomach wall muscle. Nerve plexuses in the stomach wall are responsible for such coordination.

Vomiting is due to expulsion of the gastric contents via the oesophagus. The pylorus contracts, the body of the stomach relaxes and antiperistaltis (reverse peristalsis) may occur; the diaphragm descends and the abdominal wall muscles contract so compressing the stomach. At the same time the cardiac sphincter relaxes and the gastric contents are expelled through the oesophagus. This complicated series of movements, which includes closure of the glottis to protect the lungs, is coordinated by a vomiting centre in the medulla oblongata.

Absorption in the stomach

Absorption in the stomach is limited, but small amounts of water, glucose, mineral salts, some drugs and alcohol are probably absorbed into the blood stream from the stomach.

Digestion in the small intestine

The small intestine is the principal site of digestion and absorption, the two processes taking place simultaneously. The partially digested stomach contents which pass into the duodenum are known as *chyme*. In the small intestine this chyme is mixed first with *pancreatic juice* and *bile* and then with the secretion of the intestinal glands, which is known as *succus entericus*.

The pancreatic juice enters the second part of the duodenum, together with bile, at the duodenal papilla. Pancreatic juice is rich in enzymes, containing *amylase*, *lipase*, and *protein-splitting enzymes*, as well as *mineral salts* and *bicarbonate* ions.

The functions of the pancreatic juice

1 Pancreatic juice contains a high concentration of *bicarbonate ions* which neutralize the acid of the gastric juice and provide the slightly alkaline conditions which are ideal for the pancreatic enzymes.

2 *Amylase* causes the breakdown of starch into maltose.

3 *Lipase* converts neutral fat to glycerol and fatty acids.

4 *Trypsinogen*, when mixed with enterokinase (an enzyme in the succus entericus) is converted to its active form *trypsin*, which breaks down proteins, proteoses and peptones to peptides and amino acids.

5 *Chymotrypsinogen* is the inactive precursor of *chymotrypsin* which digests proteins to small polypeptides. It also clots milk. Chymotrypsinogen is converted to chymotrypsin by the action of trypsin.

4 *Carboxypeptidase* splits peptides by a series of steps into amino acids. The enzyme is secreted by the pancreas in its inactive form, procarboxypeptidase which is converted to carboxypeptidase by enterokinase.

The control of pancreatic juice secretion

The secretion of pancreatic juice is controlled by nervous and hormonal mechanisms. Impulses passing down the vagus nerves, just after eating, stimulate pancreatic juice secretion, but this mechanism is less important than the nervous control of gastric secretion and less important than the hormonal control of pancreatic secretion. The presence of acid chyme in the duodenum and jejunum causes the release of *secretin* and *pancreozymin*, hormones which pass in the circulation to the pancreas and stimulate the production of pancreatric juice rich in bicarbonate ions and enzymes.

Bile

This is continuously secreted by the liver and concentrated by the gall bladder since the sphincter of Oddi is usually closed. Bile contains water, mineral salts, mucus, bile salts, bile pigments and cholesterol, but no enzymes.

The functions of bile

1 *Bile salts* lower surface tension and allow fats to be emulsified. This

facilitates the absorption from the small intestine of fats as well as that of fatty acids, glycerol and fat-soluble vitamins.

2   *Bile pigments* are the breakdown products of the haem portion of hae-moglobin. They have no digestive function but are responsible for the colour of bile and of the faeces.

The control of bile secretion

Like the secretion of gastric and pancreatic juice, bile secretion is controlled by nervous and hormonal mechanisms. The presence of fats and meat in the duodenum causes the liberation of a hormone which causes the gall bladder to contract. This hormone is called *cholecystokinin (CCK)* but it is identical to *pancreozymin (PZ)* and it is now usually referred to as *cholecysto-kinin-pancreozymin* or *CCK-PZ*. The vagus nerves are responsible for relaxation of the sphincter of Oddi (which usually occurs about 30 min after eating) and are also partly responsible for contraction of the gall bladder.

Brunner's glands in the duodenum produce an alkaline secretion which contains mucus and this acts to neutralize the acid of the gastric juice.

The succus entericus is secreted by the glands of the small intestine. Succus entericus contains *water, mineral salts, bicarbonate ions* and *enzymes.*

Functions of the enzymes of the succus entericus

1   *Enterokinase* converts pancreatic *trypsinogen* and *chymotrysinogen* into their active forms (see above).
2   *Erepsin* is a mixture of enzymes which break down peptones and poly-peptides to amino acids.
3   *Amylase* breaks down starches into maltose.
4   *Maltase* breaks down maltose into glucose.
5   *Lactase* breaks down lactose into glucose and galactose.
6   *Sucrase* breaks down sucrose into glucose and fructose.
7   *Lipase* breaks down neutral fat into fatty acids and glycerol.

The secretion of succus entericus

The secretion of succus entericus is controlled by local mechanical and chemical factors stimulated by the presence of food in the small intestine. Nervous and hormonal factors are not involved.

The movements of the small intestine

The acid chyme, which is expelled from the stomach into the duodenum by gastric peristalsis, lies initially in the first part of the duodenum. Almost immediately a wave of peristalsis passes round the duodenum carrying the chyme almost as far as the duodeno-jejunal junction.

In the jejunum and ileum three types of movement occur: segmentation, pendular movements and peristalsis.

1 Segmentation involves ring-like contractions of smooth muscle fibres which divide the contents of a length of intestine into a number of segments. These contractions then relax and new sites of contraction appear about midway in between the original sites. The rate of segmentation is highest in the duodenum and decreases towards the ileo-caecal valve. Segmentation is coordinated by the nerve plexuses in the wall of the small intestine. This type of movement is responsible for thorough mixing of the intestinal contents with the intestinal juice.

2 Pendular movements move the contents of a length of intestine to and fro across the surface of the mucosa. This type of movement probably aids mixing and absorption.

3 Peristalsis in the small intestine occurs in a series of rapid movements which propel the intestinal contents towards the ileo-caecal valve. The transit time of food from the stomach to the ileo-caecal valve is about three hours but varies widely in both health and in disease states. The transit time is affected by emotion and by the nature of the intestinal contents. Peristaltic movements are superimposed upon the segmental contractions.

Absorption in the small intestine

The surface area of the small intestinal mucosa is increased because it possesses a large number of minute processes, called villi. These villi, which each contain an artery, a vein, a capillary network and a lymphatic vessel (called a lacteal) constitute a vast surface which is available for absorption of digested foodstuffs. Amino acids, glucose, water-soluble vitamins, mineral salts and water are absorbed through the intestinal mucosa into the capillaries of the villi, and pass in the portal circulation to the liver. Fatty acids and fat-soluble vitamins are absorbed into the lacteals of the villi and ultimately reach the blood stream via the thoracic duct.

The absorption of the products of digestion in the small intestine is almost complete. Vitamin B12 complexed with intrinsic factor is absorbed in the terminal ileum, although the intrinsic factor itself is not absorbed. Iron absorption occurs mainly in the duodenum and upper jejunum. Folic acid is absorbed in the jejunum. Calcium absorption takes place throughout the small intestine and depends upon an adequate amount of Vitamin D.

The functions of the large intestine

Absorption. The contents of the small intestine, which pass through the ileo-caecal valve into the caecum and colon, are still fluid although most of the nutritive material has been absorbed. Only about 100 ml of water per day reach the rectum, the remainder being absorbed in the caecum and ascending colon. The colon also absorbs small quantities of mineral salts and glucose but cannot absorb protein, fat or calcium.

In spite of absorption of water the faeces are semi-solid and contain 70% water. The other constituents of faeces are cellulose, dead bacteria, some fat, epithelial cells (which have been removed from the surface of the mucosa during the passage of the intestinal contents), bile pigments and mucus. The mucus is secreted by the large intestinal goblet cells and lubricates the faeces. Celluloses cannot be digested by the enzymes in the lumen of the intestine and pass out largely unchanged in the faeces. An increase in the cellulose (fibre) content of the diet adds considerably to the bulk of the faeces and stimulates intestinal peristalsis. A high fibre content in the diet prevents constipation, may prevent certain colonic disorders such as diverticular disease and is thought to reduce the risk of carcinoma of the colon. Fibre also delays gastric emptying and increases the feeling of fullness after eating.

Actions of colonic bacteria. The large intestine contains many bacteria which are normal inhabitants of the bowel. These bacteria can synthesize Vitamin K and also partly digest dietary fibre resulting in the formation of gases (flatus) which are passed per rectum. Most of the flatus, however, consists of nitrogen due to the swallowing of air when eating.

Movements of the large intestine

Peristalsis of the type which occurs in the small intestine does not occur in the large intestine. Various types of colonic wall activity have been described, but the most important is *mass peristalsis* which pushes the colonic contents from the caecum towards the rectum. Waves of mass peristalsis occur frequently but particularly after the eating of a meal. This is referred to as the *gastrocolic reflex*.

Defaecation

The presence of faeces in the rectum gives rise to the desire to defaecate. Faeces are not driven into the rectum by mass peristalsis but accumulate at the recto-sigmoid junction, which acts as a partial sphincter, before entering the rectum. Stimulation of sensory nerves in the wall of the rectum by distension leads to relaxation of the anal sphincters coupled with contraction of the abdominal wall muscles and the diaphragm. These last two actions increase the intra-abdominal pressure. The longitudinal muscle of the distal colon and rectum contracts thus shortening and straightening the lumen of the bowel. The levator ani muscles contract and expel the faeces by pulling the anal canal upwards over the faecal mass.

During the first year of life defaecation is reflex but training leads to voluntary control of the act by higher centres. Defaecation is then only initiated when the external anal sphincter, which is under voluntary control, is allowed to relax.

Effects of radiation on the gut

Acute exposure to ionizing radiation (e.g. an industrial nuclear accident) at or above the limit of tolerance of the body, damages the haemopoietic and gastrointestinal systems. Radiation damage is greatest to actively dividing cells and these two systems have the highest cell turnover rates in the body. The cells of the small intestinal epithelium are, for example, replaced every 48 hours. The major symptoms are nausea, vomiting and diarrhoea. If such symptoms occur within two hours of exposure to radiation the outlook is grave and treatment is often ineffective.

Radiation therapy for abdominal or pelvic tumours can also damage the wall and lining of the small and large intestines. Such damage is often worse if radiation and chemotherapy and both used in the same treatment. The wall of the gut becomes inflamed, mitosis in the epithelial cells is depressed and the villi become shortened. Later the gut wall becomes thinner and abscesses and fistulae may form. Acute symptoms include diarrhoea with loss of blood per rectum. In the long term the small intestine may not properly perform its digestive or absorptive functions and malnutrition may result despite an adequate diet. This condition is known as *malabsorption*.

Metabolism

Metabolism is the name given to the processes which occur in the body for the provision of energy from food and for the growth and repair of body tissues. Two types of process are involved: *anabolism* and *catabolism*.

Anabolism, or synthesis, is the name given to the process by which molecules or tissues are made. *Catabolism*, or breakdown, involves destruction and is therefore the reverse of anabolism. Catabolic processes liberate energy whereas anabolism will only occur if energy is expended. The liver is the central metabolic organ and is responsible for many chemical interconversions (see below).

The metabolism of carbohydrates

The end product of carbohydrate digestion is mainly glucose. Glucose is absorbed from the alimentary canal and passes in the portal circulation to the liver. This glucose is then used in various ways:

1 It is converted, by the liver in the presence of insulin, into glycogen.

Glycogen is stored in the liver until it is needed for the provision of energy, when it is reconverted to glucose.

2 Some glucose remains in the circulating blood and is passed to the tissues where it is used for the provision of energy. This is particularly important for tissues such as the brain which have no energy stores and rely on glucose derived from the blood stream for the provision of energy.

3 Glucose from the blood is converted by muscles into glycogen which is stored in the muscles until glucose is needed for the provision of energy.

4 Glucose in excess of these requirements is converted into fat and stored in the fat depots. In addition glucose can be formed from non-carbohydrate sources, such as certain amino acids lactate and glycerol; this is known as gluconeogenesis and takes place in the liver.

The metabolism of carbohydrate and the control of the blood glucose concentration are influenced by many factors including eating and exercise, but the roles of the liver and insulin are of paramount importance. The actions of insulin are considered in the chapter on the endocrine system (p. 459).

The metabolism of fat

The end products of fat digestion are fatty acids and glycerol. Fatty acids are absorbed into the lacteals and glycerol into the villous capillaries. Fat in the blood stream is transported as lipoprotein—fat (lipid) bound to plasma proteins. After absorption, fat is metabolized in various ways:

1 It can be broken down to act as a source of energy.

2 It can be stored in fat depots. When there is insufficient carbohydrate for the provision of energy, fat is released from the fat depots and used as a source of energy.

3 It is used for structural synthesis e.g. cell membranes and the myelin sheath of medullated nerve fibres.

The metabolism of protein

The end products of protein digestion are amino acids. Essential amino acids are needed for growth and tissue repairs but cannot be synthesized in the body in the amounts needed. Amino acids are absorbed into the blood stream and transported to the liver via the portal circulation. From the liver some amino acids pass in the circulation to all parts of the body, where they are used for growth and repair. There is a continuous process of protein synthesis and catabolism and hence amino acids are continuously being used for protein synthesis and released as a result of protein catabolism. This amino acid pool is made up of amino acids derived from food as well as those liberated by protein catabolism. This 'pool' is not an anatomical but a physiological entity and consists of all the amino acids in the blood stream, in the other body fluids and in the tissues.

Amino acids leave the pool by protein synthesis and also as a result of deamination in the liver. In the process of deamination the nitrogenous portion of the amino acid is removed and converted into urea, which is excreted in the urine. The portion of the amino acid which remains after deamination is either used to provide energy or is converted via glucose to glycogen. Such amino acids are said to be glucogenic. Some amino acids are ketogenic, their non-nitrogenous portion being converted into ketone bodies rather than glucose.

Energy yield of foodstuffs

The three basic foodstuffs can all be used for the provision of energy. The energy values can be expressed in kiloJoules (kJ) or kiloCalories (kCal).

1 g of carbohydrate produces 17 kJ or 4 kCal.
1 g of protein produces 17 kJ or 4 kCal.
1 g of fat produces 37 kJ or 9 kCal.

Basal metabolic rate

This is the term used to describe the energy output of an individual when resting and fasting (at least twelve hours after a meal). It is estimated by measuring the rate at which oxygen is consumed. The basal metabolic rate is increased in children compared with adults; it is increased in hyperthyroidism and decreased in hypothyroidism and during starvation.

Metabolic and other changes during exercise

During exercise glucose is metabolized by working muscle resulting in carbon dioxide production and oxygen use. In order to meet these demands, blood flow to the muscles is increased and the depth and rate of respiration rise. Moderate exercise, such as walking or running at a steady pace, leads to an increase in pulmonary ventilation (about six-fold) and a rise in cardiac output (about eight-fold). Blood flow to the exercizing muscles increases and the oxygen extraction by the muscles is also higher, that is to say the venous blood leaving these tissues contains much less oxygen than normal. Carbon dioxide production by the muscles rises and the body temperature is also increased because of the increased metabolism. Fatigue is poorly understood but it is thought to be due to a combination of brain anoxia and muscle discomfort.

Severe exercise can only be sustained for short periods of time, e.g. sprinting distances up to 400 m. This type of exercise is characterized by the inability of the circulatory system and lungs to cope with the increased metabolic demands of the working muscles. As a result an *oxygen debt* is incurred and glucose is metabolized anaerobically (without oxygen), leading to the accummulation of lactic acid. This type of exercise is followed by increased pulmonary ventilation which may continue for up to 45 min after the event, as the increased carbon dioxide which has been produced is excreted via the lungs.

The functions of the liver

Some of the many functions of the liver have already been described (see above) but, for the sake of convenience, they are summarized here:

1 **The secretion of bile**. The main constituents of bile are:

a) *Bile pigments*, which are formed as a result of the breakdown of haemoglobin from old red blood cells. The bile pigments colour the faeces but have no digestive function.

b) *Bile salts* which are necessary for adequate digestion and absorption of fats.

c) *Electrolytes*.

d) *Cholesterol*.

2 **Storage**. The liver stores glycogen, fat, some proteins, Vitamins A, D, K and B12, iron and folic acid.

3 **Red cell formation and destruction**.

4 **Synthesis**. The liver is the site of formation of plasma proteins, fibrinogen and prothrombin, both of which are necessary for the normal clotting of blood, and heparin, which prevents the intravascular clotting of blood.

5 **Detoxication**. The liver is capable of chemically altering or completely destroying many substances, such as drugs and hormones. This action protects the body from the harmful (toxic) effects of continued high levels of the substances concerned.

6 **Metabolism**. The central role of the liver in metabolism has been described; the products of digestion reach the liver via the portal circulation and other substances are transported in the systemic circulation. In the liver they are stored, chemically converted, excreted in the bile or pass unchanged into the general circulation. Although described separately, metabolic processes are continuous and are influenced by a large number of factors, e.g. chemical, hormonal and nervous.

CHAPTER 15

The urinary system

The urinary system is composed of: the *left and right kidneys*, which excrete the urine; the *left and right ureters*, which convey the urine to the urinary bladder; *the urinary bladder* in which the urine is collected to be discharged to the exterior at intervals; and the *urethra*, through which the urine is discharged to the exterior in the act of micturition.

The kidneys

The kidneys are two organs of characteristic shape which lie on the posterior abdominal wall, beneath the peritoneum, one on either side of the abomen. Their *upper poles* lie at the level of the upper border of the 12th thoracic vertebra and their *lower poles* lie at the level of the 3rd lumbar vertebra, about 2.5 cm above the highest palpable parts of the iliac crests. The central indentation of the medial border of each kidney, which is called the *renal hilum*, lies at the level of the transpyloric plane. However, since the right kidney is slightly lower in position than the left, the left renal hilum lies just above, and the right renal hilum just below, this plane. The kidneys also lie slightly obliquely so that their upper poles lie 2.5 cm from the midline and their lower poles 7.5 cm from the midline.

The relations of the kidneys

Posteriorly: the diaphragm, the quadratus lumborum muscle, the psoas muscle and the 12th rib.

    Anteriorly: anteriorly the relations of the kidneys differ on each side.

    The right kidney is related:

1 at its upper pole to the right suprarenal gland;
2 by the upper two-thirds of its anterior surface to the right lobe of the liver;
3 at its lower pole to the hepatic flexure of the colon; and
4 at its medial border to the second part of the duodenum.

    The left kidney is related:

1 at its upper medial border to the left suprarenal gland;
2 by the upper two-thirds of its lateral border to the spleen;

403

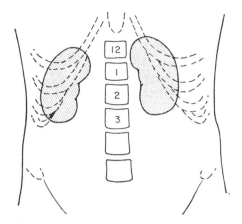

Fig. **15.1.** The position of the kidneys.

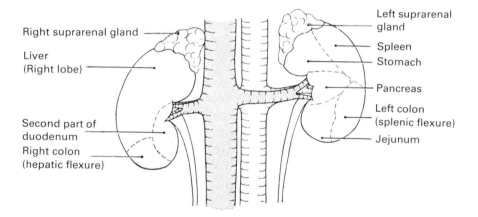

Fig. **15.2.** The anterior relations of the kidneys.

3 medially to this area to the stomach above, and
4 to the splenic vessels and pancreas below;
5 at its lower pole medially to the jejunum; and
6 laterally to the splenic flexure of the colon.

The blood supply to the kidneys

The kidneys are supplied with arterial blood by the renal arteries, which are direct branches of the abdominal aorta. Normally there is one renal artery to each kidney, but sometimes there are two on one or both sides.

The blood from the kidneys is drained by the left and right renal veins into the inferior vena cava.

The structure of the kidneys

The kidneys are enclosed in a thin capsule of fibrous tissue, which, in the undiseased kidney, is easily removed. Each kidney has an anterior and a posterior surface, a lateral convex border and a medial border which is convex above and below, but has a central indentation, the *renal hilum*, through which the renal vessels and the ureter pass.

The hilum leads into a recess within the kidney, the *renal sinus*, which contains the renal vessels and the upper expanded portion of the ureter, the *pelvis of the ureter*. The pelvis of the ureter divides into two or three branches called the *major calyces* and these further divide into several smaller branches, the *minor calyces*.

If the kidney is divided into an anterior and posterior half it is seen to consist of an outer layer, the *cortex*, and the inner layer, the *medulla*. The medulla consists of conical masses, the apices of which point towards the hilum and project into the minor calyx as a renal papilla.

The cortex of the kidney contains approximately 1 000 000 units called *nephrons*, which are the functional units of the kidney. Each nephron commences as a blind expanded end, called *Bowman's capsule*, and this is indented by a cluster of capillaries, the *glomerulus*. Bowman's capsule and the glomerulus together form a unit called a *Malpighian corpuscle*. Immediately by the Malpighian corpuscle the nephron consists of a coiled tube, called the *proximal convoluted tubule*. The next part of the nephron is like a U-tube. The first limb runs towards the hilum of the kidney into the medulla for a variable distance, then turns and the second limb runs

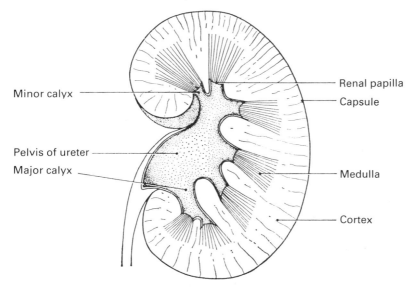

Renal papilla

Capsule

Minor calyx

Pelvis of ureter

Major calyx

Medulla

Cortex

Fig. 15.3. Cross-section of the kidney.

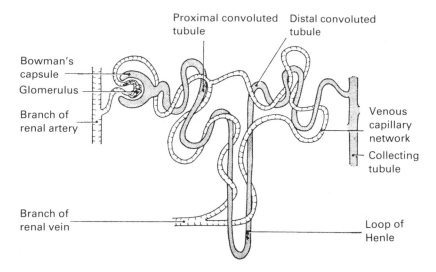

Fig. 15.4. The nephron.

parallel to it back into the cortex. This loop is called the *loop of Henle*. The nephron then again lies in a coil, called the *distal convoluted tubule*, and finally straightens out to join a collecting duct near the medulla; several collecting ducts unite to form larger ducts which finally empty into a minor calyx on the surface of a renal papilla.

The blood is brought to the glomerulus by a branch of the renal artery. After passing through the glomerular capillaries the blood enters a venous capillary network surrounding the tubules, and is finally drained into a tributary of the renal vein.

The ureters

The ureters are two muscular tubes which convey the urine, by peristaltic action, from the kidney to the urinary bladder. Each ureter is 25 cm in length and is described in three parts.

1 **The pelvis of the ureter** has already been described.

2 **The abdominal portion of the ureter** runs downwards and slightly medially, beneath the peritoneum, on the surface of the psoas muscle which separates it from the transverse processes of the lumbar vertebrae.

3 **The pelvic portion of the ureter** commences at the brim of the pelvis, crosses the common iliac vessels and then runs downwards and backwards on the posterolateral wall of the pelvis, beneath the peritoneum. At the level of the ischial spine it turns and runs medially and forwards to the base of the bladder.

The structure of the ureters

The ureters consist of three layers.

1 *An outer fibrous layer* is continuous with the capsule of the kidney.

2 *A muscular layer* consists of an outer circular layer and an inner longitudinal layer of smooth muscle fibres.

3 *An inner layer of mucous membrane* lies in folds when the ureter is empty but is smooth when the ureter is full. The mucous membrane has an inner lining or transitional epithelium.

The urinary bladder

The urinary bladder acts as a reservoir for the urine, and lies in the pelvis behind the pubic symphysis and the pubic bones. When empty it lies entirely within the pelvis, but, as it fills, it rises above the pubic symphysis into the abdominal cavity. The bladder usually holds 170−284 ml of urine. The shape of the bladder varies with the volume of its contents, but it is described as having an apex, a base, a superior surface and two inferolateral surfaces. The base of the bladder faces backwards and downwards and is roughly triangular in shape. The lowermost point or apex of the base is

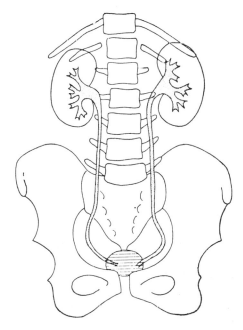

Fig. 15.5. A tracing of an intravenous pylogram to show the relationship of the ureters to the bony landmarks.

relatively fixed and is called the neck of the bladder. The neck is pierced by the internal urethral orifice. The apex of the bladder points towards the pubic symphysis.

The relations of the urinary bladder

In the male the bladder is related:
1 *anteriorly* to the pubic symphysis;
2 *superiorly* to the peritoneum, the small intestine and the sigmoid colon; and
3 *posteriorly* to the rectum, the vasa deferentia and the seminal vesicles.
The neck of the bladder rests on the prostate gland.

In the female the bladder is related:
1 *anteriorly* to the pubic symphysis;
2 *superiorly* to the peritoneum and the uterus; and
3 *posteriorly* to the vagina and cervix.
The neck of the bladder lies on the pelvic fascia, which surrounds the urethra.

The interior of the bladder

The interior of the bladder is lined by mucous membrane which, in the empty bladder, is thrown into folds, except over an area at the base where it

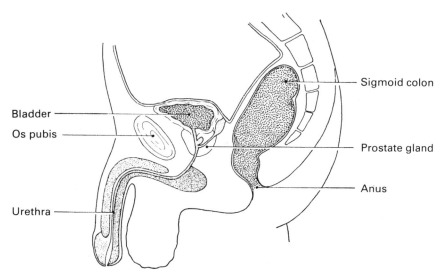

Fig. 15.6. Sagittal section of the male pelvis to show the relations of the urinary bladder.

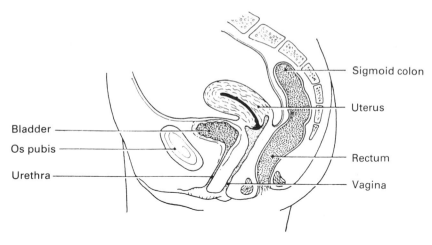

Fig. 15.7. Sagittal section of the female pelvis to show the relations of the urinary bladder.

is smooth and firmly adherent to the wall. This area is shaped like an inverted triangle and is called the *trigone*. The ureters open into the upper angles of the triangle and the urethra opens at its apex.

The structure of the bladder

The bladder is composed of three coats.

1 *An outer serous coat*, which is the peritoneum, covers the superior surface of the bladder.

2 *A middle muscular coat* consists of three layers of muscle fibres. The inner and the outer layer of fibres are arranged in a longitudinal fashion and the middle layer in a circular fashion. The circular fibres form a thickened band around the urethral outlet to form the *internal urethral sphincter*.

3 *An inner mucous coat* is pale pink in colour. It is continuous with the mucous membrane lining the ureters and the urethra and its inner lining is composed of transitional epithelium.

The blood supply to the bladder

The arterial supply to the bladder is from branches of the internal iliac arteries. The veins from the bladder enter the internal iliac veins.

The lymphatic drainage of the bladder

The lymphatic vessels from the bladder drain mainly into the external iliac lymph glands. A few vessels from the superior surface, however, may pass to the internal iliac or common iliac lymph glands.

The urethra

In the male the urethra is shared with the reproductive system and it is therefore described with that system (p. 469).

In the female the urethra is about 4 cm length. It commences at the internal urethral orifice of the bladder and runs downwards and forwards on the anterior wall of the vagina. It opens in front of the vagina, 2.5 cm below the clitoris. There are numerous small mucous glands opening into the urethra.

The structure of the female urethra

The female urethra is composed of three coats.

1   *A muscular coat* is continuous with that of the bladder and consists of an inner layer of longitudinal and an outer layer of circular muscle fibres. Near its termination the urethra is surrounded by the external urethral sphincter which is under voluntary control.

2   *A thin coat of spongy tissue* contains a plexus of veins.

3   *An inner layer of mucous membrane* is continuous above with that of the bladder and below with that of the vulva and its inner lining is composed of transitional epithelium.

The functions of the kidneys

The kidneys are concerned with:

1   the removal from the body of waste products of protein metabolism, such as urea, uric acid, creatinine, phosphates and sulphates;

2   the regulation of body water;

3   the regulation of electrolytes;

4   the maintenance of acid–base balance by the body;

5   the removal of toxic substances and drugs from the body; and

6   metabolic functions including the maintenance of blood pressure, red cell production and calcium metabolism.

The kidneys perform the first four of these functions by the production of urine. The urine consists mainly of *water* and contains *urea, uric acid, creatinine, sodium chloride, potassium, calcium, phosphates* and *sulphates*. Normally 1–2 l of urine are produced per day.

The volume depends upon the fluid intake and the amounts of fluid lost by sweating and in the stools. Increased fluid loss due to increased sweat production or gastrointestinal losses (diarrhoea or vomiting) will result in the passage of a decreased volume of concentrated urine, unless fluid intake is increased.

The processes involved in the production of urine are *filtration, reabsorption* and *secretion*.

Filtration. The blood entering the glomerulus is under high pressure and, at rest, up to 25% of the cardiac output flows through the kidneys. The fluid filtered into the Bowman's capsule is plasma minus the plasma proteins and cells.

The fluid filtered at the glomerulus is altered during its passage down the tubules by removal of some of its constituents and by the addition of some others. The processes involved are, respectively, reabsorption and secretion. Thus the fluid which enters the ureter is very different in composition from that which was filtered at the glomerulus. By the processes of reabsorption and secretion the composition of the extracellular fluid is kept constant.

Reabsorption. Some of the constituents of the fluid which is filtered at the glomerulus are reabsorbed into the blood stream. This process may be *active* or *passive*, active transport requiring energy expenditure. For example, glucose is present in the glomerular filtrate but is normally absent from the urine. The glucose is completely reabsorbed from the glomerular filtrate and returned to the blood stream by the action of the cells of the proximal convoluted tubule, i.e. it is *actively reabsorbed*. Urea on the other hand, passes out of the tubule back into the blood by diffusion, i.e. it is *passively reabsorbed*.

Sodium chloride is actively and virtually completely reabsorbed by the renal tubule, the reabsorption in the distal tubule occurring under the control of aldosterone.

Approximately 6.5 l of fluid are filtered at the glomerulus in each hour, but only 1−2 l of urine are produced every 24 h. Therefore nearly all the water filtered must be reabsorbed from the renal tubules. The reabsorption of water occurs at such a rate as to keep the osmotic pressure (osmolality) of the body fluids constant. The rate of reabsorption of water from the tubule is controlled by the secretion of antidiuretic hormone (ADH) from the posterior pituitary gland. The loop of Henle dips deep into the medulla of the kidney, where there is a high osmotic pressure due to the active transport of sodium out of the tubule at that point. ADH increases the permeability to water of the distal tubular cells and the cells lining the collecting ducts. Water therefore passes into the area of high osmotic pressure, i.e. out of the renal tubule.

Secretion. The cells of the tubules remove potassium and hydrogen ions from the venous blood and secrete them into the tubules. The secretion of hydrogen ions into the tubules causes the production of an acid urine. Since metabolic processes generate a great deal of hydrogen ions, i.e. acidity, this function of the kidney is very important in maintaining the correct pH of extracellular fluid.

Tubular secretion is the method by which the kidney rids the body of drugs such as penicillin.

The kidney and the production of red blood corpuscles

The kidney is the main source of the hormone erythropoietin, which increases the rate of red cell production. Patients with renal failure are anaemic and patients with renal cell carcinoma have an increased number of red cells.

The kidneys and the blood pressure

Cells in the region of the glomerulus produce an enzyme, renin, which converts angiotensinogen in the blood to angiotensin I. A further enzyme causes the production of angiotensin II from angiotensin I. Angiotensin II is a powerful constrictor of blood vessels and raises the arterial blood pressure by this action. It also stimulates the production of aldosterone, the sodium-retaining hormone, from the zona glomerulosa of the adrenal cortex.

The secretion of renin is stimulated by a fall in the blood pressure within the kidney or by a fall in the plasma sodium concentration.

The kidney and calcium metabolism

The kidney is the site of formation of 1,25-dihydroxycholecalciferol, the most active form of Vitamin D. (The actions of Vitamin D are described in Chapter 14 and Chapter 17.) The most important action of this renal metabolite is to increase calcium absorption from the intestine, especially to meet the demands of growth, pregnancy and lactation.

Renal clearance

The efficiency of the glomerulus may be investigated by studying the clearance of *creatinine,* a product of protein metabolism. Creatinine is filtered at the glomerulus and is then neither reabsorbed from, nor secreted into the renal tubule. Certain other substances, such as the radiological contrast agents sodium diazotrizoate (Hypaque) and iohexol (Omnipaque), are not only filtered but are also secreted into the tubules. The high iodine content of these drugs makes them radio-opaque, and this allows them to be used to visualize the renal tract on radiographs. The clearance of such substances is equal to the renal blood flow.

The control of micturition (bladder emptying)

The bladder stores urine formed by the kidneys until a moment convenient for the voiding of urine presents itself. As the bladder fills with urine the pressure within it increases, thus stimulating somatic and autonomic nerve endings in the bladder wall. If it is not convenient, or appropriate, to pass urine, impulses from the cerebral cortex cause relaxation of the bladder wall muscle. With practice, large volumes of urine can be accomodated in the bladder, especially in women. When the appropriate moment arrives, under voluntary control, the bladder wall contracts, the internal and external sphincters are relaxed and urine is expelled into the urethra. At the same time the abdominal wall muscles contract, the diaphragm contracts (and descends) both of which actions increase the intra-abdominal pressure, thus compressing the bladder. The perineal muscles are relaxed and the urine is expelled.

The desire to micturate is affected by emotion and by disease. Pre-examination 'nerves' or inflammation of the bladder mucosa both result in frequent voiding of small amounts of urine. In babies the bladder empties reflexly and micturition is not under the control of the cerebral cortex.

Dialysis

When the kidneys fail to carry out their normal excretory and regulatory functions, substances such as urea and creatinine accumulate in the blood stream. If renal function does not resume, death will result. It is possible to use the principle of dialysis to remove such potentially toxic substances from the blood. In haemodialysis the blood is circulated through a semi-permeable tube immersed in a urea-free solution, similar in composition to the blood plasma, and is then returned to the patient. Urea and creatinine pass from the blood into the urea-free fluid. Peritoneal dialysis uses the same principle but the fluid is run into the peritoneal cavity via a catheter, allowed to equilibrate with blood and then removed. This process is repeated as necessary.

CHAPTER 16

The nervous system

The ability to respond to the environment is a characteristic of all living organisms and is called *irritability*. In unicellular animals a stimulus applied to the surface of the cell may cause movement away from, or towards, the source of the stimulus depending on its nature. The one cell thus appreciates the stimulus and responds accordingly.

In the more primitive multicellular animals special cells, called *sensory receptors*, become differentiated and are usually found near the surface of the organism. These cells are specialized for the reception of stimuli. Other cells, called *effectors*, are also differentiated to form muscles or glands and these produce the response to the stimuli. The connection between these effectors and sensory receptors is formed by further specialized cells, called *neurons*, or nerve cells, which possess the property of conductivity, and are composed of a cell body from which elongated protoplasmic processes extend. Thus a mechanism called a *reflex arc* is built up. The simplest form of reflex arc is composed of a sensory receptor, two neurons

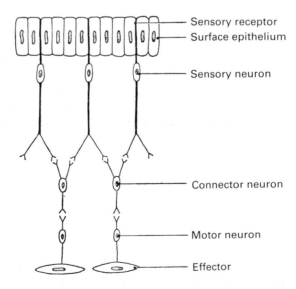

Sensory receptor
Surface epithelium
Sensory neuron
Connector neuron
Motor neuron
Effector

Fig. 16.1. A simple reflex arc.

and a muscle or a gland. The neuron which conveys the impulse away from the sensory receptor is called a *sensory neuron*, and the neuron which conveys the impulse to the effector is called a *motor neuron*. In more complicated reflex arcs several neurons, called *connector neurons*, are interposed between the sensory and the motor neurons.

In the more complex multicellular animals most of the cell bodies of the neurons become collected together near the midline of the animal, where they constitute the *central nervous system*. The cells of the sensory neurons, however, generally lie outside the central nervous system in groups, which are called *ganglia*. The cells of the motor neurons and connector neurons lie within the central nervous system forming the *grey matter*. The processes of the neurons which run through the central nervous system constitute the *white matter*. The processes of the sensory and motor neurons which run outside the central nervous system are collected into bundles, called *nerves*, and these constitute the *peripheral nervous system*.

Because of the direction of movement of animals, sensory receptors of different types are most numerous and complex at the head end, and this part of the central nervous system becomes enlarged, forming the *brain*, while the rest of the central nervous system constitutes the *spinal cord*.

In higher animals, centres which are capable of organizing behaviour develop in the brain, and many reflex arcs become modified by voluntary control. Thus, sensory impulses are then received in the brain and are assessed before the appropriate response is initiated.

The viscera of the body, that is the smooth muscles and glands, are not under voluntary control but their activities are regulated by the brain to correlate with the activities of the other parts of the body.

The nervous system is, therefore, the system of the body which enables an individual to react to the environment and which regulates and controls the activities of the other systems of the body.

The nervous system in man is divided into three parts.

The central nervous system is composed of the brain and spinal cord.

The peripheral nervous system is composed of bundles of nerves which contain sensory and motor fibres.

The autonomic nervous system is concerned with the regulation of the activities of smooth muscle and of glands.

Nervous tissue

Nervous tissue is composed of *neurons*, which are the cells responsible for the transmission of nervous impulses, and *connective tissue*. In the central nervous system the connective tissue is composed of special cells called *neuroglia*, but in the peripheral nervous system the nerve fibres are bound together by loose connective tissue.

Neurons

A neuron is composed of a cell body from which protoplasmic processes arise. There are usually two types of processes called *axons* and *dendrites*. The axons are long processes and only one axon arises from each nerve cell. The axon is usually called the nerve fibre and it conveys nervous impulses away from the cell body. The dendrites are short processes, and one or more arise from each nerve cell. The dendrites convey impulses towards the cell body.

The cell bodies of neurons vary considerably in shape and some of them are the largest cells found in the human body. Each cell body has a single nucleus but, in mature neurons, there is usually no centrosome present as neurons are unable to reproduce.

The axon is a fine thread-like process which may be as much as 100 cm in length. It is usually surrounded by a fatty sheath called the *myelin sheath*. The diameter of this sheath is constricted at intervals along its length, and these constrictions are called the *nodes of Ranvier*. Axons which possess a myelin sheath are called *myelinated* fibres, while those with no myelin sheath are called *unmyelinated* fibres. Unmyelinated fibres are found predominantly in the autonomic nervous system. All axons, whether myelinated or unmyelinated, are surrounded by a fine nucleated sheath called the *neurolemma*. The nuclei of the neurolemma are surrounded by an area of cytoplasm and are called *Schwann cells*. They lie at intervals along the length of the nerve fibre. In myelinated fibres the nuclei usually lie midway between the nodes of Ranvier. The neurolemma does not surround axons inside the central nervous system.

The dendrites are usually shorter than the axons and they do not possess myelin sheaths.

Types of neurons

Neurons are classified according to the number of the processes which they possess.

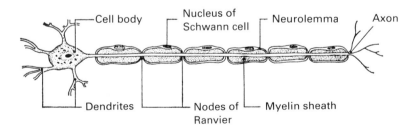

Fig. 16.2. A neuron.

Multipolar neurons have one axon and two or more dendrites. Most motor neurons and connector neurons are of this type.

Bipolar neurons are oval cells which have one axon and one dendrite, and these arise from opposite poles of the cell body. The neurons of the retina, the olfactory area of the nose and the auditory nerve are of this type.

Pseudounipolar neurons are spherical cells which possess a single process which divides, close to the cell body, to form one axon and one dendrite. Most sensory neurons are of this type, one of the processes entering the central nervous system and the other passing to the periphery to form a sensory ending.

The termination of the processes of neurons

1 Sensory neurons carry sensory impulses from the periphery towards the central nervous system. They are either pseudounipolar or bipolar neurons and their cell bodies lie outside the central nervous system in sensory ganglia. Their peripheral processes end in relationship to the skin, mucous membrane, ligaments, muscles, etc. and the structure of the ending varies according to the sensation which they transmit. A sensory fibre ending in the skin, for example, loses its myelin sheath and divides into a number of branches which end in close relationship to the cells of the epidermis.

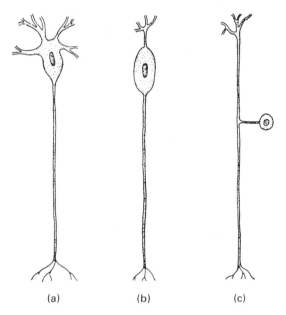

(a)   (b)   (c)

Fig. 16.3. Types of neuron: (a) multipolar, (b) bipolar, and (c) pseudounipolar.

Other fibres end in relationship to hairs or to special structures called tactile corpuscles. The centrally directed processes of sensory neurons lose their neurolemmal sheaths and enter the central nervous system to end in relationship to processes of motor or connector neurones.

2   Motor neurons carry impulses away from the central nervous system to voluntary muscles. They are multipolar neurons. The peripheral process of a motor neuron ends in relationship to one or more muscle fibres at a specialized structure called a motor end-plate. The nervous impulse travelling down the motor fibre is transferred to the muscle fibre at the motor end-plate producing contraction of the muscular fibre. The dendrites of motor neutrons end in relationship to processes of connector or sensory neurons.

3   Connector neurons, as their name suggests, form connections between the processes of neurons within the central nervous system. Apart from the connector neurons which are concerned with the autonomic nervous system, all connector neurons and their processes lie within the central nervous system.

The synapse

Impulses are passed from one nerve cell to another at junctions called *synapses*. At these junctions the axon of one neuron divides into a series of small branches, which end in minute swellings called *end feet*. These end feet are closely applied to the dendrites and cell bodies of other neurons, but there is no direct structural continuity. The axon can only carry nervous impulses away from the cell body of a neuron and, similarly, at the synapse an impulse can pass in one direction only, namely from the axon of one neuron to the dendrites and cell bodies of another.

Neuroglia

Neuroglia are the connective tissue cells of the central nervous system. They are of three types:

1   **astrocytes**, which are stellate cells possessing numerous processes;

2   **oligodendrocytes**, which are similar to astrocytes but possess fewer processes; and

3   **microglial** cells, which are small cells that have few processes; they resemble the cells of the macrophage system, since they are capable of phagocytosis and of independent movement.

The central nervous system

The central nervous system is composed of the brain and spinal cord.

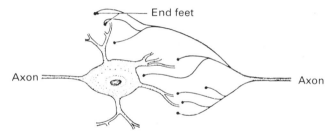

Fig. 16.4. A synapse.

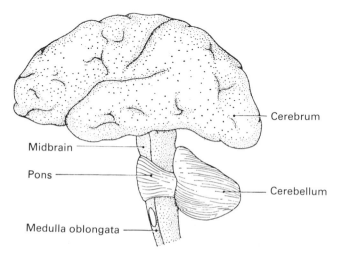

Fig. 16.5. The parts of the brain.

The brain

The brain lies within the cranial cavity and is composed of: the *cerebrum*, the *midbrain*, the *pons*, the *medulla oblongata*, and the *cerebellum*.

The cerebrum

The cerebrum is the largest part of the brain, and is composed of the right and left cerebral hemispheres, which are incompletely separated by a deep midline cleft, called the *longitudinal cerebral fissure*. A fold of dura mater, called the *falx cerebri*, projects downwards into this fissure. Below the longitudinal cerebral fissure the two cerebral hemispheres are connected by a mass of white matter, or nerve fibres, called the *corpus callosum*. Within each cerebral hemisphere is a cavity called the *lateral ventricle*.

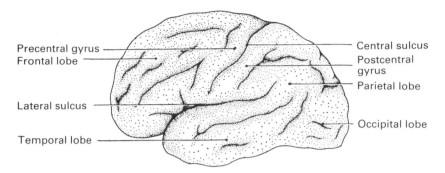

Precentral gyrus
Frontal lobe

Lateral sulcus

Temporal lobe

Central sulcus
Postcentral gyrus
Parietal lobe

Occipital lobe

Fig. **16.6.** The lateral aspect of the left cerebral hemisphere.

Each cerebral hemisphere is divided, for descriptive purposes, into four lobes, each lobe receiving its name from the bone of the cranial vault to which it is related. The four lobes are: the *frontal lobe*, the *parietal lobe*, the *temporal lobe*, and the *occipital lobe*.

The anterior end of the frontal lobe is called the *frontal pole* of the cerebral hemisphere and the posterior end of the occipital lobe is called the *occipital pole*.

Each cerebral hemisphere is composed of an outer layer of grey matter called the *cerebral cortex*. The surface of the cerebral cortex is arranged in a number of folds, called *gyri* or convolutions, which are separated by *sulci*, or fissures. The most important sulci are:

1   the central sulcus (the fissure of Rolando), which runs downwards from the upper medial surface of the cerebral cortex, and separates the frontal and parietal lobes; and

2   the lateral sulcus (the fissure of Sylvius), which lies on the inferior and lateral aspects of the cerebral hemisphere, and separates the temporal lobe from the frontal and parietal lobes.

The most important gyri are:

1   the precentral gyrus, which lies in front of the central sulcus; and

2   the postcentral gyrus, which lies behind the central sulcus.

The cerebral cortex is the most highly developed part of the brain. It initiates all voluntary movement, it appreciates all sensations and it contains the so-called higher centres which are responsible for thought, memory and intelligence. Certain functions can be ascribed to different areas of the cerebral cortex.

The motor area. The precentral gyrus is frequently called the motor cortex, and from this area all voluntary movements are initiated. The nervous impulses from the motor cortex pass to muscles of the opposite side of the body. Thus the right cerebral cortex controls the muscles of the left side of the body and vice versa.

The sensory area. The postcentral gyrus is frequently called the sensory cortex. It receives and appreciates all general sensations from the opposite side of the body. Thus the right cerebral cortex appreciates sensations from the left side of the body and vice versa.

The auditory area. The cortex of the temporal lobe immediately below the lateral sulcus is called the auditory area, and is concerned with the appreciation of impulses from the inner ear which are transmitted in the vestibulocochlear nerve.

The visual area. The cortex of the greater part of the occipital lobe is called the visual area, and it receives impulses from the retina which are transmitted in the optic nerves.

The motor speech area. The cortex of the frontal lobe, just above the anterior end of the lateral sulcus, is called Broca's area and is concerned with initiating the voluntary movements which produce speech. This area is found in the left cerebral cortex in right-handed persons and vice versa.

The parietal area. The cortex of most of the parietal lobe is thought to be associated with stereognosis, that is the recognition of objects by touch and feel without the aid of vision. The lower part of the parietal cortex is thought to be associated with the ability to interpret the written and spoken word, and is therefore called **the sensory speech area**.

The interior of the cerebral hemisphere is composed mainly of white matter, or nerve fibres. These are arranged in tracts, or groups of fibres, which are of three types:

1 **association fibres**, which connect one part of the cerebral hemisphere with another;

2 **commissural fibres**, which connect the two cerebral hemispheres and form the corpus callosum; and

3 **projection fibres**, which connect the other parts of the central nervous system with the cerebral cortex; these fibres pass through a narrow area in the lower part of the cerebral hemisphere, called the *internal capsule*.

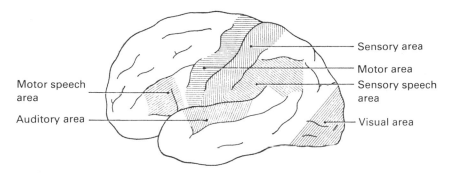

Fig. 16.7. The lateral aspect of the left cerebral hemisphere showing the functional areas.

Within the white matter of the cerebral hemisphere there are a number of collections of grey matter. The three important collections of grey matter are as follows:

1   The basal ganglia lie on the lateral wall of the anterior and central parts of the lateral ventricle.

2   The thalamus is a mass of grey matter which lies below the corpus callosum, and forms the lateral wall of the cavity known as the *third ventricle*. The lateral side of each thalamus is separated from the basal ganglia by the internal capsule. The thalamus is mainly composed of nerve cells which relay sensory impulses from the spinal cord towards the sensory cortex.

3   The hypothalamus lies below the thalamus at the base of the brain and forms the floor of the third ventricle. Projecting downwards from the lower surface of the hypothalamus is the stalk which attaches the *pituitary gland* to the base of the brain. Just in front of the stalk of the pituitary gland is the *optic chiasma* which forms a communication between the two optic nerves. The hypothalamus contains centres which are responsible for the regulation of the body temperature, the regulation of sleep and the control of the metabolism of fat and carbohydrate. The hypothalamus also plays an important part in the control of the pituitary.

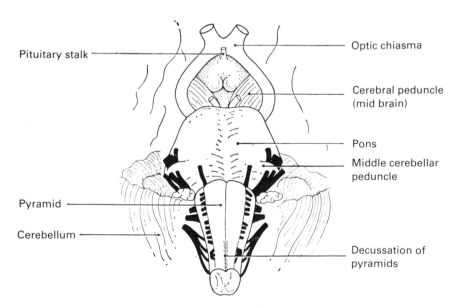

Fig. 16.8. The midbrain, the pons and the medulla oblongata.

The midbrain

The midbrain connects the cerebral hemispheres above to the pons below. Anteriorly it is composed of two cerebral peduncles which are mainly formed by fibres passing to and from the cerebral hemispheres, and posteriorly it is composed of four rounded eminences called the *quadri-geminal bodies* which are reflex centres for sight and hearing. Within the centre of the midbrain is a narrow canal, called the *cerebral aqueduct* (aqueduct of Sylvius), which connects the 3rd ventricle above with the 4th ventricle below.

The pons

The pons lies below the midbrain, in front of the cerebellum and above the medulla oblongata. It is composed mainly of nerve fibres passing between the cerebral hemispheres and the spinal cord, and of nerve fibres passing from one cerebellar hemisphere to the other. Lying deeply within the white matter are collections of grey matter which are nuclei for some of the cranial nerves.

The medulla oblongata

The medulla oblongata is continuous with the pons above and the spinal cord below. It lies just above the foramen magnum, in the posterior cranial fossa.

Anteriorly, on each side of the midline, the medulla oblongata has a rounded eminence, called the *pyramid*, which is composed of motor fibres passing downwards from the cerebral cortex. In the lower part of the medulla these motor fibres cross from one side to the other at the *decussation of the pyramids* and thus the fibres from the right cerebral cortex pass to the left side of the body. Some of the sensory fibres passing upwards towards the cerebral cortex also cross from one side to the other in the medulla. The medulla contains a number of collections of grey matter which are:

1 nuclei for certain of the cranial nerves; and
2 relay stations for some of the sensory fibres passing upwards from the spinal cord.

Also contained within the medulla are the so-called *vital centres* which are the respiratory centre, the cardiac centre and the vasomotor centre.

The cerebellum

The cerebellum lies in the posterior cranial fossa behind the pons and medulla oblongata. It is composed of two hemispheres which are joined across the midline by a narrow strip called the *vermis*. Like the cerebrum, the cerebellum is composed of an outer cortex of grey matter and has white matter in its interior. The outer surface of the cerebellum is thrown into a large number of narrow folds which are separated by deep fissures.

Nerve fibres enter the cerebellum in three pairs of bundles called the cerebellar peduncles. These are:

1   the superior cerebellar peduncles, which connect the cerebellum with the midbrain, and transmit efferent fibres from the cerebellum;

2   the middle cerebellar peduncles, which connect the cerebellum with the pons, and contain afferent fibres; and

3   the inferior cerebellar peduncles, which connect the cerebellum with the medulla oblongata and which also contain afferent fibres.

The cerebellum is concerned with the maintenance of balance and posture, and the coordination of voluntary muscular movement. The cerebellum receives from the muscles and joints sensory impulses which are used for coordinating voluntary movement, and are also used, together with impulses from the semicircular canals of the inner ear, to maintain balance. Efferent impulses from the cerebellum pass to the basal ganglia and pass from there down the spinal cord. They exert a controlling influence over voluntary impulses which are initiated in the motor cortex. Impulses which pass to the cerebellum do not give rise to conscious sensations.

The spinal cord

The spinal cord is the elongated, almost cylindrical, part of the central nervous system which lies in the upper two-thirds of the vertebral canal. It extends from the level of the upper border of the atlas vertebra, where it is continuous with the medulla oblongata, to the level of the lower border of the 1st lumbar vertebra where it ends in a pointed extremity called the *conus medullaris*. From the conus medullaris a fine fibrous ligament, called the *filum terminale*, extends downwards to be attached to the back of the body of the 1st coccygeal vertebra. The spinal cord is about 45 cm in length.

Thirty-one pairs of nerves arise from the spinal cord. Each nerve arises by two roots, an *anterior root* and a *posterior root*. The anterior (ventral) root contains efferent or motor fibres and some efferent fibres of the autonomic nervous system. The posterior (dorsal) root contains sensory fibres passing into the spinal cord. Just lateral to the spinal cord each dorsal root possesses an oval swelling, called the dorsal root ganglion, which contains the cell bodies of the pseudounipolar neurons from which the sensory fibres arise. Each cell of the dorsal root ganglion gives off a single process which branches into two. One branch passes to the periphery where it makes a

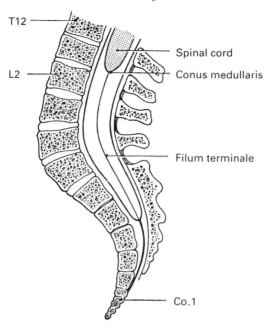

T12

L2

Spinal cord

Conus medullaris

Filum terminale

Co.1

Fig. 16.9. The lower end of the spinal cord.

sensory ending, the other branch passes into the spinal cord. The two roots fuse just lateral to the ganglion to form a spinal nerve.

The 31 spinal nerves are arranged as follows: eight *cervical*, twelve *thoracic*, five *lumbar*, five *sacral*, and one *coccygeal*.

The upper nerve roots pass horizontally to leave the vertebral canal through the appropriate intervertebral foramen, but, because the spinal cord is shorter than the canal in which it lies, the lower nerve roots pass increasingly obliquely downwards to reach the appropriate foramen. Below the conus medullaris the nerve roots pass downwards in a bundle which is called the *cauda equina*.

The spinal cord is not of uniform diameter, but shows two swellings or enlargements. The upper, or cervical, enlargement extends from the 3rd cervical to the 2nd thoracic segment of the cord and corresponds to the origin of the nerves to the upper limb. The lower, or lumbar, enlargement begins at the level of the 9th thoracic vertebra and ends at the level of the lower border of the 12th thoracic vertebra, below which the cord tapers away to the conus medullaris. The lumbar enlargement corresponds to the origin of the nerves of the lower limb.

The spinal cord is not completely cylindrical, being slightly flattened

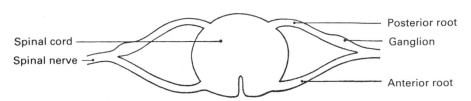

Fig. 16.10. The mode of origin of a spinal nerve from the spinal cord.

from front to back. It is incompletely divided into right and left halves by an *anterior median fissure* and a *posterior median septum*.

The interior of the spinal cord

The spinal cord is composed of white matter, grey matter and neuroglial tissue.

The grey matter lies centrally and on transverse section of the cord is seen to have an H-shaped arrangement. The two limbs of the H are symmetrical and each has an anterior and a posterior horn.

The anterior horn of the grey matter contains nerve cells, the axons of which form the motor fibres of the anterior nerve roots.

The posterior horn of the grey matter contains nerve cells of the connector type and many of the posterior root fibres form synapses with these cells.

The central limb of the grey matter is called the *grey commissure.* Running centrally through the grey commissure is the *central canal* of the

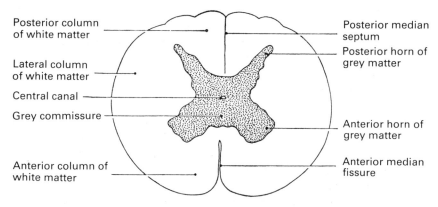

Fig. 16.11. Cross-section of the spinal cord to show the arrangement of the grey matter and the white matter.

spinal cord, which is continuous above with the lower end of the 4th ventricle of the brain and contains cerebrospinal fluid.

The white matter of the spinal cord surrounds the central mass of grey matter and is divided, by the limbs of the grey matter, into three columns called the posterior, lateral and anterior columns.

The fibres within the white matter are of three types:

1   **ascending fibres**, which carry sensory impulses to the brain;

2   **descending fibres**, which carry motor impulses from the brain to the cells of the anterior horn of the spinal cord; and

3   **intersegmental fibres**, which carry impulses from one level of the cord to another.

The motor and sensory fibres in the spinal cord run in definite pathways, or tracts, and the origin and destination of these tracts will now be described.

The sensory pathways

Sensory impulses enter the spinal cord in the posterior nerve roots. The fibres from the posterior nerve roots may take one of several courses on entering the cord.

1   They may pass directly upwards in the white matter of the cord to reach the brain.

2   They may form synapses with nerve cells in the anterior horn of the grey matter thus forming a reflex arc.

3   They may form synapses with connector neurons in the posterior horn of the grey matter. The axons of these connector neurons may, in turn, either pass upwards to the brain or they may form synapses with anterior horn cells thus forming a reflex arc.

The sensations which enter the spinal cord in the posterior nerve roots are sensations of pain, temperature, light touch, pressure or proprioception. (Proprioceptive impulses are impulses which arise in muscles, joints and ligaments and convey a sense of position.)

Pressure, pain and temperature. Fibres carrying sensations of pressure, pain and temperature on entering the spinal cord form synapses with connector neurons in the posterior horn of the grey matter. The axons of these connector neurons then cross to the opposite side of the cord and pass upwards through the lateral column of white matter to the brain. In the brain they ascend to the thalamus where another synapse is formed. From the thalamus the impulse then passes to the sensory area of the cerebral cortex where it gives rise to a conscious sensation.

Proprioception and light touch. Fibres carrying sensations of proprioception and light touch enter the spinal cord in the posterior nerve roots and then pass directly upwards, in the posterior columns of the white

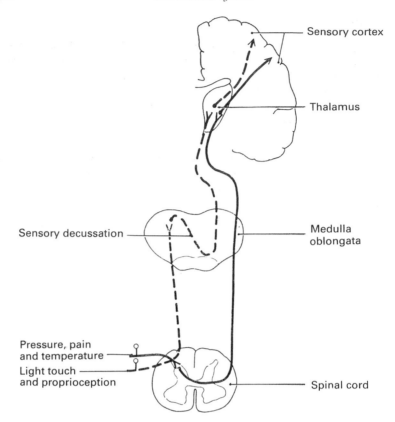

Sensory cortex

Thalamus

Sensory decussation

Medulla oblongata

Pressure, pain and temperature

Light touch and proprioception

Spinal cord

Fig. 16.12. The sensory pathways.

matter of the same side of the cord, to the medulla oblongata. In the medulla oblongata they form synapses with connector neurons, and the axons of these connector neurons then cross to the opposite side of the medulla oblongata and pass upwards to the thalamus. In the thalamus a further synapse is formed and the impulses then pass onwards to the sensory cortex where they give rise to conscious sensations.

Proprioceptive impulses which are destined for the cerebellum, where they are used for the coordination of muscular movement, follow a different course. On entering the cord the fibres carrying these impulses form synapses with connector neurons in the posterior horns. The axons of these connector neurons then pass upwards in the lateral columns of white matter on the same side of the cord and enter the cerebellum through the inferior cerebellar peduncle. The impulses which pass to the cerebellum do not give rise to conscious sensations.

The motor pathways

Motor impulses which produce contractions of voluntary muscles are initiated in the cells of the motor area of the cerebral cortex. The fibres from these cells then pass downwards through the internal capsule, the midbrain and the pons to the medulla oblongata. In the medulla oblongata they cross the midline in the decussation of the pyramids and then pass downwards in the lateral columns of the white matter of the spinal cord. The finally form synapses with anterior horn cells at the level of the cord for which they are destined. The axons of the anterior horn cells leave the cord in the anterior nerve roots and pass to the muscle concerned. The motor neuron in the cerebral cortex is frequently referred to as the upper motor neuron and that in the anterior horn as the lower motor neuron.

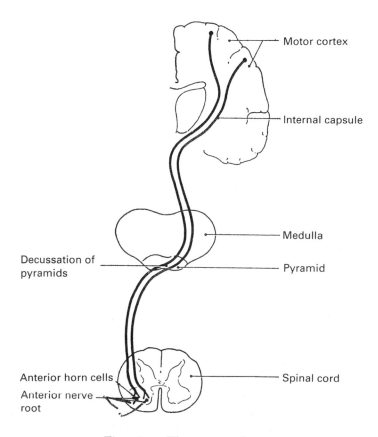

Fig. 16.13. The motor pathways.

Reflexes

Reflexes are automatic motor responses to sensory stimuli and they occur without the direct involvement of the brain. Two, or more usually three or more neurons are involved in a reflex action.

1 A sensory neuron conveys a sensory impulse to the spinal cord from the periphery. The axon of this sensory neuron forms a synapse with a connector neuron in the posterior horn of the grey matter.

2 The connector neuron then conveys an impulse to a motor neuron in the anterior horn of the grey matter of the spinal cord.

3 The anterior horn cell then conveys an impulse to a muscle producing contraction.

Usually several sensory neurons are stimulated and each forms several synapses with connector neurons so that ultimately large numbers of motor impulses pass out from the spinal cord.

Some common examples of reflex action

1 **Withdrawal from a painful stimulus**. If the fingers receive a painful stimulus reflex contraction of the flexor muscles of the fingers and arm occurs, removing the fingers from the source of the stimulus. The individual is usually consciously aware of the pain only fractionally after the limb has been withdrawn.

2 **The plantar reflex**. Scratching the sole of the foot produces reflex contraction of the flexor muscles of the toes and the toes are dorsiflexed.

3 **Tendon reflexes**. A sudden tap on the tendon of certain muscles causes an alteration in the length of the muscle and a reflex contraction of the muscle follows. Thus tapping the patellar tendon at the knee produces a reflex contraction of the quadriceps muscle.

4 The alterations in the activity of the viscera which occur to coincide with voluntary activity are technically reflex actions but these will be discussed in the section on the autonomic nervous system (p. 444).

The meninges

The brain and spinal cord are covered by three membranes called the meninges.

The dura mater is the outermost layer of the meninges and it is a thick dense membrane. The dura mater of the cranial cavity differs in some respects from the dura mater of the vertebral canal.

The dura mater of the cranial cavity is divided into two layers. The outer layer acts as the periosteum of the inner surface of the bones which form the cranial vault. The inner layer acts as a protective covering for the brain. The two layers are, in the main, closely united, but the inner layer separates from the outer layer at several sites:

1 The venous sinuses, which drain the venous blood from the brain, lie between the two layers of the dura mater.

2 The falx cerebri, which projects into the longitudinal fissure between the left and right cerebral hemispheres, is a sickle-shaped fold of the inner layer of the dura mater. The superior sagittal sinus runs in the upper margin of the falx cerebri and the inferior sagittal sinus in the lower free margin.

3 The tentorium cerebelli, which partially covers the posterior cranial fossa, is a crescentic-shaped fold of the inner layer of the dura mater. It lies between the upper surface of the cerebellum and the occipital lobes of the cerebral hemispheres. The posterior and lateral margins of the tentorium cerebelli enclose the transverse sinuses. The anterior border bears a large notch in which the midbrain lies. The posterior margin of the falx cerebri is attached to the tentorium cerebelli in the midline and the straight sinus runs backwards in the line of attachment.

4 The falx cerebelli is a small sickle-shaped fold of the inner layer of the dura mater which lies below the tentorium cerebelli and projects between the cerebellar hemispheres.

The dura mater which lines the vertebral canal is composed of one layer only and corresponds to the inner layer of the cranial dura mater, the outer layer of the cranial dura mater ending at the foramen magnum. The spinal dura mater forms a loose protective covering for the spinal cord. At the level of the 2nd sacral vertebra it constricts to cover the filum terminale and passes downwards to fuse with the periosteum of the posterior surface of the back of the coccyx. The space between the periosteum of the bones of the vertebral canal and the dura mater is called the *extradural space*.

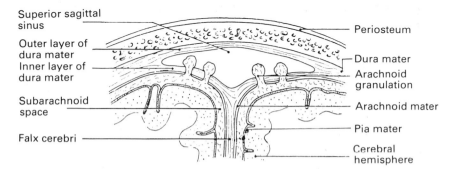

Fig. 16.14. Cross-section of the superior sagittal sinus and the falx cerebri to illustrate the arrangement of the meninges.

The arachnoid mater is the middle layer of the meninges. It is a thin delicate membrane which is separated from the dura mater by a narrow space called the *subdural space*. This space contains a thin film of fluid. The arachnoid mater lines the inner surface of the dura mater, and extends down the vertebral canal to the level of the 2nd sacral vertebra, where the dura mater comes to lie on the filum terminale. The arachnoid mater projects in small tufts into the superior sagittal and transverse sinuses to form structures called the *arachnoid granulations*.

The pia mater is the innermost layer of the meninges and it is separated from the arachnoid mater by a fluid-filled space called the *subarachnoid space*. The pia mater is a thin membrane which is composed of fine areolar tissue containing large numbers of small blood vessels. It closely invests the surface of the brain and spinal cord and dips downwards into the cerebral sulci. The pia mater covering the spinal cord is less vascular and somewhat thicker than the pia mater which covers the brain, and it extends downwards from the conus medullaris as the filum terminale.

The ventricles of the brain

The ventricles of the brain are intercommunicating, fluid-filled spaces which lie within the cerebrum, the midbrain, the pons and the medulla oblongata.

The lateral ventricles. The two lateral ventricles lie one on each side of the midline in the substance of the cerebral hemispheres. They are the largest of the ventricles, and they each have an anterior horn which extends into the frontal lobe, a posterior horn which extends into the occipital lobe and a temporal horn which extends into the temporal lobe. The posterior part of the anterior horn opens into a narrow canal called the interventricular foramen, which unites the two lateral ventricles and extends downwards to open into the upper part of the 3rd ventricle at an opening called the *foramen of Monro.*

The 3rd ventricle is roughly quadrilateral in shape and lies in the midline between the two thalami. The foramen of Monro opens into its upper anterior angle. The lower anterior angle projects downwards towards the pituitary gland and is called the *infundibular recess.* Just anterior to the infundibular recess is a small recess, called the *optic recess*, which lies over the optic chiasma. The posterior border of the 3rd ventricle is related to the pineal gland. The cerebral aqueduct opens into the lower posterior angle of the 3rd ventricle.

The cerebral aqueduct (aqueduct of Sylvius) is a narrow canal which runs through the midbrain from the lower posterior angle of the 3rd ventricle to open into the 4th ventricle.

The 4th ventricle lies in front of the cerebellum and behind the pons

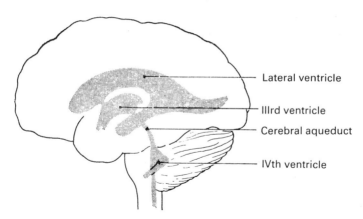

Fig. 16.15. Lateral view of the brain to illustrate the position of the ventricles.

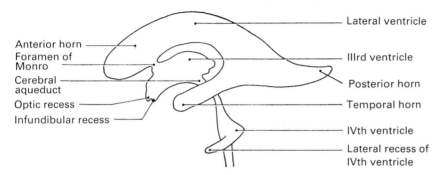

Fig. 16.16. The ventricles of the brain, lateral aspect.

and the upper part of the medulla oblongata. When seen from behind the 4th ventricle is roughly diamond-shaped. The cerebral aqueduct opens into its upper angle and the *central canal of the spinal cord* into its lower angle. On each side of the 4th ventricle a recess, called the *lateral recess*, passes laterally behind the inferior cerebellar peduncles. The roof of the 4th ventricle is tented and is formed above by the under surface of the cerebellum and below by a thin sheet of fibrous tissue.

There are three foramina in the roof of the 4th ventricle through which the ventricular system communicates with the subarachnoid space. The *foramen of Magendie* lies in the midline of the roof and the *foramina of Luschka* lie in the roofs of the lateral recesses.

The ventricular system and the central canal of the spinal cord have a lining of ciliated epithelium called the *ependyma*.

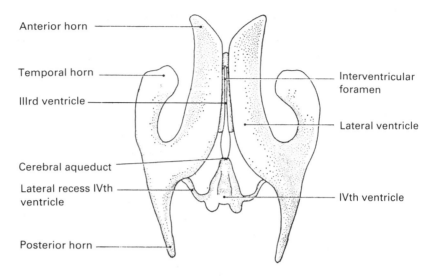

Anterior horn

Temporal horn

IIIrd ventricle

Cerebral aqueduct

Lateral recess IVth ventricle

Posterior horn

Interventricular foramen

Lateral ventricle

IVth ventricle

Fig. **16.17.** The ventricles of the brain, superior aspect.

The circulation of the cerebrospinal fluid

The subarachnoid space, the ventricles of the brain and the central canal of the spinal cord are filled with a clear colourless fluid called the cerebrospinal fluid.

Projecting into each lateral ventricle is a highly vascular fold of pia mater called the *choroid plexus*. Two similar folds project into the roof of the 3rd ventricle and the roof and lateral recesses of the 4th ventricle.

The cerebrospinal fluid is secreted by the choroid plexuses, and it flows from the lateral ventricles, through the interventricular foramen into the 3rd ventricle. From the 3rd ventricle it passes down the cerebral aqueduct into the 4th ventricle, and from the 4th ventricle it flows out through the foramen of Magendie and the foramina of Luschka into the subarachnoid space. The flow of fluid then passes upwards through the gap in the tentorium cerebelli and forwards and laterally over the surface of each cerebral hemisphere. It is reabsorbed back into the blood stream through the arachnoid granulations which project into the superior sagittal sinus.

The cerebrospinal fluid also passes down the subarachnoid space surrounding the spinal cord and down the central canal of the spinal cord. It is probably reabsorbed back into the blood stream through local veins in the vertebral canal.

The peripheral nervous system

The peripheral nervous system is composed of twelve pairs of cranial nerves which arise from the brain and 31 pairs of spinal nerves which arise from the spinal cord.

The cranial nerves

The cranial nerves are twelve pairs of nerves which arise directly from the brain. Some of the cranial nerves contain only sensory fibres, others contain only motor fibres and some contain both motor and sensory fibres and are called *mixed nerves*. The cranial nerves are referred to either by numbers or by names. They are: *I olfactory nerve, II optic nerve, III oculomotor nerve, IV trochlear nerve, V trigeminal nerve, VI abducent nerve, VII facial nerve, VIII vestibulocochlear nerve IX glossopharyngeal nerve, X vagus nerve, XI accessory nerve, and XII hypoglossal nerve.*

I The olfactory nerve (sensory) is a purely sensory nerve, and it is concerned with the sense of smell. The fibres of the nerve arise in the mucous membrane of the olfactory area of the nose and pass upwards, through *the foramina in the cribriform plate of the ethmoid bone*, to enter the

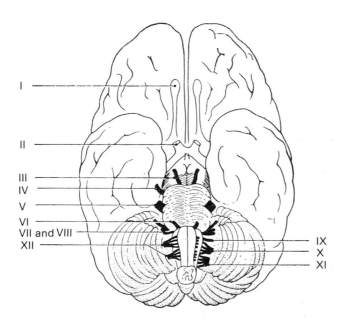

Fig. 16.18. The base of the brain showing the origins of the cranial nerves.

olfactory bulb which lies on the under surface of the frontal lobe of the cerebral hemisphere. From each bulb nerve fibres pass backwards to the cortex of the temporal lobe of the cerebral hemisphere.

II The optic nerve (sensory) is a purely sensory nerve and it is concerned with the sense of sight. The fibres originate in the retina of the eye and pass backwards from the eyeball to enter the cranial cavity through *the optic foramen*. In the cranial cavity the fibres from the nasal half of the retina cross in the optic chiasma (p. 493) to join the fibres from the temporal half of the opposite retina. The nerve then runs backwards to the visual area in the cerebral cortex of the occipital lobe.

III The oculomotor nerve (motor) is a purely motor nerve. The fibres of the nerve arise from grey matter in the midbrain, and then leave the brain to run forward to enter the orbit through *the superior orbital fissure*. The oculomotor nerve supplies all the muscles which move the eye, except the lateral rectus and superior oblique muscles (p. 495). It also supplies the ciliary muscle, the circular muscle of the iris and the muscle which elevates the upper eyelid.

IV The trochlear nerve (motor) is a purely motor nerve. The fibres of the nerve arise from grey matter in the midbrain. It also enters the orbit through *the superior orbital fissure* and it supplies the superior oblique muscle of the eye (p. 495).

V The trigeminal nerve (mixed) contains both motor and sensory fibres and it is the largest of the cranial nerves. It is the main sensory nerve to the head and its motor fibres supply the muscles of mastication. It arises by two roots, a sensory root and a motor root, from the pons. The fibres of the sensory root pass through a ganglion, called the trigeminal ganglion, which lies on the floor of the cranial cavity near the apex of the petrous temporal bone. The motor fibres arise from grey matter in the upper part of the pons. Just beyond the trigeminal ganglion the sensory fibres divide into three main branches which are called the ophthalmic division, the maxillary division and the mandibular division. The motor fibres join the mandibular division.

The ophthalmic division of the trigeminal nerve contains only sensory fibres. It enters the orbit through the *superior orbital fissure* and supplies the lacrimal gland, the conjunctiva and the skin of the upper eyelid. The residual fibres then leave the orbit through *the supraorbital foramen* to supply the skin of the forehead and the scalp as far back as the lambdoid suture.

The maxillary division of the trigeminal nerve contains only sensory fibres. It leaves the cranial cavity through *the foramen rotundum* to enter the pterygopalatine fossa and then passes into the orbit through *the inferior orbital fissure*. It soon leaves the orbit again through the *infraorbital foramen* to reach the face. It supplies the lower eyelid, the cheeks and the upper teeth and gums.

The mandibular division of the trigeminal nerve contains sensory and motor fibres. It leaves the cranial cavity through *the foramen ovale*. It supplies motor fibres to the muscles of mastication and sensory fibres to the lower lip, the teeth and gums, the chin, the posterior part of the skin of the face and the pinna.

VI The abducent nerve (motor) is a purely motor nerve. The fibres of the nerve arise from grey matter in the pons and leave the brain at the lower border of the pons. The nerve enters the orbit through *the superior orbital fissure* and supplies the lateral rectus muscle of the eye (p. 495).

VII The facial nerve (mixed) contains both sensory and motor fibres. The motor fibres which supply the muscles of facial expression, arise from grey matter in the lower part of the pons. The sensory fibres, which convey sensations of taste from the anterior two-thirds of the tongue, pass to the taste area in the lower part of the precentral gyrus of the cerebral cortex. On leaving the brain the two roots of the nerve pass into *the internal acoustic meatus* with the auditory nerve. Within the internal acoustic meatus the two roots unite and then enter a canal, called *the facial canal*, which runs through the temporal bone to open at the base of the skull at the *stylomastoid foramen*. While in the facial canal the facial nerve gives off a branch, called the chorda tympani nerve, which conveys taste fibres to the tongue. On leaving the facial canal the facial nerve passes into the substance of the parotid gland and divides into its terminal branches which supply the muscles of facial expression.

VIII The vestibulocochlear nerve (sensory) contains only sensory fibres. These sensory fibres are divided into two nerves:
1   **the cochlear nerve**, which arises from the cochlea of the inner ear and is concerned with hearing; and
2   **the vestibular nerve**, which arises from the vestibular portion of the middle ear and conveys sensations of balance to the brain.

The two nerves unite and enter the cranial cavity through *the internal acoustic meatus* to pass to the lower border of the pons. The nerve then enters the brain just below the facial nerve. The cochlear fibres pass to the auditory area in the temporal lobe of the cerebral hemisphere. The vestibular fibres pass to the cerebellum.

IX The glossopharyngeal nerve (mixed) contains both motor and sensory fibres. The motor fibres arise from grey matter in the medulla oblongata and they supply some of the muscles of the pharynx and the secretory cells of the parotid gland. The sensory fibres carry general sensations to the brain from the pharynx and tonsils and also sensations of taste from the posterior third of the tongue.

The nerve leaves the brain as three or four small roots which arise from the side of the medulla oblongata. These roots unite and the nerve then leaves the cranial cavity through *the jugular foramen* and passes to the pharynx and tongue.

X The vagus nerve (mixed) contains both sensory and motor fibres and has the most extensive distribution of all the cranial nerves. It arises mainly from grey matter in the medulla oblongata and leaves the brain as eight or more small roots from the side of the medulla oblongata. These roots unite to form one trunk and the nerve then leaves the cranial cavity through *the jugular foramen*. It passes downwards through the neck and thorax and breaks up to form a plexus in the lower part of the mediastinum. From this plexus two nerve trunks pass into the abdomen through the oesophageal opening in the diaphragm. The vagus nerve supplies branches to the larynx, the pharynx, the trachea, the heart and lungs, the oesophagus and, below the diaphragm, the gastrointestinal tract and its accessory glands.

XI The accessory nerve (motor) contains only motor fibres and is formed from two roots, one of which arises from the medulla oblongata and the other from the spinal cord.

The cranial root is the smaller and its fibres arise from a nucleus in the medulla oblongata. The fibres emerge from the brain as four or five small rootlets which unite and run laterally to the jugular foramen. Close to the jugular foramen the cranial root is joined by the spinal root.

The fibres of the **spinal root** arise from a nucleus in the grey matter of the anterior horn of the upper part of the spinal cord. They emerge from the side of the spinal cord, midway between the anterior and posterior roots of the upper cervical nerves, and unite to form a single trunk which enters the skull through the foramen magnum and runs laterally to the jugular foramen to join the cranial root. The nerve emerges from the skull through *the jugular foramen* and supplies some of the muscles of the soft palate, the trapezius and sternomastoid muscles.

XII The hypoglossal nerve (motor) contains only motor fibres. These fibres arise from grey matter in the lower part of the medulla oblongata and leave the brain as a series of small roots which arise from the side of the medulla oblongata anteriorly to the roots of the vagus nerves. These roots unite to form a single trunk which leaves the skull through *the hypoglossal canal*.

The hypoglossal nerve supplies the muscles of the tongue and the muscles around the hyoid bone.

The spinal nerves

Thirty-one pairs of nerves arise from the spinal cord and leave the vertebral canal through the intervertebral foramina. Each spinal nerve is composed of a posterior (sensory) root and an anterior (motor) root. On each posterior root there is a ganglion which contains the cell bodies of the pseudounipolar neurons of the sensory fibres. These ganglia usually lie in the invertebral foramina. The two roots fuse to form the spinal nerve just lateral to the posterior root ganglion. The spinal nerves are grouped as follow: eight *cervical*, twelve *thoracic*, five *lumbar*, five *sacral* and one *coccygeal*.

The spinal nerve, on leaving the intervertebral foramen, divides into two branches called the anterior and posterior primary rami.

The posterior primary ramus of each spinal nerve contains both motor and sensory fibres and passes backwards to supply the skin and muscles of the posterior part of the trunk.

The anterior primary ramus of each spinal nerve receives, soon after its origin, a branch from the sympathetic trunk, which is part of the autonomic nervous system (p. 444). The anterior primary rami of the thoracic and upper lumbar nerves also give off a branch which passes to the sympathetic trunk.

The anterior primary rami of the thoracic spinal nerves pass round the chest wall as the intercostal nerves and supply the intercostal muscles and the skin of the chest wall and the upper part of the abdominal wall.

The anterior primary rami of the other spinal nerves form plexuses from which nerves emerge to supply the trunk and limbs.

The cervical plexus

The cervical plexus is formed from the anterior primary rami of the upper four cervical nerves. Branches of the cervical plexus supply the muscles of the neck and the skin of the front and sides of the neck.

The most important branch of the cervical plexus is the *phrenic nerve* which passes down through the neck and mediastinum to supply the diaphragm.

The brachial plexus

The brachial plexus is formed by the anterior primary rami of the lower four cervical nerves and the anterior primary ramus of the 1st thoracic nerve. The plexus lies in the lower part of the neck and the axilla and its branches supply the upper limb.

The brachial plexus is formed in the following way.

1 The anterior primary rami of the 5th and 6th cervical nerves unite to form an *upper trunk*.

The anterior primary ramus of the 7th cervical nerve forms the *middle trunk* of the plexus.

The anterior primary rami of the 8th cervical and the 1st thoracic nerve unite to form the *lower trunk* of the plexus.

2 The three trunks all divide, after a very short course, into *anterior* and *posterior divisions*.

3 The anterior divisions of the upper and middle trunks unite to form the

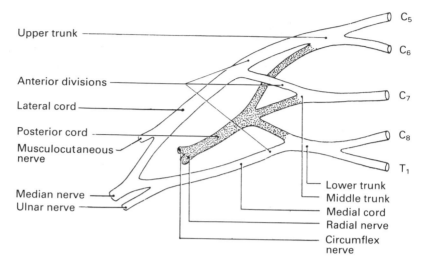

Upper trunk

Anterior divisions

Lateral cord

Posterior cord

Musculocutaneous
nerve

Median nerve
Ulnar nerve

C₅

C₆

C₇

C₈

T₁

Lower trunk
Middle trunk
Medial cord
Radial nerve
Circumflex
nerve

Fig. 16.19. The brachial plexus and its major branches.

lateral cord of the plexus. The most important nerves which arise from the lateral cord are the *musculocutaneous nerve* and the *lateral root of the median nerve*.

The anterior division of the lower trunk forms the *medial cord* of the plexus. The most important branches of the medial cord are the *ulnar nerve* and the *medial root of the median nerve*.

The posterior divisions of three trunks unite to form the *posterior cord* of the plexus. The most important branches of the posterior cord are the *circumflex nerve* and the *radial nerve*.

The circumflex nerve arises from the posterior cord of the brachial plexus and curves round the back of the surgical neck of the humerus below the shoulder joint. It supplies the deltoid muscle and the skin of the lateral side of the arm. It also gives a branch to the shoulder joint.

The musculocutaneous nerve arises from the lateral cord of the brachial plexus. It passes through the coracobrachialis muscle and then runs down the lateral side of the upper arm. It supplies the coracobrachialis, the brachialis and the biceps brachii muscles and the skin of the lateral side of the forearm.

The radial nerve arises from the posterior cord of the brachial plexus. It runs behind the brachial artery, and then curves round the back of the humerus in the spiral groove. It passes in front of the lateral epicondyle of the humerus at the elbow joint, and then runs down the forearm initially in front of, and then behind, the radius. It divides into its terminal branches at the lower end of the forearm. It supplies the triceps muscle, the extensor

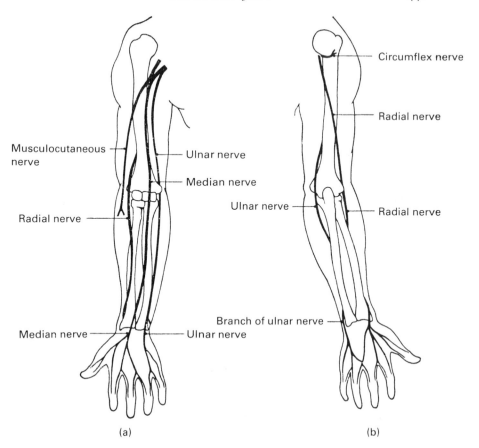

Fig. 16.20. The main nerves of the upper limb: (a) anterior aspect, and (b) posterior aspect.

muscles of the forearm and supinator muscle. At the elbow joint it gives a branch to the joint and it supplies the skin of the posterior aspect of the forearm, the skin of the radial side of dorsum of the hand and the skin of the dorsal surface of the thumb, and index finger and the radial side of the middle finger.

The ulnar nerve arises from the medial cord of the brachial plexus. It runs down the upper arm on the medial side of the brachial artery, curves backwards to pass behind the medial epicondyle of the humerus, and finally runs down the medial side of the forearm to the hand where it divides into its terminal branches. It supplies the flexor carpi ulnaris muscle, part of the flexor digitorum profundus muscle and some of the small muscles of the hand. It also supplies the skin of the ulnar border of the hand, the skin of the ulnar border of the dorsum of the little finger, the ring finger and the ulnar side of the middle finger, and the skin of the palmar surface of the little finger and the ulnar side of the ring finger.

The median nerve arises by two roots, one of which is derived from the lateral cord of the brachial plexus and the other of which is derived from the medial cord. The median nerve runs down the upper arm at first lateral to, and then medial to, the brachial artery. It then passes down the middle of the forearm and onto the palm of the hand, where it divides into its terminal branches. It supplies the muscles of the flexor aspect of the forearm, with the exception of those muscles which are supplied by the ulnar nerve, and some of the muscles of the hand. It also supplies the skin of the palmar surface of the hand and the skin of the palmar surfaces of the thumb, the index and middle fingers and the radial side of the ring finger.

The lumbar plexus

The lumbar plexus is formed from the anterior primary rami of the upper three lumbar nerves and part of the anterior primary ramus of the 4th lumbar nerve. The plexus lies within the psoas muscle. The most important nerves which arise from the lumbar plexus are the obturator nerve and the femoral nerve.

The obturator nerve emerges from the medial side of the psoas muscle, and then enters the medial side of the thigh through the obturator foramen. It supplies the obturator externus muscle, the abductor muscles and the skin on the medial side of the knee. It also gives branches to the knee and hip joints.

The femoral nerve is the largest branch of the lumbar plexus. It emerges from the lateral side of the psoas muscle and passes under the inguinal ligament, just lateral to the femoral artery. After a short course in the anterior part of the thigh it divides into its terminal branches. It supplies the quadriceps muscle, the pectineus and sartorius muscles and the skin of the anterior and medial aspects of the thigh and the medial aspect of the leg. It also gives branches to the hip and knee joints.

The sacral plexus

The sacral plexus is formed from part of the anterior primary ramus of the 4th lumbar nerve and the anterior primary rami of the 5th lumbar and upper four sacral nerves. The plexus lies on the posterior wall of the true pelvis. The main nerve which arises from the sacral plexus is the sciatic nerve.

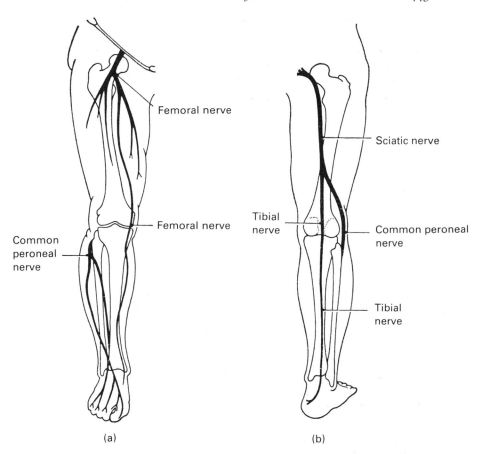

Fig. 16.21. The main nerves of the lower limb: (a) anterior aspect, and (b) posterior aspect.

The sciatic nerve

The sciatic nerve is the largest nerve in the body. It passes out of the pelvis through the greater sciatic notch under the gluteus maximus muscle. It then runs down the thigh, between the adductor magnus and the hamstring muscles, and finally divides into the common peroneal and tibial nerves just above the knee. In the thigh the sciatic nerve gives branches to the hamstring muscles and the adductor magnus muscle.

The common peroneal nerve is the smaller of the two terminal branches of the sciatic nerve. It runs down the lateral side of the popliteal fossa, curves round the neck of the fibula and the divides into its terminal branches which run downwards to the foot. It supplies the anterior and lateral muscles of the leg and the skin of the lateral side of the leg and the dorsum of the foot. It also gives branches to the knee and ankle joints.

The tibial nerve is the larger of the two terminal branches of the sciatic nerve. It runs through the popliteal fossa to enter the calf where it runs downward, finally passing behind the medial malleolus to reach the sole of the foot. It supplies the muscles of the calf, the small muscles of the foot and the skin of the posterior aspect of the leg and the sole of the foot. It also gives branches to the knee and ankle joints.

The autonomic nervous system

The autonomic nervous system controls the activities of the heart, smooth muscle and the glands. It is not under conscious control but its activities are closely related to those of the organs which are under voluntary control so that the viscera can meet the physiological demands imposed on them by voluntary activity.

The autonomic nervous system is under the control of the brain and it consists mainly of efferent fibres which pass out from certain parts of the brain and spinal cord. The efferent fibres of the autonomic nervous system all form synapses with nerve cells outside the central nervous system. These nerve cells lie in groups called ganglia. The fibres which pass out from the spinal cord are myelinated and are called *preganglionic fibres*. The fibres which pass from the ganglia to the organs which they innervate are usually unmyelinated and are called *postganglionic fibres*.

The afferent fibres of the autonomic nervous system are relatively few in number. They pass from the viscera to the brain and spinal cord and their impulses give rise to such sensations as hunger, nausea and pain.

The autonomic nervous system is divided into two parts: the *sympathetic nervous system*, and the *parasympathetic nervous system*.

The sympathetic nervous system

The preganglionic fibres of the sympathetic nervous system pass out from the grey matter of the spinal cord in the spinal nerves of the whole of the thoracic region and the upper part of the lumbar region. These fibres then pass in branches of the spinal nerves to the *sympathetic trunk* which is a chain of ganglia lying at the side of the vertebral column. The sympathetic trunks, one on each side of the vertebral column, extend from the upper part of the neck down into the pelvis where they meet in the midline at the front of the coccyx. The preganglionic fibres from the spinal cord either form synapses in the ganglia of the trunk or pass onwards to form synapses with nerve cells which lie in ganglia in one of the three main sympathetic nerve plexuses. The main sympathetic plexuses are:

1 **the cardiac plexus**, which lies behind the heart and supplies the heart and lungs;

2 **the coeliac plexus**, which lies in front of the abdominal aorta around the origins of the coeliac axis and the superior mesenteric artery; it supplies the abdominal organs; and

3 **the inferior hypogastric plexus**, which lies in the pelvis and supplies the pelvic organs.

Postganglionic fibres from the sympathetic trunks pass to join each spinal nerve and are distributed to the smooth muscle in the walls of the blood vessels, the arrectores pilorum of the skin (p. 489) and the sebaceous glands and sweat glands in the skin.

The functions of the sympathetic nervous system

The sympathetic nervous system is antagonistic to the parasympathetic nervous system. Its actions are:

1 it increases the rate and force of the heartbeat;

2 it causes dilation of the coronary arteries and the arteries of supply to voluntary muscle;

3 it causes constriction of the vessels of the skin and the intestine;

4 it causes contraction of the arrectores pilorum producing the appearance of goose flesh and it stimulates the sweat glands, to cause sweating;

5 it stimulates the adrenal medulla to produce adrenaline;

6 it produces dilatation of the bronchi;

7 it causes dilatation of the pupil of the eye;

8 it causes relaxation of the muscle of the urinary bladder; and

9 it decreases the motility of the stomach and intestine.

All these actions are similar to those produced by adrenaline which is secreted by the adrenal medulla (p. 464), and they prepare the body to meet a situation of excitement or stress.

The parasympathetic nervous system

The preganglionic fibres of the parasympathetic nervous system pass out from the brain in certain of the cranial nerves and from the spinal cord in the 2nd, 3rd and 4th sacral nerves. This limited outflow is often referred to as the *craniosacral outflow*. These preganglionic fibres pass to ganglia which are situated close to the organs which they supply.

The oculomotor nerve contains preganglionic parasympathetic fibres which pass through a ganglion to the ciliary muscle and the sphincter muscle of the iris of the eye. They produce accommodation and constriction of the pupil.

The facial nerve contains preganglionic parasympathetic fibres which pass via ganglia to the lacrimal gland and the submandibular and sublingual salivary glands. They stimulate the production of the secretions of these glands.

The glossopharyngeal nerve contains preganglionic parasympathetic fibres which pass through a ganglion to the parotid salivary gland. They stimulate the production of saliva by the gland.

The vagus nerve contains very large numbers of preganglionic parasympathetic fibres. They supply:

1   the heart, where they cause slowing of the heart beat and constriction of the coronary arteries;
2   the lungs, where they cause constriction of the bronchi; and
3   the stomach, the small intestine and the proximal large intestine, together with the pancreas and liver.

They cause increased motor activity, relaxation of the sphincters and increased secretion of the digestive juices.

The preganglionic parasympathetic fibres which pass out from the spinal cord in the sacral region pass to the descending colon, the pelvic colon, the rectum and the urinary bladder. They stimulate contraction of the smooth muscle in the walls of these organs and cause relaxation of the sphincters thus producing defaecation and micturition.

CHAPTER 17

The endocrine system

The endocrine system consists of a number of glands which are widely distributed throughout the body. They are known as *ductless glands* because their secretions, which are called *hormones*, pass directly into the blood stream. A **hormone** is defined as a substance which is produced in one gland or tissue and is secreted into the blood stream by which it reaches the target organs or tissues on which it exerts its effects.

The glands of the endocrine system have important controlling functions on development and growth as well as influencing the many metabolic activities of the body. There are parallels between the activities of the endocrine system and the nervous system in that they both exert effects distant from the point of origin of the stimulus, but whereas the nervous system uses well-defined anatomical pathways, the endocrine system uses the blood stream as its means of communication. However, the two systems may act together and there is a close link between the two systems at the hypothalamus which, through its links with the pituitary gland, integrates their actions. Most hormonal changes are slow, whereas the majority of changes effected by the nervous system occur swiftly. The action of the two systems in concert is essential for the maintenance of a stable internal environment.

The glands of the endocrine system are:
1 *the pituitary gland*;
2 *the thyroid gland*;
3 *the parathyroid glands*;
4 *the islets of Langerhans* in the pancreas;
5 *the adrenal glands*;
6 *the gonads* (the testes in the male, the ovaries in the female); and
7 *the pineal gland*.

The *hypothalamus* must also be considered to be a part of the endocrine system as must certain cells in the gastrointestinal tract which secrete hormones. The thymus gland is concerned with the regulation of lymphocyte function rather than endocrine function and it is considered in the chapter on the lymphatic system (Chapter 12).

Endocrine disorders are of two main types: those characterized by overproduction of hormone (*hyperfunction*) and those in which hormone production is defective or totally absent (*hypofunction*). Most endocrine disorders are of slow onset and progress over months, or even years, before a diagnosis is made. Hyperfunction may be due to a functional tumour, benign or malignant, of the gland concerned, or it may be due to hyperstimulation of the normal control mechanism. Hypofunction is caused by removal of the gland during a surgical operation or by destruction of the gland by inflammation, infiltration or neoplasia. More rarely, the picture of underactivity may be caused not by undersecretion but by resistance of the target organ, gland or tissue to normal circulating concentrations of the hormone. Diagnosis of these disorders depends upon confirmation of disturbed endocrine gland function by measuring the level of hormones, or of their metabolic products, in the circulation, or in the urine.

The pituitary gland

Although the endocrine glands are widely separated they are, to a large extent, interdependent. The pituitary gland, in particular, exerts an important controlling effect, either directly or indirectly, on several of the other endocrine glands. In turn the pituitary gland is subject to control by the hypothalamus, and via the hypothalamus is influenced by the central nervous system, which integrates many external stimuli.

The pituitary gland lies in the *sella turcica*, or pituitary fossa, of the sphenoid bone. The gland is reddish-brown in colour and is roughly oval-shaped. It is about 12 mm across and 8 mm in length. It is attached to the hypothalamus of the brain by a thin stalk. The optic chiasma, which is a communicating tract between the two optic nerves, lies just in front of the pituitary stalk. In front of, and below, the pituitary fossa is the sphenoid sinus. On each side of the fossa run the cavernous sinuses, which are venous channels containing the internal carotid artery, the IIIrd, the IVth, the upper two divisions of the Vth, and the VIth cranial nerves. The roof of the pituitary fossa is formed by a layer of dura mater, called the *diaphragma sellae*, through which the pituitary stalk passes.

Tumours of the pituitary gland may:
1 *enlarge within the pituitary fossa*, eroding its margins and causing enlargement of the fossa which becomes visible on plain radiographs of the skull;
2 *enlarge upwards* causing pressure on the optic chiasma producing visual field losses; and
3 *enlarge laterally* encroaching on the cavernous sinus causing symptoms and signs due to pressure on the cranial nerves lying within the sinus.

Blood supply. Branches of the internal carotid and the posterior cerebral arteries supply arterial blood to the posterior pituitary gland, to the pituitary stalk and to parts of the hypothalamus. Veins draining the hypothalamus form the hypophysial portal system which supplies blood to the

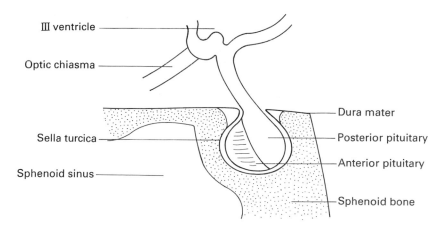

III ventricle

Optic chiasma

Sella turcica

Sphenoid sinus

Dura mater

Posterior pituitary

Anterior pituitary

Sphenoid bone

Fig. 17.1. The pituitary gland.

anterior pituitary gland. Releasing hormones, which are produced in the hypothalamus (see below), pass in this portal circulation to reach the anterior pituitary and control its functions. The anterior pituitary receives no direct arterial blood supply.

The structure and functions of the pituitary gland

The pituitary gland consists of two major parts. The anterior pituitary arises in fetal life as an upward growth from the pharynx, and the posterior pituitary as a downward projection from the region of the hypothalamus of the brain.

The anterior pituitary gland

The anterior pituitary contains three main types of hormone-secreting cell which differ in size and shape and also in the way in which they take up dyes when stained. The main cell types are *acidophil, basophil* and *chromophobe* cells. The anterior pituitary produces *growth hormone, prolactin, thyroid stimulating hormone (TSH), adrenocorticotrophic hormone (ACTH)*, and *gonadotrophic hormones (LH and FSH)*. Growth hormone and prolactin are secreted by the acidophil cells, while the basophil cells secrete TSH, ACTH, and the gonadotrophic hormones. The chromophobe cells are thought to be non-secreting, although some may to be capable of secreting ACTH.

1 *Growth hormone* is a protein hormone which stimulates bone growth in particular, but it is also needed for protein synthesis generally. The secretion

of growth hormone by the anterior pituitary is controlled by the hypothalamus which produces two hormones, one of which stimulates, and the other of which inhibits, the release of growth hormone.

Overproduction of growth hormone in childhood, prior to the closure of the epiphyses, leads to *gigantism* and due to this condition some patients may reach a height of 2.4 m (8 ft). In **adults** a tumour of the anterior pituitary which secretes growth hormone produces *acromegaly* (enlargement of the extremities). Since, at this age, the bones cannot grow in length, thickening occurs, particularly affecting the fingers and toes. The lower jaw increases in size, so that it protrudes beyond the upper jaw and the teeth become separated. There is also an increase in the size of the frontal sinuses and overgrowth of the supraorbital ridges. Acromegaly also leads to an increase in the size of other organs and, if the condition is not treated, the patient usually dies prematurely of heart disease. In addition to the endocrine effects of acromegaly the causative tumour may produce symptoms from its expansion within, or without, the pituitary fossa.

Undersecretion of growth hormone in childhood leads to dwarfism. The impairment of growth hormone secretion may be an isolated disorder or may occur in conjuction with other pituitary hormone deficiencies. In **adults** growth hormone deficiency produces no obvious effects.

2 *Prolactin* is a protein hormone which is very similar to growth hormone. Together with oestrogen it is secreted in increased amounts during pregnancy and is responsible for the breast changes associated with lactation. Prolactin secretion is normally inhibited by the secretion of an inhibitory hormone by the hypothalamus.

Overproduction of prolactin is due either to tumours of the anterior pituitary, which secrete prolactin on its own or with growth hormone, or to drugs which interfere with the production or action of the hypothalamic inhibitory hormone. High prolactin levels inhibit gonadotrophin secretion leading to amenorrhoea and infertility in women, and erectile impotence and lack of sperm production in men. Milk production from the breast (*galactorrhoea*) occurs in women but it is rare in men because the oestrogen levels are low. Prolactin deficiency is rare but it can occur due to necrosis of the pituitary following childbirth which is complicated by severe uterine blood loss, and it then leads to failure of lactation

3 *Thyroid stimulating hormone (TSH)* is a glycoprotein (protein molecule containing carbohydrate) which controls the rate of hormone production by the thyroid gland. In its turn TSH secretion is governed by a hypothalamic hormone which causes release of TSH from the anterior pituitary. There is an inverse relationship between thyroid hormone concentrations in the blood and TSH levels. When thyroid hormone levels fall as a result of destruction or removal of the thyroid gland the TSH secretion is increased. Conversely, increased secretion of thyroid hormone suppresses the production of TSH. This type of control mechanism is known as a *negative feedback*.

Overproduction of TSH by a pituitary tumour is rare, but increased TSH secretion does occur in patients with hypothyroidism. **Undersecretion of TSH** usually arises as part of a generalized decline in pituitary function (*hypopituitarism*) and it leads to underactivity of the thyroid gland.

4 *Adrenocorticotrophic hormone (ACTH)* is a polypeptide hormone, containing 39 amino acids, which stimulates the adrenal cortex to secrete cortisol and adrenal androgens. In its turn ACTH secretion is stimulated by a hypothalamic releasing hormone. Cortisol exerts a negative feedback control on both ACTH and its hypothalamic releasing hormone. ACTH itself also exerts negative feedback control on its releasing hormone.

Overproduction of ACTH can occur in response to destruction of the adrenal cortex and also as a result of hypothalamic or pituitary disease. The latter produces *Cushing's syndrome*, the features of which will be described in the section on the adrenal cortex. **ACTH deficiency**, like growth hormone deficiency, can either occur as an isolated hormone lack, or in combination with other pituitary hormone deficiencies. The features of ACTH deficiency are due to underproduction of cortisol by the adrenal cortex and consist of vomiting, weight loss, low blood pressure (*hypotension*) and low blood glucose levels (*hypoglycaemia*). There are, however, no disorders of sodium and potassium because the secretion of aldosterone by the adrenal cortex is not affected (see below).

5 *The gonadotrophic hormones* are, like TSH, glycoproteins and have a similar molecular structure. There are two gonadotrophic hormones, *follicle stimulating hormone (FSH)* and *luteinizing hormone (LH)*, which in the male is called *interstitial cell stimulating hormone (ICSH)*. The secretion of FSH and LH by the anterior pituitary are both under the control of the same hypothalamic releasing hormone. In both men and women the sex steroids, androgens and oestrogens, exert negative feedback control on both FSH and LH secretion, but in women positive feedback control occurs during the follicular phase of the menstrual cycle. This is further described in the chapter on the reproductive system (Chapter 18). FSH, in the female, stimulates the development and maturation of the ovarian follicle, and in the male, stimulates the production of spermatozoa. LH, in the female, stimulates the development of the corpus luteum of the ovary after ovulation. In the male LH (ICSH) stimulates the interstitial cells of the testis to secrete testosterone.

Overproduction of FSH or LH is rarely due to pituitary tumours but occurs after castration (removal of the gonads), and after the menopause in women because of failure of ovarian oestrogen secretion and operation of the normal feedback control. Selective **deficiencies of FSH** and **LH** are rare, but gonadotrophin deficiency is common in destructive lesions of the pituitary, such as pituitary tumours, and leads to failure of gonadal function.

The posterior pituitary gland

The posterior pituitary contains thin, non-myelinated nerve fibres, which are the termination of the modified nerve tracts whose cell bodies lie within the hypothalamus, and also specialized glial cells called *pituicytes*. Vasopressin, also called antidiuretic hormone (ADH), and oxytocin are produced in the hypothalamus and are transported along the nerve fibres to the posterior pituitary by a process called neurosecretion. They are stored in the posterior pituitary from which they are released into the circulation.

1 *Antidiuretic hormone (ADH)* is a peptide containing eight amino acids. It increases the renal tubular reabsorption of water and decreases urine flow. It can also stimulate the contraction of smooth muscle in the walls of blood vessels and in the walls of the gut, the urinary bladder and the gall bladder but these actions are not thought to be physiologically important. ADH secretion is stimulated by an increase in the plasma osmolality and a decrease in blood volume. There are cells in the region of the hypothalamus which are sensitive to changes in the blood osmolality, while cells in the walls of the atria of the heart and of the arteries sense changes in blood volume. The vagus nerve is thought to act as an afferent pathway linking these pressure stimuli to the hypothalamus.

Deficient production of ADH leads to failure of water reabsorption by the renal tubule and the consequent passage of large volumes of dilute urine. This condition is called *diabetes insipidus* and it is caused by damage to the posterior pituitary and the neighbouring hypothalamus by pituitary tumours, surgery or inflammatory destruction. In some patients the posterior pituitary and hypothalamus are normal but the ADH produced is ineffective because of an abnormality in the renal tubules which fail to respond to the ADH. This condition is known as nephrogenic diabetes insipidus to distinguish it from cranial diabetes insipidus in which ADH is deficient.

Overproduction of ADH leads to dilution of the blood (haemodilution) and to symptoms of drowsiness, weakness and confusion which are due to a low plasma sodium level (hyponatraemia). This condition may be associated with a number of disorders.

2 *Oxytocin*, which is also a peptide containing eight amino acids, causes contraction of the smooth muscle in the uterus and in the breast. In the human the role and physiological importance of oxytocin are uncertain. There are disease processes associated with oxytocin overproduction or

deficiency. The uterine muscle is very sensitive to oxytocin at the end of pregnancy and oxytocin is, therefore, sometimes used to induce labour.

The hypothalamus

Many of the functions of the hypothalamus have already been described in the section on the pituitary gland. The hypothalamus acts as an integrator, linking the short-term neurological responses, and especially those of the autonomic nervous system, with the more leisurely, longer-term changes in endocrine function. In addition to controlling the hormone secretions of the anterior pituitary, and acting as the producer of posterior pituitary hormones, the hypothalamus plays a part in the regulation of body temperature, appetite and the onset of puberty. The cyclical activity of the hypothalamus is responsible for the circadian rhythm of certain pituitary hormone secretions, for example the secretion of ACTH which is at its lowest between 10 p.m. and 2 a.m. and at its highest between 6 a.m. and 9 a.m. The hypothalamus is also responsible for the cyclical gonadotrophin production of the normal menstrual cycle.

The pineal gland

The pineal gland is a small, reddish-grey body which lies in the midline of the brain below the corpus callosum and behind the 3rd ventricle. It consists of special cells with processes, called *pinealocytes*, neuroglial cells and blood vessels. The secretions of the pineal are stimulated by darkness and inhibited by light exposure and they are probably involved in the regulation of sexual function and development. The precise function of the pineal gland in human physiology is not yet certain. Pineal tumours and tumours in the pineal region in children cause precocious puberty but this is probably due to pressure effects on the hypothalamus.

The thyroid gland

The thyroid gland lies in the lower anterior part of the neck. It consists of two lobes, one lying on either side of the trachea, which are joined across the midline by a narrow isthmus. The upper border of the thyroid gland lies at the level of the 5th cervical vertebra and the lower border at the level of the 1st thoracic vertebra. Each lobe is roughly conical in shape and is about 5 cm long and 2—3 cm across at its widest point. The apex of each lobe reaches the side of the thyroid cartilage and the base lies at the level of the 4th or 5th tracheal ring. The isthmus usually covers the 2nd and 3rd rings of the trachea. Occasionally a small, conical, pyramidal lobe arises from the upper border of the isthmus. The posterolateral surface of each

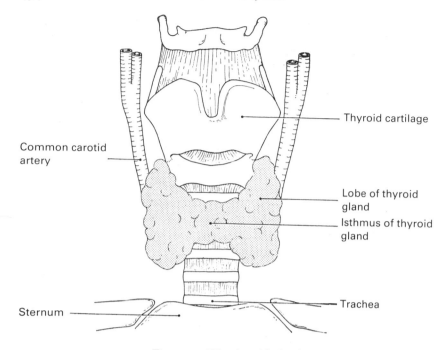

Fig. 17.2. The thyroid gland.

lobe is related to the common carotid artery, and the posterior surface is related to the parathyroid glands. The convex, anterior surface of each lobe is covered by the superficial muscles of the neck.

Blood supply. The thyroid gland is supplied by two arteries on each side:

1 the *superior thyroid artery*, which is a branch of the external carotid artery; and
2 the *inferior thyroid artery*, which is a branch of the subclavian artery.

The veins draining the thyroid gland form a network on its surface and then join to form the superior, middle and inferior thyroid veins which drain into the internal jugular veins and the left brachiocephalic vein.

The structure of the thyroid gland

The thyroid gland is covered by a thin capsule of connective tissue. Microscopically the gland is seen to consist of a large number of spherical follicles which are lined with cuboidal epithelium and which contain *colloid*. The colloid consists of a glycoprotein, called *thyroglobulin*, which is the storage form of the thyroid hormones (*thyroxine* and *triiodothyronine*). In a

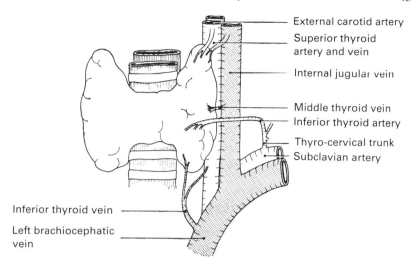

External carotid artery

Superior thyroid
artery and vein

Internal jugular vein

Middle thyroid vein

Inferior thyroid artery

Thyro-cervical trunk

Subclavian artery

Inferior thyroid vein

Left brachiocephatic
vein

Fig. 17.3. The blood supply to the thyroid gland.

very active thyroid gland the stores of colloid are small and the follicular epithelial cells become hypertrophied and columnar.

There is also a second type of cell, the *parafollicular cell*, which is present in smaller numbers than the follicular cells. This cell secretes calcitonin.

Functions of the thyroid gland

The thyroid gland takes up inorganic iodine from the blood stream and converts it into the thyroid hormones by a complex series of reactions each of which is controlled by a different enzyme. When the thyroid gland is stimulated by TSH the outputs of thyroxine and triiodothyronine are increased as a result of increased activity in each of the steps.

Thyroid hormones have the following actions:

1 They increase the basal metabolic rate and stimulate oxygen consumption; and

2 they are necessary for normal growth and development and for the maturation of brain tissue in the new-born.

Abnormalities of thyroid function

Undersecretion (*Hypothyroidism*). *During infancy* thyroid underactivity leads to *cretinism*. The features of cretinism are:

1 retarded physical and mental development;

2 a typical appearance consisting of a large tongue, an expressionless face, a protuberant abdomen and a dry skin; and

3 lethargy and a low temperature.

Without instant recognition of the condition and treatment with thyroxine, the eventual IQ of many of these children is impaired. Currently many centres screen all neonates (newly-born children) for hypothyroidism.

During later childhood hypothyroidism causes growth retardation and at the normal age of puberty there is a delay in the development of the secondary sexual characteristics.

During adult life hypothyroidism, which is then called *myxoedema* because of the swollen appearance of the skin and subcutaneous tissues, causes:

1 mental and physical sluggishness;

2 a dislike of cold weather;

3 a tendency to gain weight, but this is not very marked;

4 dryness of the skin and brittle hair;

5 constipation;

6 a hoarse voice; and

7 a low basal metabolic rate.

Hypothyroidism is usually due to a disease of, or destruction of, the thyroid gland. The commonest disease affecting the thyroid is *autoimmune thyroiditis* (p. 47), a condition which is characterized by the presence of circulating antibodies to the thyroid gland and consequent inflammatory destruction of the thyroid gland tissue. Surgical removal of thyroid tissue, and radioiodine therapy for thyrotoxicosis (see below), can both cause thyroid hormone undersecretion. Less commonly hypothyroidism is due to TSH deficiency, which is usually part of generalized hypopituitarism.

Oversecretion (*hyperthyroidism* or *thyrotoxicosis*). Oversecretion most commonly affects young women. The following features are common:

1 there is weight loss despite a normal, or even increased, appetite;

2 the patient is intolerant of heat;

3 diarrhoea;

4 tremor, increased excitability and nervousness;

5 a rapid heart rate;

6 prominence of the eyes; and

7 an increased basal metabolic rate.

Hyperthyroidism is most often due to thyroid stimulation by an immunoglobulin, not TSH. In some cases thyroid hormone overproduction is associated with a nodular thyroid gland and most commonly with a multinodular thyroid. Occassionally, however, a single hyperfunctioning nodule, or '*toxic adenoma*', is present.

Goitre is a general term used to describe enlargement of the thyroid gland. It does not imply any particular cause, nor any disorder of thyroid function. When iodine intake is insufficient, which is usually due to a diminished amount in the drinking water, thyroid enlargement occurs. This condition is known as endemic goitre and is particularly common in the Alps, the Himalayas and the Andes, areas where the iodine content of the water, the food and the soil is low. The addition of small quantities of potassium iodide to salt and foods has made endemic goitre rare in developed countries, but it is still a problem in the Third World countries.

Goitre may be associated with disturbances of thyroid function (e.g. hyperthyroidism), with autoimmune thyroiditis, with defects in thyroid hormone biosynthesis (dyshormonogenetic goitre or goitrous cretinism), and can occur in the absence of any disease or disorder of thyroid function, when it is called *simple*, or *colloid*, *goitre*.

Calcitonin is secreted by the parafollicular cells of the thyroid gland. It inhibits osteoclastic bone resorption and it is important for maintaining normal bone in some species. In man the physiological importance of calcitonin is uncertain but circulating concentrations are increased in patients with medullary carcinomas of the thyroid, a malignant tumour of the parafollicular cells.

The parathyroid glands

There are usually four parathyroid glands which lie on the posterior surface of the thyroid lobes, two one each side. They are small, yellowish-brown, oval glands which measure 6 mm, by 3 mm, by 2 mm, and weigh about 50 mg, each. The parathyroid glands, however, may vary in number from three to six or more, and the anatomical location of the two inferior parathyroid glands may vary.

Blood supply. The *inferior thyroid artery* supplies the parathyroid glands. Their venous return is via the thyroid veins.

Structure of the parathyroid glands

The parathyroid glands have a thin connective tissue capsule. The cells of the glands are known as *chief cells*, and they are arranged in columns. According to the staining characteristics of their cytoplasm the cells are called *light, dark* and *clear cells*. These cells secrete *parathyroid hormone*. There is also a second type of cell, the *oxyphil cell*, which is inactive in terms of hormone synthesis.

Functions of the parathyroid glands

The parathyroid glands secrete an 84 amino acid polypeptide, parathyroid hormone, which is concerned with the metabolism of calcium and phosphate.

The secretion of parathyroid hormone is not controlled by the pituitary hypothalamic unit but is stimulated by a fall in the calcium concentration of the blood. Parathyroid hormone has the following actions:

1   it increases bone destruction, liberating calcium into the blood stream;

2   it increases the excretion of phosphate by the kidneys; and

3   it increases the formation, in the kidney, of active Vitamin D (see the chapters on the skeleton and the digestive system).

Actions 1 and 2 cause a rise in the blood calcium concentration (hypercalcaemia) and a fall in the blood phosphate concentration (hypophosphataemia), while action 3 leads to an increased absorption of calcium from the small intestine because of the increased level of active Vitamin D.

Abnormalities of parathyroid function

Oversecretion of parathyroid hormone is called *hyperparathyroidism*. This condition is characterized by hypercalcaemia and hypophosphataemia. Mild cases may be virtually without symptoms but, if the rate of bone destruction is high, bone changes may become obvious on X-rays. Areas of bone destruction are called *osteitis fibrosa cystica*. In some cases the increased urinary excretion of calcium leads to urinary stone formation. Most cases of hyperparathyroidism are due to a single benign parathyroid adenoma. Occasionally chronic hypocalcaemia, (due to renal disease or gastrointestinal disorders) may lead to hyperplasia of all four parathyroid glands producing *secondary hyperparathyroidism*.

Undersecretion of parathyroid hormone is called *hypoparathyroidism*. This condition can be due to accidental removal of, or damage to the blood supply of, the parathyroid glands during surgery of the thyroid gland. Less frequently it is due to autoimmune destruction of the parathyroid glands, and, more rarely still, to tissue resistance to parathyroid hormone. Hypoparathyroidism leads to lowered blood calcium levels (hypocalcaemia) and raised blood phosphate levels (hyperphosphataemia). Hypocalcaemia causes increased excitablity of nerves and muscles, causing *tetany*—pins and needles in the hands and feet which are sometimes associated with muscular stiffness and cramps.

The islets of Langerhans

The islets of Langerhans are collections of endocrine cells interspersed between the exocrine cells of the pancreas. These spheroidal collections of cells, which may number up to one million, are most numerous in the tail of the pancreas. They have a rich blood supply and they have a nerve supply from the sympathetic autonomic nervous system, and through the vagal nerve fibres, from the parasympathetic nervous system as well. Different cell types within each islet secrete *insulin, glucagon, somatostatin* and also other hormones whose precise physiological role is not yet clear.

Insulin is a polypeptide hormone containing 51 amino acids. It is secreted by the *B-cells* of the islets of Langerhans and, like other pancreatic hormones, it enters the hepatic portal circulation. It is important in controlling not only the blood level of glucose but it also has a profound effect on the metabolism of fat and protein. Insulin increases the uptake and use of glucose by most body tissues but its most important actions occur within the liver. There it increases glycogen formation and lowers glucose output. Fat formation is increased, and fat breakdown is decreased, by insulin. The hormone also stimulates protein synthesis. The secretion of insulin is stimulated by a rising concentration of blood glucose. Although not directly under anterior pituitary control, insulin secretion can be stimulated by both growth hormone and ACTH. In addition, certain hormones secreted by the gastrointestinal tract can stimulate insulin release and are probably important in generating the insulin response to food. Stimulation of the vagus nerve also causes secretion of insulin.

Abnormalities of function of the islets of Langerhans

Undersecretion of insulin causes *diabetes mellitus* which is invariably characterized by a raised blood glucose level (hyperglycaemia). If the insulin lack is mild then hyperglycaemia and the resultant glycosuria (the presence of glucose in the urine) may be the only metabolic abnormalities. However, if insulin lack is severe the diabetes is characterized by excessive lipolysis (fat breakdown) and decreased protein synthesis as well as hyperglycaemia. This type of diabetes usually occurs in children and young adults whereas the milder type is common in older people.

Oversecretion of insulin may be due either to an insulinoma, which is an insulin-secreting tumour of the pancreas which is usually benign, or to diffuse hyperplasia of the B-cells of the islets of Langerhans. The excessive secretion of insulin causes a low blood glucose level (hypoglycaemia) which leads to feelings of hunger, weakness, sweating, palpitation and headache. These symptoms are due to adrenaline release but, in addition, the brain, which has no glycogen stores, is also affected. The effects on the brain lead to mental confusion and sometimes to unconciousness and epileptic fits.

Glucagon is, like insulin, a polypeptide but contains 29 amino acids. It is secreted by the *A-cells* of the islets of Langerhans and its major effect is to raise the blood level of glucose by causing the breakdown of liver glycogen to glucose. It also stimulates the secretion of insulin. Glucagon secretion is stimulated by a fall in the level of glucose in the blood. Disorders of glucagon secretion rarely cause identifiable disorders of glucose metabolism.

Glucagonomas, however, are tumours of the islets of Langerhans which secrete glucagon, but the resultant diabetes is mild.

Somatostatin is a peptide made up of 14 amino acids. Although it was originally isolated from the hypothalamus, from where it is secreted as growth hormone release inhibiting hormone, it was subsequently found in the islets, the gut and nervous tissue. Its name, somatostatin, derives from its first discovered action of inhibiting growth hormone release by the anterior pituitary (growth hormone is also known as somatotrophin). However, it has a multiplicity of actions which depend on the tissue from which it is released, and also on which of its multiple targets is affected. In addition to inhibiting growth hormone release it also inhibits insulin, glucagon and TSH secretion and affects gut secretions. Since its effects on other islet hormones and gut cells do not depend upon secretion into the blood stream, but are direct effects on neighbouring cells, they are known as *paracrine* rather than endocrine actions. Somatostatin excess, due to a tumour of the pancreas or gut which secretes the hormone, has been found in some patients but it is extremely rare. The symptoms include mild diabetes in the case of the pancreatic—somatostatin secreting tumours.

The adrenal glands

The adrenal (suprarenal) glands are found on either side of the vertebral column, in the retroperitoneal space, on the upper pole of each kidney. Like the kidneys they are surrounded by adipose tissue and they lie within the renal fascia.

The right adrenal gland is roughly pyramidal in shape and the left is crescentic. Each gland is yellowish in colour, is about 5 cm long and 3 cm thick, and weighs about 5 g.

Blood supply. Each gland is supplied by three arteries which are branches of the inferior phrenic artery, the abdominal aorta and the renal artery. There is only one suprarenal vein on each side. The left suprarenal vein drains into the left renal vein and the right suprarenal vein drains directly into the inferior vena cava.

The structure of the adrenal glands

The adrenal glands consist of an outer cortex and an inner medulla. The cortex makes up about 90% of the glands' total weight. Whereas the cortex is yellowish in colour the medulla is either grey, or dark red, and is functionally separate.

The adrenal cortex consists of three layers:
1 an outer layer, called the *zona glomerulosa*, which is composed of small polyhedral cells arranged in clumps;

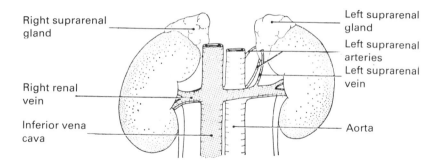

Right suprarenal gland

Right renal vein

Inferior vena cava

Left suprarenal gland

Left suprarenal arteries

Left suprarenal vein

Aorta

Fig. 17.4. The adrenal (suprarenal) glands.

2   a middle layer, called the *zona fasciculata*, which is composed of large polyhedral cells arranged in columns; and
3   an inner layer, called the *zona reticularis*, which is composed of a network of rounded cells.

The adrenal medulla is composed of groups and columns of *chromaffin cells*, which are so-called because they can be stained with chromium salts. These cells are derived from cells which form the sympathetic nervous system (the neural ectoderm). This type of cell is found in a number of sites in relation to the sympathetic nervous system. Scattered among the chromaffin cells are nerve cells.

Accessory adrenal glands are sometimes found in the fat surrounding the adrenal glands or in the renal cortex.

The functions of the adrenal glands

The adrenal cortex secretes a large number of steroid hormones but these can be considered in three main groups.

The glucocorticoids are mainly concerned with carbohydrate metabolism and, acting in concert with other metabolic and endocrine mechanisms, are responsible for maintenance of a normal blood glucose level. The main glucocorticoid is *hydrocortisone* (cortisol). Glucocorticoids are produced by the zona fasciculata under the influence of ACTH.

The mineralocorticoids are secreted by the zona glomerulosa which is not under the control of ACTH but of the renin-angiotensin system (see Chapter 15). *Aldosterone* is the principal mineralocorticoid and acts on the renal tubule, leading to retention of sodium and loss of potassium. Retention of sodium is accompanied by secretion of antidiuretic hormone because of a rise in plasma osmolality.

The adrenal androgens are secreted mainly by the zona reticularis, which also secretes small amounts of oestrogen and progesterone. Secretion of these hormones is under the control of ACTH.

The adrenal cortex is essential to life and complete removal leads to death, unless glucocorticoid and mineralocorticoid replacements are given to the patient.

The adrenal medulla functions independently of the adrenal cortex and produces two hormones, *adrenaline* and *noradrenaline*. The functions of the adrenal medulla and the sympathetic nervous system are closely inter-related. Most adrenaline is produced by the adrenal medulla but noradre-naline is produced principally by the sympathetic nervous system at its nerve endings.

The actions of these two hormones prepare the body for sudden action when faced with fear or pain. Their actions are frequently described as the 'fight or flight' reaction and, although their actions are similar in many respects, they are not identical.

Adrenaline causes:

1   an increase in the heart rate;
2   an increase in the blood glucose level; and
3   a relaxation of smooth muscle in some small arteries leading to a rise in blood flow to skeletal muscle and some viscera.

Noradrenaline causes more generalized constriction of small arteries leading to a considerable rise in both the systolic and the diastolic blood pressure. It has less effect on the heart rate than adrenaline.

Abnormalities of adrenal function

The cortex

Undersecretion of the hormones of the adrenal cortex can be due either to diseases of the adrenal cortex or to impaired output of ACTH due to pituitary disease. When ACTH deficiency occurs only cortisol (glucocorti-coid) and adrenal androgen secretions fail. Diminished glucocorticoid secre-tion leads to hypoglycaemia but mineralocorticoid (aldosterone) secretion continues and electrolyte disturbance does not occur. Destruction of the adrenal cortex may be caused by autoimmune disease or by tuberculosis. Although the adrenal medulla is also destroyed, no symptoms arise from the lack of this part of the gland due to the output of adrenaline and noradrenaline by the sympathetic nervous system. The condition of adrenal gland destruction is called *Addison's disease*. The clinical features of this disease are:

1   weakness and tiredness;

2 weight loss;

3 low blood pressure especially on standing; and

4 brown pigmentation in scars, in areas exposed to sunlight, in the mouth and in the axillae and groins.

Occasionally the disease presents acutely with low blood pressure, vomiting and extreme dehydration. Unless it is rapidly treated with hydrocortisone and intravenous fluids death may result.

The features of Addison's disease are due to combined failure of mineralocorticoid (aldosterone) and glucocorticoid secretion. The pigmentation is caused by a gradual onset of adrenal insufficiency with persistently raised ACTH secretion which is due to the feedback mechanism (see p. 451). ACTH in large quantities stimulates melanin production in the skin. The lack of adrenal androgens in women with Addison's disease causes loss of secondary sexual hair (axillary and pubic hair). This effect of the disease is not noticeable in men because of the testicular secretion of androgens.

Oversecretion of adrenal cortical hormones may be considered under three headings: oversecretion of cortisol, mineralocorticoid or adrenal androgens.

Oversecretion of cortisol causes *Cushing's syndrome.* The excess cortisol production is caused either by a tumour of the adrenal cortex, which may be benign or malignant, or by increased ACTH secretion. The increased ACTH secretion may be produced from the anterior pituitary, due to hypothalamic pituitary disease, or from a non-endocrine malignant tumour such as a bronchial carcinoma. Cushing's syndrome can also be seen in patients given corticosteroid drugs, or ACTH, for the treatment of nonendocrine disease, such as bronchial asthma or rheumatoid disease. The features of Cushing's syndrome are:

1 obesity of the trunk with the deposition of excessive subcutaneous fat;

2 high blood pressure;

3 purple stretch marks (striae) in the skin, particularly over the abdominal wall;

4 thin arms and legs, with wasting and weakness of the muscles, particularly those of the proximal limb girdle muscles;

5 excess hair growth (hirsutism); and

6 hyperglycaemia and glycosuria (diabetes mellitus) in some patients.

These features are frequently not all found in patients with Cushing's syndrome secondary to a malignant non-endocrine tumour as the patients do not live long enough to develop the typical picture described above. The ACTH hypersecretion in these patients causes pigmentation; the weight loss due to the carcinoma is obvious and the muscle weakness is usually due to a low plasma potassium concentration.

Oversecretion of mineralocorticoid (*primary aldosteronism*) is a rare condition which is due to a tumour of the adrenal cortex which secretes excess aldosterone and not the other adrenocortical hormones. The resulting retention of sodium and water leads to high blood pressure with a low plasma potassium.

Oversecretion of adrenal androgens (*congenital adrenal hyperplasia*). This condition is due to an inherited defect of one of the enzymes involved in adrenal steroid synthesis. Glucocorticoids, mineralocorticoids and adrenal androgens all have the same basic chemical structure and are derived originally from cholesterol. The chemical pathways leading from cholesterol to these three main types of adrenal steroids are complex and interrelated. When an enzyme defect occurs there is a block in the pathway to the formation of one or more of these three main classes of adrenal hormone. Although there are various types of congenital adrenal hyperplasia, because of the number of enzymes involved in steroid biosynthesis the commonest picture is that of cortisol lack with excessive androgen production. The cortisol lack stimulates ACTH secretion which stimulates hyperplasia of the adrenal cortex with a further rise in adrenal androgens. In female infants the high circulating concentration of adrenal androgens produces masculinization of the genitalia because the process begins during intrauterine development. Although an uncommon disorder in clinical practice congenital adrenal hyperplasia is one of the commonest inborn errors of metabolism.

The Medulla

Oversecretion. Rarely increased amounts of adrenaline and noradrenaline are secreted by a tumour of the adrenal medulla known as a *phaeochromocytoma*. The result is high blood pressure, which may be sustained or intermittent. When the high blood pressure is intermittent the patient may experience paroxysms of a fast heart rate (tachycardia), sweating and pallor or flushing. The exact pattern of symptoms depends upon the proportions of adrenaline and noradrenaline produced by the phaeochromocytoma.

Undersecretion of adrenaline and noradrenaline produces no definite symptoms.

Gastrointestinal hormones

Throughout the gut there are found cells which secrete hormones. If combined together these cells would form the largest endocrine gland. Most of these cells are found in the stomach, the duodenum and the small intestine. These cells produce a variety of polypeptide hormones which are responsible for coordinating the motility of the gut and controlling its secretions. Some of these hormones are released into the circulation in true endocrine fashion while others exert local effects on neighbouring cells in the paracrine fashion (see somatostatin).

CHAPTER 18

The reproductive system

The male reproductive system

The male reproductive system is composed of the following organs: two *testes* and two *epididymes* which lie in the scrotum, two *seminal ducts*, two *seminal vesicles*, two *ejaculatory ducts*, the *prostate gland*, and the *penis*.

The scrotum

The scrotum is a pouch of pigmented skin which is thrown into numerous folds by a thin subcutaneous sheet of unstriped muscle fibres, called the *dartos muscle*. The scrotum lies below the pubic symphysis, in the front of the thighs. The dartos muscle gives off a muscular septum which divides the scrotum into right and left compartments, in which are contained the right and left testes. Deep to the dartos muscle is the *cremasteric fascia*, which is a layer of fibrous tissue containing some unstriped muscle fibres.

The testes

The testes are ellipsoidal organs which hang in the scrotum suspended by the spermatic cords. Each testis is about 4.5 cm in length, 2.5 cm in breadth and 3 cm thick. Three membranes invest the testes.
1   **The tunica vaginalis**, which is the outer covering, is a closed sac of serous membrane. The tunica vaginalis is thus composed of an outer layer called the *parietal layer* and an inner layer called the *visceral layer*. The two layers are separated by a thin film of serous fluid.
   The testes develop in the abdominal cavity and, during the seventh month of intrauterine life, start to move downwards towards the scrotum carrying with them a pouch of peritoneum which forms the tunica vaginalis. By birth they usually lie in the scrotum and the communication between the main peritoneal cavity and the tunica vaginalis is gradually obliterated.
2   **The tunica albuginea**, which is the middle membrane, is a dense, bluish, sheet of fibrous tissue. The testis is divided into a number of lobules by incomplete fibrous septa which are attached to the tunica albuginea.
3   **The tunica vasculosa**, which is the inner membrane, is a layer of thin

areolar tissue which contains a plexus of blood vessels. It lines the tunica albuginea and the fibrous septa of the testis.

The structure of the testis

Each testis consists of 200–300 lobules and each lobule is composed of several minute tortuous tubules, called the *convoluted seminiferous tubules*. These tubules are connected by a loose layer of connective tissue which contains a number of cells called the *interstitial cells*. Each tubule is composed of a basement membrane lined by three layers of epithelial cells called the *germinal epithelium*. The tubules finally straighten out and pass to the upper part of the testis where they end in a small number of ducts, called *efferent ducts*. These ducts pass through the tunica albuginea to enter the epididymis.

The epididymis

The epididymis is composed of a tortuous canal which constitutes the first part of the efferent duct of the testis. The canal is about 5.5 m in length but it is folded and tightly packed to form the elongated epididymis. The epididymis has an expanded *head*, which lies at the upper pole of the testis, a *body*, which lies on the posterior aspect of the testis, and a *tail*, which lies at the lower pole of the testis. The tail of the epididymis is continuous with the seminal duct.

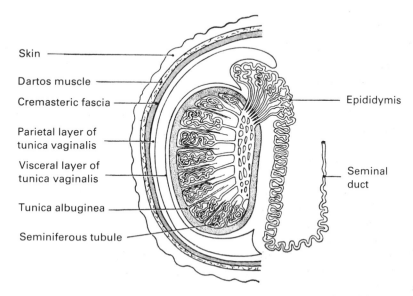

Skin

Dartos muscle

Cremasteric fascia

Parietal layer of tunica vaginalis

Visceral layer of tunica vaginalis

Tunica albuginea

Seminiferous tubule

Epididymis

Seminal duct

Fig. 18.1. Diagrammatic cross-section of the scrotum, testis and epididymis.

The seminal duct

The seminal duct, or vas deferens, is the continuation of the canal of the epididymis. It is at first tortuous, but soon straightens out and passes upwards in the inguinal canal to enter the pelvis. In the pelvis the seminal duct passes medially towards the base of the bladder, where it is joined by the duct of the seminal vesicle to form the ejaculatory duct.

The spermatic cord

The spermatic cord suspends the testis in the scrotum. It is composed of: the *seminal duct*, the *testicular artery*, the *testicular vein, lymphatic vessels*, and *nerves*.

These structures are bound together by loose connective tissue and are surrounded by a thin layer of fibrous connective tissue.

The blood supply to the testis. The arterial supply to the testis is via the testicular artery, which arises from the abdominal aorta just below the renal artery. It passes downwards, from this level, to enter the spermatic cord.

The venous return from the testis is through the testicular vein which passes upwards to the level of the renal veins. The right renal vein enters the inferior vena cava and the left testicular vein enters the left renal vein.

The lymphatic drainage of the testis. The lymphatic vessels from the testis ascend in the spermatic cord and then pass upwards with the testicular vessels to end in the lateral aortic and preaortic lymph nodes.

The lymphatic vessels from the seminal duct, however, pass to the external iliac lymph nodes.

The seminal vesicles

The seminal vesicles are two sacs, about 5 cm in length, which lie on the posterior aspect of the base of the bladder. Each vesicle is composed of a single coiled tube from which arise several diverticula. At its medial end it combines with the seminal duct to form the ejaculatory duct.

The seminal vesicles are each composed of three coats:

1   *an outer coat* of areolar tissue;
2   *a middle coat* of smooth muscle fibres; and
3   *an inner layer* of mucous membrane which is lined with columnar epithelium. The diverticula contain goblet cells which produce a large proportion of the seminal fluid.

The ejaculatory ducts

Each ejaculatory duct is formed by the union of the seminal duct and the seminal vesicle, and is about 2 cm in length. They pass downwards and forwards to enter the prostate gland where they open into the prostatic part of the urethra.

They are thin-walled but are composed of tissue similar to the seminal vesicles.

The prostate gland

The prostate gland is a chestnut-shaped gland which lies at the base of the bladder, behind the pubic symphysis, and surrounds the first part of the urethra. It has a base, which faces upwards, and an apex, which is downwardly directed. Its posterior surface is in close relationship to the anterior wall of the rectum.

The structure of the prostate gland

The prostate gland has an outer capsule of fibrous tissue.

The substance of the gland is composed of fibromuscular tissue and glandular tissue. The glandular tissue is composed of numerous follicles which are lined with columnar epithelium. Ducts from the glandular portion open into the prostatic urethra.

The lymphatic drainage of the prostate gland

The lymphatic vessels from the prostate gland, together with those from the seminal vesicles, drain into the external and internal iliac groups of lymph nodes.

The male urethra

The urethra in the male forms the outflow pathway for the urine and for the seminal fluid.

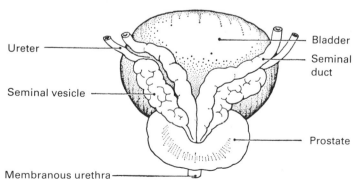

Fig. 18.2. The base of the male bladder to show the seminal vesicles and the prostate gland.

The male urethra is 18–20 cm in length and extends from the internal urethral orifice at the base of the bladder to the external urethral orifice at the tip of the penis. It describes a double curve and is divided into three parts, the prostatic part, the membranous part and the spongy part.

The prostatic part is the widest portion of the urethra. It is about 3 cm in length and runs through the prostate gland, from its base to its apex. The ejaculatory ducts and the small prostatic ducts open into the prostatic part of the urethra.

The membranous part is the shortest and narrowest portion of the urethra. It runs downwards for about 2 cm to pierce the fascia of the perineum. It is surrounded by the external urethral sphincter muscle.

The spongy part of the urethra is about 15 cm long and commences below the membranous part. It runs through the penis to the external urethral orifice.

The penis

The penis is composed of a *root*, which lies in the perineum, and a *body*, which is free and is covered with skin. The penis is composed of three columns of erectile tissue. Two of these columns lie anteriorly and are called the *corpora cavernosa*. The third column, which surrounds the spongy part of the urethra, lies posteriorly, between the corpora cavernosa, and is called the *corpus spongiosum*. At the end of the penis the corpora cavernosa

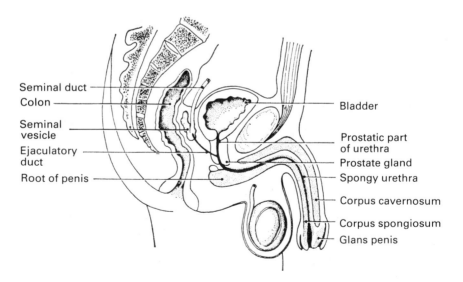

Fig. 18.3. Cross-section of the male pelvis.

is expanded to form the *glans penis*. A fold of skin, called the *prepuce* or foreskin, covers the glans penis.

The functions of the male reproductive organs

At puberty the testes increase rapidly in size, the interstitial cells begin to secrete *testosterone* and the seminiferous tubules start to produce spermatozoa. The precise mechanism by which puberty is initiated is disputed, but the result is that the gonadotrophins, LH and FSH, are secreted from the anterior pituitary in increased amounts.

FSH stimulates spermatogenesis in the seminiferous tubules. Each spermatozoon consists of a head, containing the nucleus and the genetic material, and a tail. The tail contains energy stores and contractile axial filaments, which are responsible for the movement of the spermatozoon. Spermatozoa, after formation, are stored in the epididymis, where they mature under the influence of the testosterone in the epididymal fluid. The seminal vesicles secrete fluid into the ejaculatory duct. This fluid contains fructose which is the energy source for the spermatozoa.

LH stimulates the interstitial cells of the testis to secrete testosterone, the male hormone. *Testosterone* has the following actions:
1 It stimulates the growth of accessory reproductive organs—i.e. the penis, the prostate gland, the seminal vesicle and the epididymis.
2 It causes the physical and psychological changes which occur at puberty—i.e. the growth of secondary hair, the enlargement of the larynx, the muscular development and the male drive and aggression.

Vasectomy

Ligation and cutting of the vas deferens (spermatic duct) is an effective means of sterilization in the male. This operation causes some atrophy of the seminiferous tubules but testosterone secretion remains unimpaired. For a few weeks after the operation motile spermatozoa may remain in the distal end of the vas deferens.

The female reproductive organs

The female reproductive organs are divided into an internal group of organs which lie in the true pelvis, and an external group which lie below and in front of the pubic arch.

The internal organs are: two *ovaries*, two *uterine (Fallopian) tubes*, the *uterus*, and the *vagina*.

The external organs are collectively called the *vulva*.

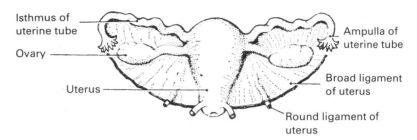

Isthmus of uterine tube

Ovary

Uterus

Ampulla of uterine tube

Broad ligament of uterus

Round ligament of uterus

Fig. 18.4. The ovaries and uterine tubes.

The ovaries

The ovaries are the female sex glands and they lie, one on each side of the uterus, on the lateral wall of the true pelvis. Each is attached to the upper layer of a peritoneal fold, called the *broad ligament of the uterus,* and lie below and behind the lateral end of the uterine tube. Each ovary is about 3 cm in length, 1.5 cm in width and 1 cm thick. The ovaries are almond-shaped and greyish pink in colour. Each ovary is attached to the broad ligament by a fold of peritoneum called the *mesovarium.* Between the two layers of this fold vessels and nerves pass to the hilum of the ovary.

The structure of the ovaries

Each ovary is composed of an outer zone called the cortex and an inner zone, called the medulla.

The medulla is composed of connective tissue which contains large numbers of elastic fibres and many blood vessels.

The cortex of the ovary has an outer layer of cuboidal epithelium called the *germinal epithelium.* Beneath this is a connective tissue framework, or *stroma,* which contains the ovarian follicles and corpora lutea. The thickness of the cortex varies according to age. Before puberty the cortex is relatively thin and contains numerous *primary ovarian follicles.* After puberty one, or occasionally more, of these follicles matures and releases an ovum during each menstrual cycle.

The primary follicle is composed of a large central cell, called the *oogonium,* surrounded by a single layer of cuboidal cells, called the *follicular cells.* When development of this primary follicle begins, the follicular cells increase in number to form several layers and fluid collects between the cells to form a cavity. The follicular cells are now divided into two groups, an outer layer of cells called the *membrana granulosa* and an inner layer of cells which contains the ovum. While these changes are occurring the oogonium matures to form the ovum. The mature follicle, now called a *Graafian follicle*, approaches the surface of the ovary and ruptures, releasing the ovum into the peritoneal cavity, whence it enters the uterine tube.

After the release of the ovum the remaining cells of the Graafian follicle increase in size and a pigment, called *lutein*, becomes deposited in their cytoplasm. This new structure, called the *corpus luteum*, becomes vascularized and, if fertilization of the ovum does not occur, remains like this for 13—14 days. After this it gradually degenerates and is replaced by fibrous tissue. This fibrous remnant is called the *corpus albicans*. If fertilization does occur, however, the corpus luteum remains intact throughout pregnancy.

The blood supply of the ovaries

The ovaries are supplied with arterial blood by the ovarian arteries which arise from the abdominal aorta just below the renal arteries.

The venous blood from the ovary is collected into a plexus of veins in the broad ligament of the uterus and is drained from there by the ovarian veins. The right ovarian vein passes upwards to enter the inferior vena cava just below the right renal vein. The left ovarian vein passes upwards to enter the left renal vein.

The lymphatic drainage of the ovary

The lymphatic vessels from the ovary pass upwards, alongside the ovarian vessels, and drain into the lateral aortic and preaortic lymph nodes.

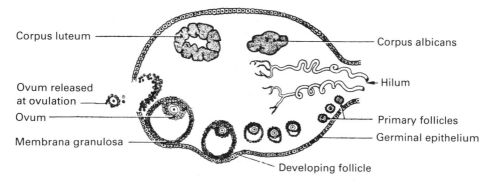

Fig. 18.5. Diagrammatic cross-section of the ovary to illustrate the stages of development of the ovarian follicle and the corpus luteum.

The uterine tubes

The uterine (Fallopian) tubes lie in the upper margins of the broad ligaments of the uterus. They are each about 10 cm in length and they serve to transmit the ovum from the ovary to the uterus. Each runs medially from the lateral wall of the pelvis to the upper part of the uterus. They are divided, for descriptive purposes, into four parts.

1   **The infundibulum** is the outer expanded portion of the uterine tube which opens into the peritoneal cavity. The mouth of the opening is surrounded by a number of finger-like processes called the *fimbriae*.

2   **The ampulla** is the widest portion of the uterine tube. It runs a tortuous course and is about 5 cm in length.

3   **The isthmus** is the medial cord-like portion of the tube which lies between the ampulla and the uterus.

4   **The uterine portion** is about 1 cm in length and is the part of the tube which runs through the uterine wall.

The structure of the uterine tubes

The uterine tubes are composed of three coats:

1   *an outer serous coat*, which is the peritoneal covering;

2   *a middle muscular coat*, which is composed of an outer layer of longitudinal and an inner layer of circular smooth muscle fibres;

3   *an inner layer of mucous membrane*, which lies in longitudinal folds and is covered by a layer of ciliated epithelium.

The uterus

The uterus is a hollow, pear-shaped muscular organ which is approximately 7.5 cm in length, 5 cm in width and 2.5 cm thick. It is divided, for descriptive purposes, into two parts: the *body of the uterus* and the *cervix of the uterus*.

On the lower part of the uterus there is a slight constriction which marks the site of a narrowing of the cavity of the uterus called the *internal os*. The portion of the uterus above this constriction is the body of the uterus and the portion below is the cervix.

The rounded upper part of the body of the uterus, above the site of entry of the uterine tubes, is called the *fundus of the uterus*. Below this portion the body gradually narrows down to its junction with the cervix. The cavity of the body of the uterus is roughly triangular in shape. The apex of the triangle lies at the internal os and the lateral angles at the site of entry of the uterine tubes.

The cervix of the uterus is about 2.5 cm in length. It is narrower than the body of the uterus and is roughly cylindrical in shape. The cavity of the

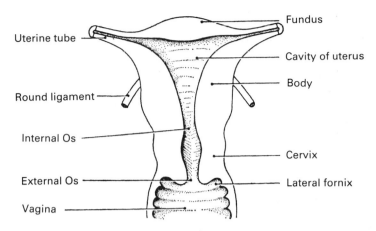

Fig. **18.6.** Cross-section of the uterus.

cervix uteri is wider in its middle part than above or below. The opening of the cavity of the cervix into the uterus is called the *internal os* and the opening into the vagina is called the *external os*. The lower part of the cervix protrudes into the anterior wall of the vagina.

The long axis of the uterus shows a slight curve, which is concave anteriorly. This is called *anterflexion*. The uterus is usually turned forward and lies on the superior surface of the bladder. This is called *anteversion*.

The structure of the uterus

The wall of the uterus is composed of three coats.

1 *An outer serous,* or *peritoneal,* coat covers the body of the uterus and the posterior surface of the portion of the cervix which lies above the vagina.

2 *A middle muscular coat,* called the *myometrium,* forms the main bulk of the uterus. It is composed of bundles of smooth muscle fibres. During pregnancy this muscular coat becomes greatly thickened.

3 *An inner mucous coat,* called the *endometrium,* is continuous with the mucous membrane of the uterine tubes above and with the mucous membrane of the vagina below. In the body of the uterus the mucous membrane is lined with columnar epithelium and contains numerous glands, called the *uterine glands.*

The ligaments of the uterus

The uterus is supported by several ligaments:

1 **The broad ligaments** are folds of peritoneum which pass from each side of the uterus to the lateral walls of the pelvis. The uterine tube lies in the free edge of the broad ligament.

2   The round ligaments are two narrow bands of fibrous tissue which are attached to the uterus just below the uterine tubes. They run laterally, between the layers of the broad ligament, to the lateral walls of the pelvis. Having reached the lateral pelvic walls, they pass forwards through the inguinal canals to the vulva, where they blend with the adjacent areolar tissue.

3   The anterior and posterior ligaments are folds of the peritoneum which are attached to the bladder and the rectum respectively. Two bands of fibrous tissue run backwards on either side of the posterior ligament, and attach the cervix to the sacrum.

The blood supply to the uterus

The uterus receives arterial blood from the uterine arteries, which are branches of the internal iliac arteries, and from branches of the ovarian arteries.

The venous blood from the uterus drains into the ovarian veins and the uterine veins.

The lymphatic drainage of the uterus

The lymphatic drainage of the uterus is described in Chapter 12.

The vagina

The vagina is a fibromuscular canal, about 7.5 cm in length, which extends from the vulva to the uterus. It runs upwards and backwards, behind the bladder and urethra and in front of the rectum and anal canal. The long axis of the vagina forms an angle of 90° or more with the long axis of the uterus. Its anterior and posterior walls are normally in contact. The lower part of the cervix uteri protrudes into the anterior wall of the vagina. The recess of the vagina behind the cervix uteri is called the *posterior fornix* and the smaller recesses in front and at the sides are called the *anterior and lateral fornices* respectively.

The structure of the vagina

The vagina is composed of three coats:
1   *an outer coat* of loose areolar tissue, which contains vessels and nerves;
2   *a middle coat* of smooth muscle fibres; and

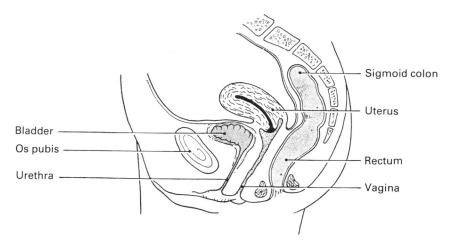

Sigmoid colon

Uterus

Bladder

Os pubis

Urethra

Rectum

Vagina

Fig. **18.7.** Cross-section of the female pelvis.

3 *an inner coat* of mucous membrane, which firmly attached to the muscular coat. The surface of the mucosa is thrown into a number of folds, called rugae, and the inner lining is of stratified squamous epithelium.

The blood supply to the vagina

The vagina is supplied with arterial blood by branches of the internal iliac arteries. The veins of the vagina drain into the internal iliac veins.

The lymphatic drainage of the vagina

The lymphatic vessels from the vagina pass mainly to the internal and external iliac groups of lymphatic nodes. The lymphatic vessels from the lower end of the vagina, however, pass to the superficial inguinal lymph nodes.

The vulva

The vulva is composed of: the *mons pubis*, the *labia majora* and *minora*, the *clitoris*, the *vestibule of the vagina*, the *hymen* and the *greater vestibular glands*.

The mons pubis is a rounded eminence, in front of the pubic symphysis. It is formed by a collection of adipose tissue beneath the skin.

The labia majora are two longitudinal folds of skin which extend downwards from the mons pubis. They form the lateral boundaries of the

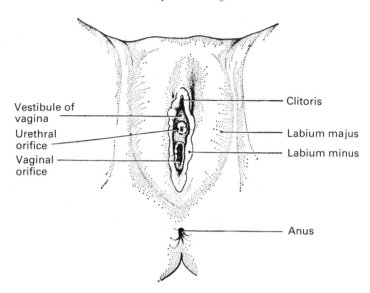

Vestibule of vagina

Urethral orifice

Vaginal orifice

Clitoris

Labium majus

Labium minus

Anus

Fig. 18.8. The vulva.

vulva and they fuse posteriorly, about 3 cm in front of the anal canal. They possess large numbers of sebaceous glands on their inner surfaces.

The labia minora are two smaller longitudinal folds of skin which extend downwards and backwards on either side of the vagina. They contain large numbers of sebaceous glands on their surfaces. The labia minora each divide anteriorly into two folds which enclose the clitoris.

The clitoris is equivalent to the male penis and contains erectile tissue. It is suspended by a ligament from the pubic symphysis and is enclosed in the upper portions of the labia minora.

The urethra, the vagina and the ducts of the greater vestibular glands open into the **vestibule**.

The hymen is a thin fold of mucous membrane which, in the virgin state, partially occludes the orifice of the vagina.

The greater vestibular glands are two oval shaped glands which lie one on either side of the vagina. They secrete mucus which lubricates the vagina.

The blood supply to the vulva

The vulva is supplied with arterial blood by branches of the internal iliac and femoral arteries.

The veins from the vulva drain into the internal iliac veins.

The lymphatic drainage of the vulva

The lymphatic vessels from the vulva drain into the superficial inguinal lymph nodes.

The functions of the female reproductive organs

The ovaries at birth contain oocytes which are the forerunners of the ova. After birth no new oocytes are formed. This is in contrast to spermatogenesis in the male which does not begin until puberty.

The onset of menstruation (the menarche) is caused by hypothalamic stimulation of the anterior pituitary to secrete LH and FSH in a cyclical fashion. The stimulation of the ovary by these two hormones causes the development of the female reproductive organs and initiates menstruation. Menstruation then continues at regular intervals until the menopause. As in the male, the precise mechanism causing the hormonal changes which result in puberty remains obscure.

The menstrual cycle

The menstrual cycle consists of a series of changes which occur in the female throughout the reproductive years and which generally result in ovulation. During each cycle one ovum (or occasionally two ova) is released from the ovary and changes occur in the uterine lining, the endometrium, in preparation for pregnancy should fertilization of the ovum occur. The cycle lasts on average 28 days and is divided into three phases: the *proliferative* or *follicular phase*, the *secretory* or *luteal phase*, and *menstruation*. The changes of the menstrual cycle are due to a series of intricate and reciprocal hormonal changes brought about by the ovary and the pituitary.

The first day of the menstrual cycle is the day on which menstrual blood loss begins, but for ease of description the follicular phase will be considered first.

1   **The follicular phase**. During this part of the menstrual cycle pituitary secretion of FSH and LH stimulate the development of an ovarian (Graafian) follicle which secretes oestrogen in increasing amounts. The rising oestrogen concentration stimulates proliferation of the endometrium. From the 6th to the 14th days of the cycle, the glands, blood vessels and stroma increase causing the endometrium to double in thickness. On the 14th day, shortly after a sharp peak of LH and FSH secretion, ovulation occurs. The ovarian follicle ruptures onto the surface of the ovary and the mature ovum is released briefly into the peritoneal cavity before it is rapidly swept into the ampulla of the uterine duct.

2   **The luteal phase**. The luteal phase starts at the 15th day of the cycle.

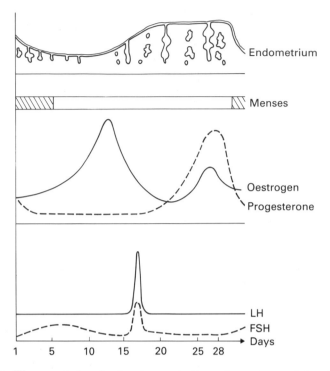

Fig. 18.9. The menstral cycle: the changes in the endometrium and the changes in the plasma levels of sex steroids and gonadotrophins.

The high circulating oestrogen concentration suppresses the secretion of FSH thus allowing only one follicle to develop. (In women treated for infertility by ovulation induction with gonadotrophic hormones this does not occur, and multiple follicles may develop resulting in a multiple pregnancy.) The corpus luteum secretes progesterone, which causes a further increase in endometrial thickness due to growth of the glands, which start to secrete. The arteries of the endometrium become tortuous and exudation of fluid and blood into the endometrium occurs. These changes are in preparation for implantation of a fertilized ovum.

3   Menstruation. If fertilization of the ovum does not occur, the corpus luteum regresses, the levels of oestrogen and progesterone fall, LH and FSH levels start to rise again in preparation for a further cycle, and the superficial part of the endometrium is shed. The flow of blood, mucus and cast-off endometrium from the uterus is called the menstrual flow (menses) and lasts about four to six days.

Feedback control of pituitary gonadotrophin secretion by the ovary is complex during the menstrual cycle. The sharp increase in oestrogen secretion during the follicular phase stimulates the LH peak. This is a positive feedback effect, whereas during the luteal phase rising progesterone and oestrogen concentrations suppress gonadotrophin secretion by negative feedback. The complex and intricate interrelationships of the ovary and hypothalamic–pituitary unit which result in the changes of the normal menstrual cycle are easily disturbed by emotion and excessive weight loss or gain. Not all cycles are associated with ovulation even though menstruation occurs.

Pregnancy

If the ovum is fertilized, which usually occurs in the ampulla of the uterine tube, the corpus luteum does not degenerate and the endometrium is not shed. Cell division of the fertilized ovum proceeds and after about seven days, when the blastocyst (see Chapter 1) has grown to 200 cells, it becomes implanted into the intact endometrium. After implantation the trophoblast secretes *chorionic gonadotrophin*, which is very similar to LH, and this prolongs the life of the corpus luteum. This causes continued ovarian secretion of oestrogen and progesterone which are necessary for the maintenance of pregnancy. The placenta later becomes the main site of progesterone production which continues until the end of pregnancy, when it declines prior to the onset of labour. After about ten weeks the placenta starts to secrete *placental lactogen*, a hormone which resembles growth hormone and prolactin. The secretion of this hormone increases steadily until term is reached and is necessary for the growth and development of the breasts and of the fetus. The secretion of chorionic gonadotrophin, however, rises to a peak at ten weeks and then falls rapidly to a steady level at which its secretion continues until term. The presence of this hormone in urine forms the basis for all diagnostic tests of pregnancy.

Physiological changes during pregnancy

The following changes occur in the mother during pregnancy:
1 *The uterus* increases greatly in size, its weight rising 20-fold.
2 *Blood.* Blood volume rises to a maximum at about 32 weeks. Because the increase is mainly in plasma volume, the haemoglobin falls slightly.
3 *Cardiac output.* The cardiac output increases, but *arterial blood pressure*, both systolic and diastolic, falls slightly.
4 *Ventilation* of the lungs increases and the *metabolic rate* rises.
5 *Gastric acid* production decreases and *gastric contractions* decrease. Nausea and vomiting frequently occur during early pregnancy.

6 *The kidneys.* Renal blood flow and glomerular filtration both increase probably due to the increased cardiac output.

7 *Endocrine glands.* The thyroid gland and adrenal cortex both increase in size but the increased hormone output by these glands is offset by the increase in the concentrations of hormone binding proteins. Thus the free concentration of thyroid hormone and cortisol are normal and hyperthyroidism and Cushing's syndrome do not result.

Labour

The changes which initiate the rhythmic, powerful uterine contractions of normal labour remain unknown. It is probable that hormonal changes in the fetus, not the mother, act as the trigger and increase the sensitivity of the uterine muscle to oxytocin.

Oral contraceptives

These are of two sorts: a combined preparation which contains both a progesterone-type substance (progestogen) and an oestrogen, and a progestogen-only pill (the so-called 'mini-pill'). The combined preparation is a more efficient contraceptive, acting by inhibiting ovulation, altering cervical mucus, thus making it resistant to spermatozoa, and altering the endometrium, thus preventing implantation. The progestogen-only oral contraceptives have less effect in inhibiting ovulation and are effective mainly by their actions on cervical mucus and on the endometrium.

The '10 day rule'

Exposure of the embryo or fetus to radiation during the early months of pregnancy can cause malformation or stillbirth. X-rays of the abdomen and pelvis at this time should be avoided. These include barium meals, barium enemas, IVUs and X-rays of the lumbar spine and pelvis. In women of the child-bearing age, X-rays of the abdomen and pelvis should only be done within 10 days of the onset of the last menstrual period, that is before conception (fertilization) has occurred.

The menopause

Menstruation usually ceases between the ages of 45 and 50 years. This is caused by failure of ovarian activity—both ovulation and hormone secretion. As a result of the fall in oestrogen secretion and operation of the negative feedback control, LH and FSH concentrations rise. The diminished oestrogen secretion also causes a number of physical and psychological changes.

These changes include:

1 Atrophy of the genitalia.

2 Hot flushes.

3 Osteoporosis. The diminished bone mass results in an increased fracture rate, especially of the femoral neck, lower forearm bones and bodies of the lumbar vertebrae.

4 Nervousness, irritability and depressive symptoms.

The female breast

The female breasts, or mammary glands, are accessory organs of the female reproductive system. The breasts develop considerably after puberty but they only reach their full functional state of development in the latter part of pregnancy. Breasts are present in an undeveloped form in the male.

After puberty each female breast forms a rounded eminence on the anterior and lateral walls of the chest, on the surface of the pectoralis major muscle. They extend from the 2nd to the 6th rib and from the lateral border of the sternum to the mid axillary line. The upper outer part of the breast extends upwards into the axilla and is known as the axillary tail of the breast. The main bulk of the breast is composed of fatty tissue and there is, therefore, considerable variation in the size in different individuals. Just below the centre of the breast the *nipple* projects anteriorly. It usually lies at the level of the space between the 4th and 5th ribs. The nipple is surrounded by a circular area of pinkish pigmented skin, which is called

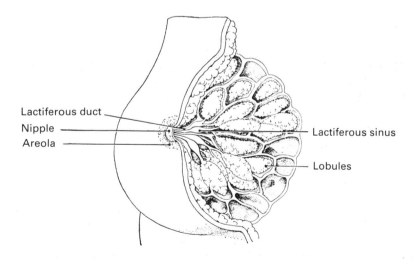

Fig. 18.10. The female breast.

the *areola*. During the first pregnancy, however, the areola becomes dark brown in colour and never subsequently regains its original hue.

The structure of the breast

The breast is composed of glandular tissue, fibrous tissue and fatty tissue. The glandular tissue is composed of 15 to 20 *lobes* each of which is divided into numerous small *lobules*. Each lobule is composed of large numbers of *secretory alveoli* which open into branches of the *lactiferous ducts*. One lactiferous duct drains each lobe of the breast. The lactiferous ducts pass upwards to the areola where they form dilated sinuses which serve as reservoirs for the milk during lactation. Beyond these sinuses the ducts pass upwards to open, by individual orifices, onto the surface of the nipple.

The surface of the breast is covered by the subcutaneous fascia from the inner surface of which numerous fibrous septa pass into the gland to support the lobules. Fibrous strands also pass from the subcutaneous fascia to the nipple and areola.

The fatty tissue lies over the surface of the gland and also lies between the lobes of the gland.

The blood supply to the breast

The breast receives its arterial supply from branches of the axillary arteries, the intercostal arteries and the internal thoracic arteries.

The veins which drain the breast form a venous network under the nipple. This network then drains into the axillary and internal mammary veins.

The lymphatic drainage of the breast

The lymphatic drainage of the breast is described in Chapter 12.

The functions of the female breast

At puberty the increased oestrogen secretion by the ovaries causes growth of the lactiferous ducts and progesterone stimulates alveolar growth. Prolactin secretion by the anterior pituitary is also necessary as are the gonadotrophins and growth hormone. During each menstrual cycle the gland undergoes some degree of proliferation followed by regression.

During pregnancy there is considerable duct proliferation and alveolar development under the influence of the greatly increased concentrations of *oestrogen*, *progesterone* and *placental lactogen*. After the birth of the child the concentrations of these hormones fall but prolactin secretion continues and increases if breast feeding is established. For continued lactation normal secretion of other hormones is needed. These include insulin, thyroid hormone and the steroid hormones of the ovary and adrenal cortex.

Suckling is associated with further temporary rises in plasma prolactin concentration and the secretion of oxytocin from the posterior pituitary. Oxytocin causes milk ejection from the alveoli into the ducts. Suckling is an important stimulus to the continued secretion of milk.

CHAPTER 19

The skin and the organs of special sense

Sensations which are appreciated by the body can be divided, for descriptive purposes, into general sensations and special sensations.

General sensations are the sensations of pain, touch, hot, cold and proprioception.

Special sensations are the sensations of sight, hearing, smell and taste.

The skin

The skin covers the surface of the body and has a protective function. It contains a large number of sensory organs which appreciate touch, pain, hot and cold and it can thus be described as an organ of general sensation. The skin also plays a part in temperature regulation and to a small extent in excretion.

The skin is divided into two layers: the superficial layer or *epidermis* and the deep layer or *dermis*.

The epidermis

The epidermis consists of stratified epithelium and is nonvascular. The thickness of the epidermis varies in different parts of the body. It is particularly thick and horny on the palms of the hands and the soles of the feet. This is partly due to pressure on the skin, but these areas of epidermis are thicker than other areas before birth so that pressure is not the only factor involved.

The surface of the epidermis is marked by a series of furrows which correspond to folds in the surface of the dermis. These folds are called the *papillae* of the skin and are particularly prominent around joints, on the palmar surface of the hands and fingers and on the soles of the feet. They are different in each individual and the furrows of the fingers are used as a method of identification (finger prints).

The epidermis is divided into two main layers: the *germinative zone*, which is the inner layer, and the *horny zone*, which is the outer layer.

The germinative zone is composed of two layers of cells, an inner layer, called the *basal cell layer*, and an outer layer, called the *prickle cell layer*.

The basal cell layer is composed of columnar epithelial cells which have oblong-shaped nuclei. These cells divide rapidly and are constantly replacing the more superficial cells. Among these cells are a number of branched cells called *dendritic cells*, or epidermal melanoblasts. These cells produce a pigment called *melanin* which gives the colour to the skin. Some of the skin colour, however, is due to the blood vessels in the dermis.

The prickle cell layer is composed of polyhedral-shaped cells which are joined together by fine processes.

The horny zone is composed of three layers of cells, an inner layer called the *stratum granulosum*, a middle layer called the *stratum lucidum* and an outer layer called the *stratum corneum*.

The stratum granulosum consists of several layers of fusiform cells which contain large numbers of granules.

The stratum lucidum appears as a clear layer and is composed of tightly packed flattened cells with clear cytoplasm and only in a few of these cells are nuclei present.

The stratum corneum consists of flattened horny epithelial cells in which the nuclei are absent. The cytoplasm of these cells has been converted into a substance called *keratin*. This layer is continuously being eroded by friction and replaced by cells from the layers below.

The dermis

The dermis is composed of connective tissue which contains a variable number of elastic fibres. It is a tough, flexible layer. The surface of the

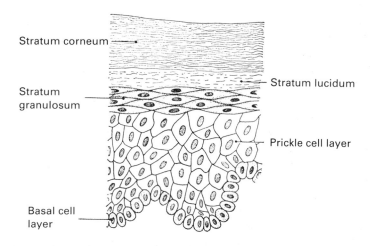

Fig. 19.1. The layers of the epidermis.

dermis is marked by a number of folds called papillae. The following structures are present in the dermis: *blood vessels* and *lymphatic vessels, sensory nerve endings, hair follicles* and *hairs, arrectores pilorum, sebaceous glands* and *sweat glands.*

The blood vessels and lymphatic vessels. The arteries supplying the skin form a network of arterioles in the dermis, and branches run from these to the hairs, the sebaceous glands, the sweat glands and the papillae. The lymphatic vessels form a network throughout the dermis.

The sensory nerve endings. There are large number of sensory nerve endings in the dermis in association with the hairs and the papillae. The appearance of these nerve endings differ according to the type of sensation to which they respond. There is thought to be different nerve endings for appreciating touch, cold, hot and pain. The nerve endings when stimulated, transmit impulses to the brain.

Hair follicles and hair. Hairs are present on almost the whole of the surface of the body, with the exception of the palms of the hands and the soles of the feet. The hairs vary in colour, diameter and length. The hairs consist of a *root*, which lies in the dermis, and a *shaft* which projects above the skin. The deep part of the hair root is expanded to form the *hair bulb* and is embedded in a structure called the *hair follicle* which lies in the dermis.

The hair follicle is an invagination of the epidermis and is composed of epithelial tissue and fibrous tissue. It runs obliquely downwards into the

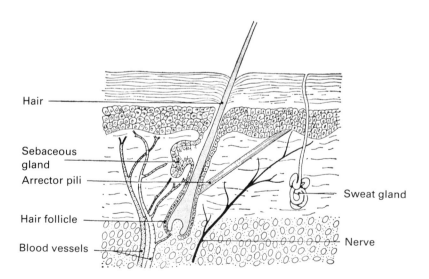

Fig. 19.2. Diagrammatic cross-section of the skin.

dermis and is expanded at its distal end to surround the hair bulb. Beneath the bulb the follicle is indented by a papilla. The hair which is composed of keratin, is budded off from the cells at the base of the papilla and the rate of growth varies from 1.5 to 2.5 mm per week. Each hair contains a variable amount of the pigment melanin which is responsible for its colour. With advancing age the melanin is replaced by minute air bubbles which give the appearance of 'grey hair'.

The arrectores pilorum are small bundles of involuntary muscle fibres which are attached to the hair follicles. They arise from the superficial layers of the dermis. When these muscle fibres contract they tend to lift the hair upright; at the same time the skin around the hair is elevated giving the appearance of 'gooseflesh'.

The sebaceous glands are small compound saccular glands situated in the dermis. They are composed of a number of alveoli and open by a single duct into the hair follicle. They vary considerably in number from area to area in the skin but, like the hairs, they are not present in the palms of the hands and the soles of the feet.

The sweat glands are coiled tubular glands which are found in the dermis in most areas of the body. They are particularly numerous in the palms of the hands and the soles of the feet. Each gland is composed of a coiled portion, called the body of the gland, and a duct, which opens onto the surface of the skin at a funnel-shaped opening called a *pore*.

The nails

The nails are flattened horny structures located on the distal ends of the fingers and toes and are equivalent to the claws of animals. The proximal part of the nail is called the *root* and it is embedded in a fold of skin called the *cuticle*. The exposed portion of the nail is called the *body* of the nail and the distal portion is called the *free edge*. The nail is equivalent to the stratum lucidum and beneath the nail the germinative layer and the dermis constitute the nail bed.

The functions of the skin

1 **Protection.** The skin is important for the protection of the deeper tissues of the body. It acts as an effective barrier against the invasion of bacteria and prevents injury to the underlying organs. It is important in preventing loss of fluid from the tissues.

2 **Sensation.** The skin is an important organ of general sensation.

3 **The secretion of sebum.** The skin secretes sebum which lubricates the hairs and keeps the skin soft and pliable. Sebum is also important for protection as it has a bactericidal action.

4 The formation of Vitamin D. The skin contains a substance called **ergosterol** which is converted into Vitamin D by the action of direct sunlight.

5 Excretion. The sweat glands play a small part in excretion, removing water, certain inorganic salts and a small amount of other waste products from the body.

6 Temperature regulation. The skin is important in the regulation of the body temperature. Heat is produced in the body as a result of the metabolic processes. A large proportion of this heat is produced in the muscles. Heat is lost from the body mainly from the skin, but also from the lungs in expiration and from the intestine and urinary bladder in the excreta during micturition and defaecation.

The body temperature is controlled by a group of cells in the hypothalamus of the brain. These cells are called the *heat regulation centre*. The heat regulation centre is sensitive to changes in body temperature and when this alters it stimulates another group of cells in the medulla oblongata. Impulses pass from these cells to the sweat glands and to the blood vessels of the skin.

The skin controls the body temperature by the amount of sweat produced and the amount of blood flowing through the skin vessels.

1 *The sweat glands*. Under normal circumstances the sweat glands continuously produce sweat but this evaporates as fast as it is formed thus cooling the body. This is called insensible sweating. If conditions call for a greater loss of heat the sweat glands are stimulated to produce more sweat, and this gathers on the surface of the body as minute beads of moisture which again evaporate. This is called sensible sweating.

2 *The blood vessels*. The skin also loses heat by convection and conduction. Alterations in the blood flow through the skin alter the rate at which heat is lost. When the body temperature rises the blood vessels in the dermis become dilated, and an increased amount of heat is lost to the surroundings. In cold conditions the blood vessels are constricted, and there is a reduction in the heat lost.

The eye

The eye is the organ of sight. It is situated in the anterior part of the orbital cavity of the skull, embedded in a pad of fat, and it is thus well protected from injury since only its anterior aspect is not surrounded by bone. The eye is supplied by the 2nd cranial (optic) nerve.

The structure of the eye

The eyeball is composed of three coats and within these coats are the contents of the eyeball.

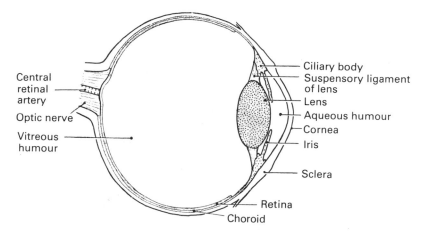

Fig. 19.3. The structure of the eye.

The coats of the eye are:
1 an outer fibrous coat, the *sclera* and the *cornea*;
2 a middle vascular coat, the *choroid*, the *ciliary body* and the *iris*; and
3 an inner nervous coat, the *retina*.

The contents of the eyeball are:
1 *the aqueous humour*;
2 *the lens*; and
3 *the vitreous body*.

The coats of the eye

1 The fibrous coat is made up of the sclera, which is opaque, and which forms the posterior 5/6ths of the outer coat, and the cornea which is translucent and which forms the anterior 1/6th of the outer coat.

The sclera is composed of strong white fibrous tissue which has a few yellow elastic fibres scattered throughout it. It has a poor blood supply. Its external surface is white and can be seen anteriorly as the 'white' of the eye. Its surface is smooth except at the sites of attachment of the muscles which move the eye (extrinsic muscles of the eye). Posteriorly the optic nerve pierces the sclera and the sclera is reflected on to the nerve to form a fibrous sheath which becomes continuous with the dura mater of the skull. Anteriorly the sclera is continuous with the cornea.

The cornea is a clear transparent circular structure which projects anteriorly and is slightly more convex than the sclera. It is composed of an outer epithelial coat, which is continuous at the margins of the cornea with the conjuctiva, and an inner coat of modified connective tissue. The cornea contains no blood vessels.

2   The vascular coat is the middle coat of the eye and is made up of the choroid, the ciliary body and the iris.

The choroid is a thin pigmented vascular membrane, dark brown in colour, which lines the inner surface of the sclera. It is only loosely attached to the sclera except at the point where it is pierced by the optic nerve, where it is firmly attached. The choroid consists mainly of a dense capillary network and pigment cells.

The ciliary body. The choroid is continuous anteriorly with the ciliary body. The ciliary body is a circular structure which lies just behind the junction of the cornea with the sclera. It contains large numbers of involuntary muscle fibres and attached to its inner surface is the *suspensory ligament of the lens* of the eye. The ciliary muscle is an intrinsic muscle of the eye and by pulling on the suspensory ligament it can alter the convexity of the lens to focus rays of light on to the retina.

The iris is a thin pigmented membrane which lies in the aqueous humour between the cornea and the lens. It is attached at its outer margin to the anterior margin of the ciliary body. The iris is the second intrinsic muscle of the eye. It contains two groups of involuntary muscle fibres, one group is arranged in a circular manner, the other in a radial manner. The central aperture of the iris is called the *pupil of the eye* and variations in the size of the aperture occur with different intensities of light. Thus in bright light the circular muscle constricts the pupil to reduce the amount of light entering the eye and in dim light the radial fibres contract to dilate the pupil to increase the amount of light entering the eye.

3   **The retina** is the innermost coat of the eye. It lines the choroid and extends as far forwards as the ciliary body. It is a delicate membrane composed of nerve cells and fibres and, on its choroidal surface, special structures called *rods* and *cones*. The rods and cones are the actual light receptors of the eye and when light strikes them impulses are initiated which pass through the optic nerve to the brain.

At about the centre of the posterior part of the retina there is an oval rather yellowish area called the *macula*. In the centre of this area is a small depression called the *fovea*. The fovea contains only cones and is the point of most accurate vision. At this point each cone is attached to one bipolar nerve cell, whereas in more peripheral parts of the retina a number of rods and cones are attached to one nerve cell thus diminishing the accuracy of vision.

Just medial to the macula is the *optic disc* at which point the nerve fibres pass backwards to enter the optic nerve. The ophthalmic artery and vein also pass through the optic disc. The optic disc contains no rods or cones and is often called the 'blind spot'.

The contents of the eye

1   The aqueous humour is a clear watery fluid situated between the cornea and the lens. This region is divided into an *anterior chamber* in front of the iris and a *posterior chamber* behind the iris. The aqueous humour is secreted in to the posterior chamber by capillaries in the ciliary body and it passes through the pupil into the anterior chamber where it drains into the ciliary veins. (Occasionally the drainage is impaired and the pressure of the aqueous humour rises producing the condition of *glaucoma*.)

2   The lens is a biconvex transparent structure, which is enclosed in a capsule and is suspended from the ciliary body by a ligament which is attached to its margins. The capsule and the lens are elastic and can be altered in shape by contractions of the ciliary body.

3   The vitreous body is a jelly-like transparent substance which fills the posterior 4/5ths of the eyeball which lies behind the lens. Its pressure helps to maintain the shape of the eyeball and to support the retina against the choroid.

The optic nerves

Each optic nerve is formed at the posteromedial side of each eyeball where the nerve fibres from the retina pierce the optic disc. The optic nerve then leaves the orbital cavity through the optic foramen to enter the middle cranial fossa. Just in front of the sella turcica there is a communication between the two optic nerves called the *optic chiasma*.

In the optic chiasma there is a changeover of nerve fibres. The nerve fibres from the nasal side of each retina cross through the optic chiasma to join the fibres from the temporal side of the opposite eye (Fig. 19.4). This

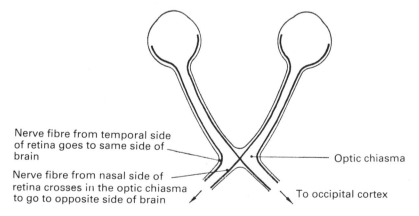

Nerve fibre from temporal side of retina goes to same side of brain

Nerve fibre from nasal side of retina crosses in the optic chiasma to go to opposite side of brain

Optic chiasma

To occipital cortex

Fig. **19.4.** The optic chiasma.

means that fibres from the right side of each retina pass to the right side of the brain, and fibres from the left side of each retina pass to the left of the brain.

Beyond the optic chiasma the fibres, now called the *optic tracts*, pass to the visual area of the cerebral cortex, which lies in the occipital lobe.

The accessory organs of sight

Associated with the eye are several accessory organs which are essential to the proper functioning of the eye. They are: the *eyebrows*, the *eyelids*, the *conjunctiva*, the *lacrimal apparatus* and the *extrinsic muscles of the eye*.

The eyebrows are the hairs on the skin covering the supraorbital ridges of the frontal bone. They serve to prevent sweat running downwards into the eye.

The eyelids are two moveable folds, an upper and a lower, which together form the anterior protection of the eyeball. The margins of the eyelids carry a number of hairs, the *eyelashes*, which are protective in function.

The eyelids have an outer covering of skin and contain within a central plate of dense fibrous tissue, called the tarsal plate. Between the tarsal plate and the skin is a small amount of areolar tissue and voluntary muscle fibres arranged in a circular manner. These muscle fibres form the *orbicularis oculi* which is the sphincter muscle of the eye. The upper lid also has a small muscle attached to it which is responsible for elevating the upper lid. The inner margin of the eyelid is lined by the conjunctiva.

The eyelids serve to protect the eye anteriorly and close frequently in the act of blinking. If an object approaches the eye or actually touches the cornea the eyelids close instantly. This is called the *corneal reflex*.

The conjunctiva is a delicate mucous membrane which covers the anterior surface of the eye and is reflected above and below on to the inner surface of the eyelids. It is closely adherent over the surface of the cornea of which it forms the outer epithelial covering. The conjunctiva is frequently referred to as the conjunctival sac, but it is only a closed sac when the eyelids are closed.

The lacrimal apparatus is composed of: the *lacrimal gland and its ducts*, the *lacrimal canaliculi*, the *lacrimal sac* and the *nasolacrimal duct*.

The lacrimal gland lies in the orbital cavity in a recess in the lateral part of the roof, just behind the supraorbital margin. The lacrimal gland secretes the tears which pass down a number of small ducts into the upper outer angle of the conjunctival sac. The tears then pass across the anterior surface of the eye. The action of blinking helps to move the tears toward the medial side of the conjunctival sac.

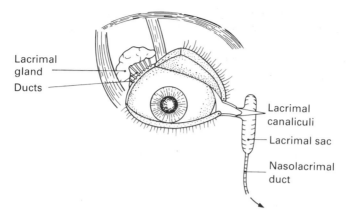

Fig. **19.5.** The lacrimal apparatus.

The lacrimal canaliculi are two small ducts which open at the medial angle of the eye into each eyelid. The tears pass through these ducts to enter the lacrimal sac.

The lacrimal sac lies in a depression on the surface of the lacrimal bone.

The nasolacrimal duct drains the tears from the lacrimal sac into the inferior meatus of the nose.

The tears are continuously secreted by the lacrimal gland and serve to moisten and clean the anterior surface of the eye. In crying, due to an emotional reaction, there is an excess of tears produced and these overflow from the conjunctival sac.

The extrinsic muscles of the eye are responsible for the coordinated movements of the eyeballs. Each eye has four straight muscles and two oblique muscles attached to it. Their names and actions are:

1 the *superior rectus*, which turns the eye upwards;
2 the *inferior rectus*, which turns the eye downwards;
3 the *medial rectus*, which turns the eye inwards;
4 the *lateral rectus*, which turns the eye outwards;
5 the *inferior oblique*, which turns the eye upwards and outwards; and
6 the *superior oblique*, which turns the eye downwards and outwards.

The muscles arise from the bony walls of the orbit and are attached to the sclera. They are voluntary muscles and are supplied by the following cranial nerves:

The oculomotor nerve supplies the *superior rectus*, the *inferior rectus*, the *medial rectus*, and the *inferior oblique*.

The trochlear nerve supplies the *superior oblique*.

The abducens nerve supplies the *lateral rectus*.

The physiology of vision

Light falling on the retina initiates impulses which pass to the occipital cortex of the brain, where a visual image is assembled from the pattern and intensity of the impulses. The eye thus serves to focus and control light from the surroundings so that a sharp image is formed on the retina. The eye can rotate round to bring into its field items of importance or interest.

The focusing of the light on to the retina is the function of the cornea and the lens. The focusing power of the cornea is fixed, but that of the lens can be altered.

Accommodation. When light from a distant object strikes the eye the rays of light are virtually parallel and the cornea and the lens focus them on to the retina at the fovea. Rays of light from a close object are diverging when they strike the eye and have to be bent more than parallel rays in order to bring them into focus on the retina. This is accomplished by an increase in the convexity of the lens which is brought about by relaxation of the ciliary body and thus relaxation of the suspensory ligament of the lens. More light reaches the eye from a near object than from a distant object and so, as the ciliary body relaxes, the pupil constricts to reduce the amount of light entering the eye. This process of alterations in the convexity of the lens and the diameter of the pupil is called accommodation. When an object is very close to the eye the eyes turn inwards, or *converge*, to ensure that the image falls on the fovea in each eye.

Binocular vision. Both eyes move together so that images strike the same part of the retina in each eye. This is achieved by coordinated movements of the extrinsic muscles of both eyes and is called binocular vision. In babies with a squint the images strike different parts of the retina in each eye and unless this is corrected early the child will not achieve binocular vision, but double vision will not occur as the cerebral cortex suppresses one image.

The use of both eyes means that every object is viewed from two slightly different angles and thus three dimensional, or stereoscopic, vision is achieved. With stereoscopic vision a much clearer assessment of distance and depth can be obtained.

The function of the retina. The retina is the light sensitive part of the eye. Two different mechanisms are involved in the stimulation of the retina. One mechanism, called *photopic vision*, operates in normal light, the other mechanism, called *scotopic vision*, operates only in dim light.

Photopic vision involves the appreciation of colour and is the responsibility of the cones. The cones are thought to be of three types, each of which are sensitive to light of different primary colours. At the area of most sensitive vision, the fovea, only cones are present.

Scotopic vision is the responsibility of the rods. As the light intensity decreases the rods take over from the cones. The rods contain a pigment called the rhodopsin or *visual purple*. In the presence of light this pigment splits into two substances, one of which initiates the nervous impulse. In light of very low intensity the visual purple is regenerated. Thus in bright light all the visual purple is dissociated and when the light intensity falls it is gradually regenerated. This regeneration takes about 20 minutes and thus vision in light of low intensity gradually improves over this period. This is called *dark adaptation*.

The ear

The ear is the organ of hearing and is located mainly in the temporal bone. Intimately associated with part of the ear are structures called the semicircular canals which are important organs for the maintenance of balance but which have no auditory function.

The ear, including the semicircular canals, is supplied by the vestibulocochlear nerve (VIIIth cranial nerve).

The structure of the ear

The ear is divided into three main parts: the *external ear*, the *middle ear* and the *inner ear*.

The external ear comprises the *pinna* or auricle (this is the part referred to by the lay term 'ear'), and the external acoustic meatus.

The pinna is attached to the side of the head at about the midpoint between the forehead and the occiput. It is made up of an upper portion composed of yellow elastic fibrocartilage, and a lower portion, called the *lobe*, which is composed of adipose and connective tissue covered with skin.

The external acoustic meatus is a slightly S-shaped tube which extends from the pinna to the *tympanic membrane*, or eardrum. The external acoustic meatus is 4 cm in length, the outer third is cartilaginous and the inner two-thirds has bony walls and lies within the temporal bone. The external acoustic meatus is lined with skin which contains *ceruminous glands*, which secrete wax.

The external acoustic meatus is separated from the middle ear by the tympanic membrane.

The tympanic membrane is an oval-shaped membrane which lies obliquely to form an angle of about 55° with the floor of the external acoustic meatus. Its greatest diameter is 9–10 mm.

It is composed of three coats:

1 *an outer coat* of stratified epithelium which is continuous with the epidermis of the external acoustic meatus;

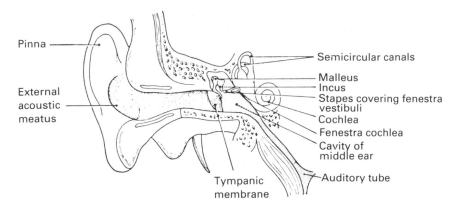

Pinna

External acoustic meatus

Semicircular canals
Malleus
Incus
Stapes covering fenestra vestibuli
Cochlea
Fenestra cochlea
Cavity of middle ear
Auditory tube

Tympanic membrane

Fig. 19.6. General arrangement of the ear.

2 *a middle coat* of fibrous tissue; and

3 *an inner coat* of cuboidal epithelium which is continuous with the lining of the middle ear.

The middle ear is an irregularly-shaped cavity situated within the petrous portion of the temporal bone. It contains three small bones, called *ossicles*, which articulate to form a chain which transmits the vibrations of the tympanic membrane to the inner ear. The middle ear communicates with the pharynx through the *auditory (Eustachian) tube.*

The middle ear is roughly shaped like a narrow box and has a roof, a floor, and anterior, posterior, lateral and medial walls.

The roof of the middle ear is a thin plate of bone which separates it from the cranial cavity.

The floor of the middle ear is a thin plate of bone which separates it from the jugular vein.

The auditory tube opens into the *anterior wall.*

In the *posterior wall* there is an opening through which the middle ear communicates with the air cells in the mastoid portion of the temporal bone.

The lateral wall is formed by the tympanic membrane.

The medial wall contains two openings:

1 **The fenestra vestibuli** (the oval window) through which the middle ear communicates with the vestibule of the inner ear; the oval window is covered by the footplate of the stapes.

2 **The fenestra cochlea** (round window) through which the middle ear communicates with the cochlea; the round window is covered by a sheet of fibrous tissue.

The middle ear is lined by mucous membrane.

The ossicles of the middle ear. The ossicles, called the malleus, the incus and the stapes, stretch across the middle ear from the tympanic membrane to the fenestra vestibuli.

The malleus is roughly hammer-shaped, and bears a handle, a neck and a head. The handle is fixed to the tympanic membrane and the head articulates with the incus.

The incus is said to resemble an anvil, and is the middle of the ossicles. It is composed of a body, a long process and a short process. The body articulates with the malleus and the long process with the stapes. The short process is connected to the roof of the middle ear by ligaments.

The stapes is shaped like a stirrup and has a head, two limbs and a footplate or base. The head articulates with the long process of the incus and the base is connected by a membrane to the oval window.

The auditory tube is about 4 cm in length and extends from the posterior wall of the nasopharynx to the anterior wall of the middle ear. It is lined by ciliated epithelium and serves to keep the pressures on either side of the tympanic membrane equal.

The inner ear is composed of two parts: the **bony labyrinth**, which is a series of cavities lying within the petrous portion of the temporal bone, and the **membranous labyrinth**, which is a series of membranous sacs lying within the bony labyrinth.

The bony labyrinth is composed of three parts, the vestibule, the cochlea and the semicircular canals. These all contain a clear fluid called *perilymph* and within them lies the membranous labyrinth.

The vestibule is the portion of the inner ear which communicates with the middle ear through the fenestra vestibuli. Opening into the vestibule are four canals, the cochlea and the three semicircular canals.

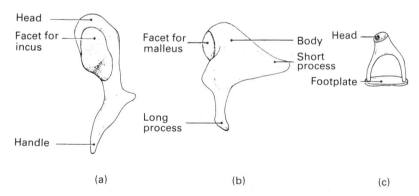

(a) (b) (c)

Fig. 19.7. The auditory ossicles: (a) the malleus; (b) the incus; and (c) the stapes.

The cochlea is shaped like a snail's shell. It lies in front of the vestibule. It is composed of a canal which describes two and a half turns round a central pillar of bone, called the *modiolus*. The cochlea is wide at its origin from the vestibule, but gradually narrows throughout its turns. Just at the point of origin of the cochlea from the vestibule the *fenestra cochlea* forms a membrane covered communication between the cochlea and the middle ear.

The three semicircular canals are described as the superior, the lateral and the posterior semicircular canals. They lie above and behind the vestibule with which they communicate at their extremities by five openings. Two of the canals have a common opening at one end.

The membranous labyrinth lies within the bony labyrinth. Within the vestibule are two small membranous sacs called the utricle and the saccule. The **utricle** is the larger of the two and opening into it are the semicircular canals. The point of origin of each semicircular duct from the utricle is widened to form the *ampulla*. The utricle communicates with the **saccule** from which arises the membranous cochlear duct which lies within the cochlea. The utricle, the saccule and the semicircular ducts are lined with epithelial cells which possess fine processes on their inner surfaces. These cells are called *hair cells*. Running between these cells are fibres of the 8th cranial nerve.

Arising from the saccule is the **cochlear duct** which spirals round inside the cochlea. The cochlear duct almost completely divides the cochlea into an upper portion called the *scala vestibuli* and a lower portion called the *scala tympani*. On the interior surface of the cochlear duct is a structure called the *organ of Corti* which is the end organ of hearing. It is supplied by

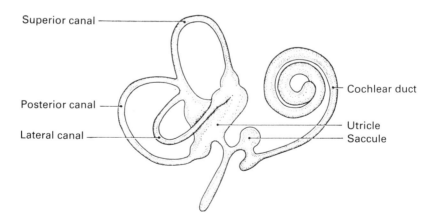

Fig. 19.8. The membranous labyrinth.

the cochlear fibres of the vestibulocochlear nerve which unite with those from the semicircular ducts to leave the petrous portion of the temporal bone to enter the cranial cavity through the internal acoustic meatus.

The whole of the membranous labyrinth is filled with a fluid called *endolymph*.

The physiology of hearing

The function of the ear is to receive sound waves and transmit them as nervous impulses to the temporal lobe of the cerebral cortex where they are interpreted. Sound waves reaching the external ear set up vibrations in the tympanic membrane. The ossicles of the middle ear transmit the vibrations of the tympanic membrane to the fenestra vestibuli. The vibrations of the fenestra vestibuli cause pressure changes within the perilymph. These pressure changes in turn cause pressure changes in the endolymph within the membranous cochlear duct. These pressure changes stimulate the organ of Corti, and this initiates nervous impulses which pass through the cochlear fibres of the vestibulocochlear nerve to the temporal lobe of the cerebral cortex, where they are interpreted.

The functions of the semicircular canals

The semicircular canals are concerned with balance and the position of the head. Movements of the head cause movements of the perilymph and

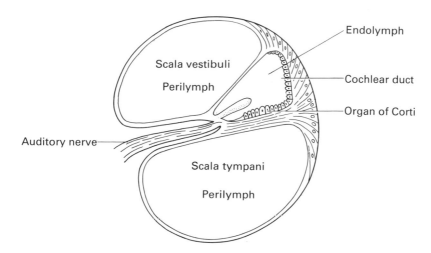

Fig. 19.9. Cross-section of the cochlea to show the arrangement of the membranous cochlear duct.

endolymph within the semicircular canals. These fluid movements stimulate
the hair cells in the utricle, saccule and ampullae. The impulses produced
travel up the vestibular fibres of the vestibulocochlear nerve to the cerebellum where they are used for the maintenance of balance.

The nose

The nose contains the organs of the sense of smell and is supplied by the
olfactory nerve. The anatomy of the nose has already been described. The
area of the nose in which smell is appreciated is called the *olfactory region*.
The olfactory region is confined to the superior nasal conchae and the area
of the nasal septum opposite the superior conchae. The nasal cavity is lined
by mucous membrane but in the olfactory region lying in between the cells
of the mucous membrane are specialized cells, called *olfactory cells*. These
are bipolar nerve cells. One fibre leads to the surface of the mucous
membrane and has fine hair-like processes attached to it. The other fibre
joins with fibres from other olfactory cells and groups of fibres pass
through the cribriform plates of the ethmoid bone to enter the olfactory
bulb. The fibres from the olfactory bulb then pass to the temporal lobe of
the cerebral cortex.

The physiology of the sense of smell

During normal breathing only small eddies of air reach the olfactory area.
The main stream of air entering and leaving the nose passes below the
superior nasal concha. Thus smells are first apparent in small amounts.
The act of sniffing, however, increases the amount of air reaching the
olfactory area and increases the appreciation of a given smell. It is not fully
known how smells stimulate the olfactory cells, but it is thought to be
mainly chemical stimulation. Particles of the substance being smelt probably
combine with receptors on the nasal cilia and then stimulate the olfactory
cells.

The sense of smell is more acute than the sense of taste and in fact the
sense of smell is important for the appreciation of 'taste'.

The tongue

The tongue is the organ of taste. Sensations of taste are carried to the
brain from the anterior two-thirds of the tongue by the facial nerve and
from the posterior third of the tongue by the glossopharyngeal nerve. The
tongue is also sensitive to the general sensations of temperature, pain and
touch. These sensations are carried to the brain from the anterior two-

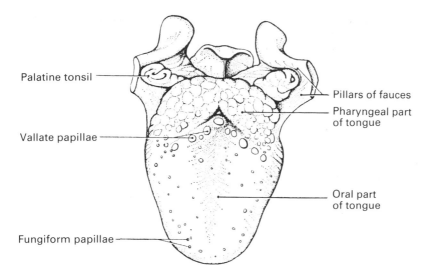

Palatine tonsil

Vallate papillae

Fungiform papillae

Pillars of fauces

Pharyngeal part
of tongue

Oral part
of tongue

Fig. 19.10. The dorsum of the tongue showing the papillae.

thirds of the tongue by the trigeminal nerve and from the posterior third of
the tongue by the glossopharyngeal nerve.

The anatomy of the tongue has already been described (p. 361). The
surface of the tongue is covered by stratified epithelium and possesses large
numbers of small projections called papillae. It will be remembered that
these are of three types, *filiform papillae*, which are found mainly on the
anterior two-thirds of the tongue, *fungiform papillae*, which are found mainly
on the tips and sides of the tongue, and *circumvallate papillae*, which are
found mainly on the posterior part of the tongue.

Associated with the papillae are structures called *taste buds*. The taste
buds are oval structures composed of slender cells each one of which has a
minute hair-like process which projects outwards through the opening of
the taste bud. It is the taste buds which are responsible for the appreciation
of taste.

The physiology of the sense of taste

Four basic tastes can be appreciated by the tongue, namely sweet, salt, sour
and bitter. All the other tastes which can be appreciated are produced by a
combination of these tastes supplemented by the sense of smell and general
sensations which a given substance evokes on the surface of the tongue.

Sweet tastes are mainly appreciated at the tip of the tongue.

Salt tastes are mainly appreciated at the sides of the anterior part of the tongue.

Sour tastes are mainly appreciated at the posterior parts of the sides of the tongue.

Bitter tastes are mainly appreciated at the back of the tongue.

The exact mechanism by which any substance stimulates the taste buds is not known.

Index